T0227507

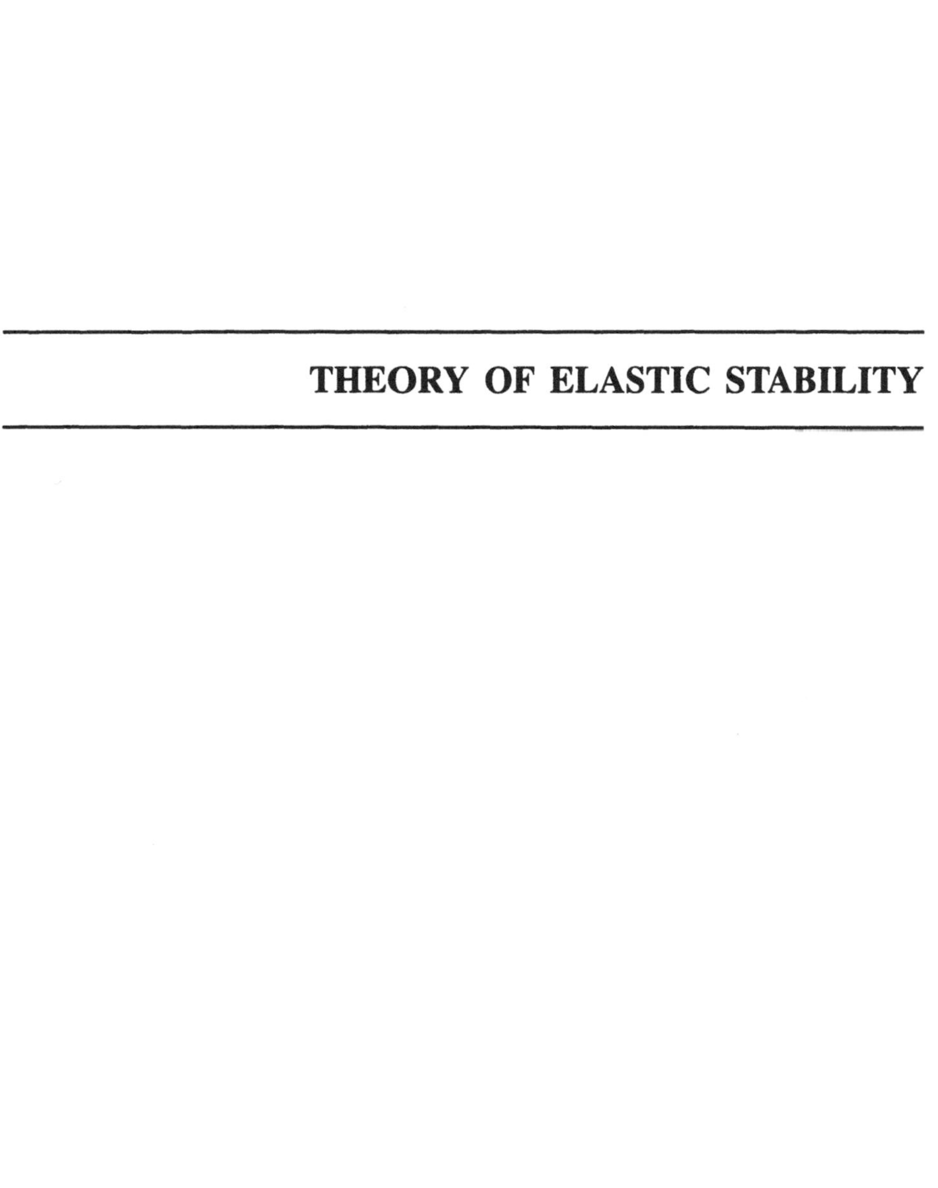

THEORY OF ELASTIC STABILITY

THEORY OF ELASTIC STABILITY
Analysis and Sensitivity

Luis A. Godoy
Department of Civil Engineering
University of Puerto Rico
Mayaguez, Puerto Rico, USA

CRC Press is an imprint of the
Taylor & Francis Group, an **informa** business
A TAYLOR & FRANCIS BOOK

THEORY OF ELASTIC STABILITY: Analysis and Sensitivity

First published 2000 by Taylor & Francis

Published 2019 by CRC Press
Taylor & Francis Group
6000 Broken Sound Parkway NW, Suite 300
Boca Raton, FL 33487-2742

ISBN 13: 978-1-56032-857-5 (hbk)

Visit the Taylor & Francis Web site at
http://www.taylorandfrancis.com

and the CRC Press Web site at
http://www.crcpress.com

Cover design by Carolyn O'Brien.

A CIP catalog record for this book is available from the British Library.

Library of Congress Cataloging-in-Publication Data

Godoy, Luis A. (Luis Augusto)
Theory of elastic stability : analysis and sensitivity / Luis A. Godoy.
p. cm.
Includes bibliographical references (p.).
ISBN 1-56032-857-6 (alk. paper)
1. Elasticity. 2. Stability. I. Title.
QA931.G582 1999
531'.382--dc21 99-41797
CIP

CONTENTS

PREFACE

The approach of the book is theoretical, as indicated in the title, but the emphasis is on the formulation of engineering problems, rather than on the proof of the fundamental principles. The complete formulation from Chapters 2 to 15 employs the total potential energy of the discrete or discretized system. An advantage of this presentation is that the same energy is the basis of equilibrium, stability, postcritical states, design sensitivity, and imperfection sensitivity. A further advantage is that the energy can be computed using the finite element method, which has become the standard tool for the solution of most engineering problems.

A second feature of the book is the use of perturbation techniques as part of the formulation and also of the analysis. This reduces the effort necessary to understand the solution of nonlinear systems. For the computation of equilibrium paths with large displacements and strains, other techniques may be more efficient; however, the emphasis here is on the discovery of a number of different problems associated with the stability of structural systems, and for that perturbation techniques are sufficient.

CONTENTS OF THIS BOOK

Chapters 2 and 3 are introductory and contain material that is usually covered in other texts. Chapter 2 reviews basic concepts of the theory of nonlinear elasticity. Perturbation theory is used throughout the book, so it was considered convenient to introduce the techniques of perturbation at an early stage (Chapter 3). The reader who is familiar with perturbation techniques can skip this chapter without any loss of comprehension of the main body of the book.

Chapters 4 to 7 deal with the problem of stability in multidegree-of-freedom systems with one load parameter. Systems are considered as perfect, meaning that only one initial geometric configuration is considered.

The concept of stability is presented in Chapter 4. Stability criteria can be viewed as tests applied to an equilibrium state to investigate a property (stability). The heart of this chapter is devoted to the criterion employed in this book and is based on the total potential energy V. In this energy criterion, stability is introduced in an axiomatic way. There are several manners in which this can be implemented in practice, and four such implementations are discussed in this chapter. In addition to stability of equilibrium states, the question of stability of a path is also addressed. Other stability criteria are presented in this chapter for the sake of completeness, including the static and the dynamic approaches.

Critical states are the subject of Chapter 5. The conditions for the presence of a critical state are formulated in terms of the energy (in fact, it is written in terms of second derivatives of the energy). Again, there are several ways to implement the computation of critical states along an equilibrium path. In the last part of the chapter we discuss the stability of the critical state and show that stability depends on the values of third- and fourth-order derivatives of the energy. Chapter 5 also contains examples of simple structural components for which the critical states are computed. The examples are columns and struts, rings, arches, frames, and plates, which are modeled with less than five degrees of freedom. The examples illustrate the behavior under specific load situations but are generalized by reference to other solutions in the literature. This is not intended to constitute a handbook of solutions but rather to illustrate concepts and algorithms.

The limit point is studied in Chapter 6 as one type of critical point. First, the general perturbation equations valid for any critical state are given, and then the contracted form of these equations is also obtained. Next, the postbuckling path is written in an asymptotic expansion, and the coefficients are computed from the perturbation equations and the contracted forms. The solutions in this chapter follow a set of assumptions that are identified as a limit point behavior.

Bifurcation points are the subject of Chapters 7 and 13. In Chapter 7 the analysis is carried out using the energy in terms of the original generalized coordinates. The perturbation equations derived in Chapter 6 are valid, and their solution under a new set of assumptions permits the identification of a family of behaviors called bifurcations. They include symmetric and asymmetric bifurcations, and in the first case the distinction between stable and unstable symmetric bifurcations is possible. Algorithms are presented for the computation of the coefficients in the asymptotic expansions. Examples of postcritical behavior are also presented in Chapter 7. Some examples are a continuation of those investigated in Chapter 5. Many theoretically inclined readers can skip the examples, but they are highly recommended in graduate courses.

Definitions of imperfections and design parameters are given in Chapter 8, together with examples. Techniques to model imperfections and damage in terms of parameters are also given. Sensitivity of limit points is investigated in Chapter 8, considering imperfections and design parameters. The study is performed by computing the complete path for the imperfect configuration and looking at the critical state.

All chapters in this book assume an elastic material; however, Chapter 9 deals with the influence of plasticity (or a discontinuity surface) in metal and composite thin-walled structures. Depending on the system, one may find plastic buckling, elastoplastic buckling, or plasticity in the postbuckling path.

Sensitivity of bifurcation points to imperfection parameters is considered in Chapter 10. An imperfection here plays the role of breaking the bifurcation and transforming the critical point. The study is carried out within the context of bifurcation theory, so that a new parameter is included. In this case it is necessary to do perturbations of the equilibrium and the stability condition together, leading to a regular (nonsingular) problem. It is shown that there are no topological differences between the response to design and to imperfection parameters. The solution of the asymptotic problem in terms of the imperfection parameter is one of singular perturbations.

Chapter 11 deals with sensitivity of bifurcation points to changes in design parameters. This is a relatively new field and includes sensitivity to geometrical imperfections and sensitivity to changes in design parameters that are relevant in optimization, stochastic analysis, etc. In this case the bifurcation is not destroyed by the presence of the new parameter, and the analysis is different from that presented in the previous chapter. Perturbation of the fundamental path as well as the eigenproblem is necessary, and two methods of solution are discussed: the direct approach and the adjoint method. Sensitivity of postcritical behavior with respect to changes in design parameters is also the subject of Chapter 12. This is a new field that is required if optimization studies considering buckling as well as postbuckling constraints are attempted. The derivatives of the load along the postbuckling path are subject to a perturbation analysis, and an asymptotic form is obtained.

The approach followed in Chapter 13 is convenient for linear fundamental paths, in which a translation of coordinates is performed, and the energy is written in terms of sliding coordinates (sliding with respect to the primary path). A new energy is computed under the name of W, and its properties are obtained. Then the postcritical path is written in terms of sliding coordinates and the equations to obtain the asymptotic coefficients are derived. The analysis is entirely similar to that presented in Chapter 7, but the resulting equations are simpler. The chapter ends with some remarks on shell problems and the limitations of the isolated mode analysis developed up to this point.

Mode interaction is explained in Chapter 14, and new modes are shown to appear as a consequence of the coincidence or near coincidence of critical loads associated with different directions of instability (in more mathematical terms, this is the coincidence of eigenvalues associated with different eigenmodes). The energy W is rewritten in terms of the amplitudes of the participating modes in the interaction, and an analysis in terms of modal amplitudes is developed. The chapter specialized the formulation to two and three interacting modes. Examples of interaction are presented, including a thin-walled I-composite column under axial load, in which the modes are set to coincide by design. Other examples are the classical Augusti column and results obtained by other authors for thin shells.

The sensitivity to imperfections of problems with mode interaction is discussed in Chapter 15. A parameter is included to control the closeness of the interacting modes, and the analysis is again carried out using perturbations.

NEW ASPECTS PRESENTED IN THIS BOOK

Some of the basic elements of elastic stability covered here have not changed since the first books on the subject were published in the 1970s.

There are, however, some new features presented in this book. The formulation based on the original form of the energy V is fully developed. Not only limit points, but also bifurcation points, are presented in this way, so that the sense of one general theory is stressed. Other books jump into the W-formulation from the beginning of bifurcation studies. The author believes that it is convenient for the student to learn just one theory instead of specialized theories for each critical state.

A similar approach based on the V-energy is employed to study imperfection sensitivity of bifurcation states.

The area of design sensitivity analysis in buckling states has been incorporated in the text. This is a new area and has great importance in view of its applications in optimization, reliability, stochastic analysis, and inverse problems. Very recent work of the author has been included in this topic.

Finally, sensitivity of postcritical states is a completely new field of current developments, and a chapter discussing this is part of the book. It is expected that research will be done in this area in the next few years, because this information is important for optimization using postbuckling as a design constraint.

The author has also tried to follow the accepted terminology and nomenclature: For example, the notation of derivatives of the energy and derivatives in the perturbation parameters is consistent with that of previous books by other authors. The same notation is employed throughout the book, and often the same examples are continued from one chapter to the next as new information is provided. It is expected that this will help the reader in following the subject.

STRUCTURING A COURSE BASED ON THIS BOOK

Chapters 2 through 9 are the basis of a one-semester (three-credit) graduate course. This course is usually called "Theory of Elastic Stability" in most universities in the United States and Canada, in civil, mechanical, or aeronautical engineering departments.

To follow the analysis presented in this book, the reader needs to be familiar with the concepts of geometric nonlinearity in structures and total potential energy. The usual prerequisites for the course are

- Theory of Elasticity and
- Energy and Variational Methods in Applied Mechanics.

To solve the problems presented and proposed at the end of each chapter it is extremely useful to have some skills using symbolic manipulators, such as Maple, Mathematica, etc.

Chapters 13 to 15 cover the areas of sensitivity in buckling problems and mode interaction. They are relevant for those developing a project in this field and could be the basis of an advanced course on stability of structures.

ACKNOWLEDGMENTS

I am indebted to many people from whom I received the enthusiasm for the field of buckling of structures. I was a doctoral student at University College London at the time there was a tremendous research program to develop a general theory of elastic stability of discrete structural systems. This program lasted with great strength for about 20 years. I was very lucky to be in contact with Professor James G. A. Croll, my Ph.D. thesis advisor, who managed to merge systematic and intuitive approaches to understand buckling, including both theory and experiments. I also enjoyed being in the University College environment, where leading researchers on stability and catastrophe theory frequently lectured. I also owe much to Professor W. Koiter, who taught us refined courses on stability in Rio de Janeiro in 1983.

Most of my research in this field has been done in cooperation with doctoral students, who helped me understand specific aspects as much as I helped them to understand the general picture of elastic stability. Such former students are Dr. L. Almanzar, Dr. E. Banchio, Dr. T. Brewer, Dr. F. Flores, Dr. A. Mirasso, Dr. I. Raftoyiannis, S. Raichman, and Dr. J. Ronda. I am also indebted to my students in the graduate courses, who were the driving force to try to explain complex ideas in simple terms. I had the input coming from teaching courses on the theory of elastic stability at the University of Puerto Rico (Mayaguez), National University of Cordoba (Argentina), West Virginia University (Morgantown), and National University of Rosario (Argentina). For those courses I started using research books based on the energy approach, until I discovered that the number of pages I wrote to make a bridge between a research book and my students had about the same number of pages as those books themselves. The two venues, my research interest and my teaching needs, put me to work on this project about four years ago.

Many other colleagues have helped to shape this book in various ways, and I acknowledge the work done in cooperation with Dr. E. Barbero, Professor R. Batista, Dr. R. Feijoo, Dr. R. Lopez-Anido, Professor C. Prato, Professor V. Souza, and Dr. E. Taroco. Fruitful discussions were possible during visits to other centers, mainly with Professor P. Ballesteros, Dr. C. Ellinas, Professor E. Onate, Professor J. Roorda, and Dr. G. Zintillis. Professor Ismael Pagan-Trinidad, head of the Department of Civil Engineering at the University of Puerto Rico, was a major supporter of this project, and I am now glad that he encouraged me so much to complete this book. The drawings of the book were prepared by L. Cortés. My most sincere thanks to all of them. Of course, they are not to be blamed for the limitations of this book, which are my entire responsibility.

My work on stability was founded by several institutions throughout the years, including the Science and Technology Research Council of Argentina (CONICET), the Science and Technology Research Council of Cordoba (CONICOR), the National Science Foundation-EPSCoR, the University of Puerto Rico, and the National University of Cordoba.

Most importantly, I wish to thank my wife Nora, who gave me the support and love so that I could write this book in a warm environment.

Luis A. Godoy, Mayaguez, January 1999

CHAPTER
ONE

INTRODUCTION

1.1 BASIC CONCEPTS ABOUT STABILITY

The photograph in Figure 1.1 was taken in the Metropolitan Museum of Art, New York, an unlikely place to start a study on stability of structures. It is the partial view of an outer (and the corresponding inner) coffin of Dynasty XXI in Egypt, about 1000 B.C. A civil servant in the Place-of-Truth, *Khonsu* was buried there, the son of *Sennedjem* and *Iynaferty*. His mummy, covered by a mummy's mask and laid in the wooden inner coffin, indicated that he was between 15 and 16 years old at his death. This coffin shows the deceased wearing a tripartite striated wig and holding in his right hand the *djed* symbol for stability, and in his left the *tyet* symbol for protection.

The *djed* was the symbol for stability and durability in the Egypt of the Pharaohs, and was incorporated as a hieroglyphic representing a column of trimmed papyrus stalks tied together. Two such representations are shown in Figure 1.2. Later the Egyptians identified it with the backbone of a god, *Osiris*. It was the sign for stability and was used by kings and queens to ensure a long and peaceful reign.[1]

In the late twentieth century we also regard the concept of stability as something valuable.

An interesting, more technical, definition of stability from *Webster's Dictionary* refers to "the strength to stand or endure without alteration of position or without material change." In broad terms, the stability of a body is defined by this dictionary as "the property of a body that causes it when disturbed from a condition of equilibrium or steady motion to develop forces or moments that restore the original condition."

[1] *Hieroglyphs, the Writing of Ancient Egypt*, N. J. Katan and B. Mintz, Atheneum, New York, 1981.

Figure 1.1 Coffin of Khonsu, who died about 1000 B.C. in Egypt. (Photograph by the author.)

Figure 1.2 The djed symbol of stability, Egypt.

Let us identify some important aspects of this definition:

(a) It is assumed that the body is at an **equilibrium state** (or at a steady motion) before this property is tested.
(b) There is a **disturbance** applied to the body in the original state.
(c) There is a **response** of the body to the disturbance (in terms of moments and forces).
(d) There is an **assessment** of the behavior of the body following the disturbance (i.e., the original condition is restored).

In the previous definition and comments, if you replace the word ''body'' by ''economy,'' ''political situation,'' or ''cultural system'' (assuming cases in which these terms have a specific meaning), it would also be possible to assess stability. This occurs not only at the level of broad definitions, but also the tools employed in the assessment of stability may be general enough to be extended to other fields.

This book is concerned with the stability of **static elastic structures**. In this context, the term **buckling** designates a loss of stability. There are two main properties that make a structure withstand loads: the constitutive material and the geometric shape. Every structure is designed with a specific shape, and it is expected that it should retain this shape during its service life. The consequences of buckling are basically geometric: There may be large displacements in the structure, to such an extent that the shape changes. The load level for which such change in geometry occurs is called the buckling load. The change in the shape can occur in a slow or in a violent way, leading to what is called **snap buckling** in the latter case. Engineers try to avoid the occurrence of buckling, and the computation of buckling loads is an important aspect in the design of many structures.

The process that occurs following buckling of a structure is called **postbuckling**. There are structures with a load reserve in their postbuckling behavior, which can adjust to the change in the shape and take more load after buckling (for example, plates); but other structures do not have postbuckling states for increasing load, and the buckling load is the maximum that the system can attain.

The phenomenon of buckling belongs to a family of physical behaviors at a critical state in which two regimes are clearly separated: One is precritical and the other is postcritical. The boiling of water is one such example. Other examples include the change of flow patterns from laminar to turbulent, a critical velocity separating subsonic and supersonic flows, etc.

Buckling loads are not always relevant. Thick-walled and massive structures do not depend on their shape as a critical factor of resistance, and buckling loads may be several orders of magnitude higher than any practical load for that particular structure. For example, the pyramid of *Cheops* has proven to be stable for several thousand years, and only cosmetic damage has occurred even when it has been subject to heavy environmental actions.

Thin-walled structures, on the other hand, are liable to buckling. The metal tank from the island of Puerto Rico, shown in Figure 1.3, buckled in 1995 under hurricane Georges. The structure cannot be used in this state and should be rebuilt, with considerably loss of time and money.

Figure 1.3 Buckling of a tank in Puerto Rico under wind load.

In the case of Figure 1.3, we can see a frozen image of buckling, because the constitutive material (steel) started the buckling process with elastic properties and underwent plasticity as deformations grew. In other cases, plasticity occurs before buckling; we then speak of **plastic buckling**. Finally, there are cases in which buckling is associated with the actual **collapse** of a structure, in the sense that parts of the structure should be found on the ground. But even if the structure does not collapse, buckling means that the stiffness has been considerably reduced and extensive appraisal or demolition has to be undertaken.

The term buckling is usually employed for the instability of a real structure. Thus, we say that steel tanks buckled in the island of St. Thomas during hurricane Marilyn, in 1995, with pressures associated with wind velocities of 120 mph. We can also say that the tests carried out in the laboratory showed that buckling occurred for an axial load of $170N$.

But one can also compute estimates of buckling using theoretical or numerical models. In this case we identify critical states of the model of the structure and say that lowest critical load should be the load causing buckling. But there are other higher loads for which we also satisfy the conditions of critical state; then we shall refer to them as critical loads and say that the lowest critical load is an estimate of the buckling load.

Buckling is sometimes a desired aspect in a structural component. For example, the keys in a computer keyboard work by elastic buckling of a shallow shell under the pressure of a finger. In cases like this, buckling is not something to be avoided, but rather something essential to the normal work of a component.

Buckling may also be a mechanism of normal behavior in living systems. In the echinoderm *Eucidaris tribuoides*,[2] for example, the ligament that surrounds the articulation joining the primary spines to the test is a collagenous structure. Any movement imparted to the spine stretches half of the ligament, while the other half is compressed. This results in the buckling (by folding or wrinkling) of the compressed surface of the ligament during a very short time.[3]

The subject of this book is elastic buckling and postbuckling of static structural systems. The systems investigated are described in terms of a finite number of generalized coordinates (the displacements of the structure at given points) and a single load parameter (this means that all the loads of the structure are increased by the same factor each time). An alternative way to study this field is by using a continuous approach in which the equations of solid mechanics are employed in the theory before discretization of the problem.

Notice that nonequilibrium states are of no interest to the question of stability in mechanics of solids, and this was the same for other fields of science until recently. In the last 15 years, interest in nonequilibrium states has spread in thermodynamics, and the stability of nonequilibrium states is currently being investigated.[4]

Before we engage ourselves any further in the study of stability, it is important to learn about the major landmarks in this field.

1.2 SOME HISTORICAL REMARKS ABOUT DEVELOPMENTS IN STABILITY

Perhaps the first to investigate structural stability using theoretical tools were the Greek masters between 400 B.C. and 200 B.C. *Aristotle* (384 B.C.–323 B.C.) employed kinematic concepts to study changes in stationary systems; and *Archimedes* (287 B.C.–212 B.C.) used geometric methods to assess the stability of floating bodies.

We do not have any strong evidence that theoretical studies of stability were carried out until the middle of the eighteenth century. Arab scientists were mainly concerned with statics: For example, *al-Khazini* (who lived in the first half of the twelfth century) reviewed most of the work done in statics by earlier scholars,[5] but to the knowledge of the author there is no information on stability. *Galileo Galilei* (1564–

[2]This sea animal feeds using a catch mechanism. In oysters and clams, muscles close the shell during catch; this echinoderm, on the other hand, keeps a given posture for long periods of time, at a very reduced energy expenditure, as compared with muscle activity alone. The ligament has the shape of a hollowed truncated cone that buckles during catch.

[3]N. Perez, J. del Castillo, and D. S. Smith, *Transient buckling upon compression of a self-adjusting collagenous structure*, VI EPSCOR Annual Conference, San Juan, PR, 6–7 May 1994, p. 46.

[4]I. Prigogine, *From Being to Becoming*, W. H. Freeman, San Francisco, 1980.

[5]See M. Rozhanskaya, On a mathematical problem in al-Khazini's Book of the Balance of Wisdom, in *From Deferent to Equant, A Volume of Studies in the Theory of Science in the Ancient and Medieval Near East*, vol. 500, The New York Academy of Sciences, New York, 1987.

1642) did not tackle stability problems either, but his contributions to the rotation of cantilever beams under self-weight[6] influenced most of the research that led to the theory of elasticity.

We tend to associate the field of stability of structures with the name of Euler. *Leonard Euler* (1707–1783) studied stability before the theory of elasticity was well established; he used valuable tools on the elastica developed by James and Daniel Bernoulli. The problem that attracted the attention of *James Bernoulli* (1654–1705) was the deflection of an elastic rod in bending, and he found that the resistance was provided by the extension and contraction of longitudinal fibers. His theoretical developments established a link between moments and the curvature of the deflected rod.[7] *Daniel Bernoulli* (1700–1782) suggested to Euler how to obtain the differential equation of the rod; Euler did that and went further to classify the solutions.[8] He found that a short column under self-weight or an applied load P at the top was fully in compression, but in a long column there was bending. The limit P_{cr} at which the behavior of the column changed was identified as

$$P_{cr} = EI\left(\frac{\pi}{2L}\right)^2,$$

where E is the modulus of elasticity, I the moment of inertia of the cross section, and L the length of the column. *Lagrange* (1736–1813) also studied the problem of the elastica.[9]

The building materials at the time of Euler were timber and stone, with low tensile strength, so that it was necessary to build relatively massive structures. Stability was not a relevant problem for those structures, and Euler's solution did not have any practical application for a century. The construction of steel railway bridges started in 1850, and the question of stability became then a relevant issue for the safety of the structure.

During the second half of the nineteenth century the studies of Euler for a column were extended to other structural forms. In 1859 *Bresse* published the first study of the buckling of an elastic ring under uniform radial pressure.[10] Rectangular frames were considered in 1893 for the first time.[11] Important experiments were also carried out: For example, experimental evidence of the buckling of tubes was reported as early as 1849.[12]

[6]Galileo Galilei, *Discorsi e Dimostrazioni Matematiche*, Leiden, 1638.

[7]James Bernoulli, Veritable hypothese de la resistance des solides, avec la demonstration de la courbure des corps qui font ressort, in *Collected Works of Jean Bernoulli*, vol. 2, Geneva, 1744.

[8]L. Euler, De curvis elasticis, in *Methudus Inveniendi lineas curvas maximi minimive proprietate gaudentes*, Laussane, 1744.

[9]Lagrange, *Miscelanea Taurinensia*, vol. 5, 1773.

[10]Bresse, *Cours de Mecanique Appliquee, Premiere Partie*, Paris, 1859.

[11]F. Engesser, *Die Zusatzkrafte und Nebenspannungen eiserner Fachwerkbrucken*, Berlin, 1893.

[12]W. Fairbairn, *An Account of the Construction of the Britannia and Conway Tubular Bridges*, London, 1849.

The first general theory of elastic stability was published in 1889 by *Bryan*.[13] Bryan found that the theorem of uniqueness of the solution in elasticity is not valid in two cases: The first case is when there are large relative displacements with small strains (for example, in thin plates and slender columns); the second case occurs when there is a displacement field that is similar to a rigid body motion. For this second case, Bryan gives the example of a spherical shell that is compressed within a circular ring of slightly smaller diameter. According to Bryan, whenever there is more than one equilibrium mode, the criterion to decide which one will be followed by the structure is given by the minimum energy.

The theory of bifurcation was developed by *Henri Poincare* (1854–1912),[14] and although it did not have any impact on the stability studies of his contemporaries, it became the basis upon which modern stability theory was built. The other contribution of equal importance was the rigorous mathematical definition of stability presented by *Lyapunov* (1857–1918). He employed a dynamic criterion,[15] which even today is considered a general criterion.

An account of the developments in the theory of stability until the end of the last century may be found in the work of some of the historians of the theory of elasticity.[16,17]

During the first half of the twentieth century, many problems of structures that undergo buckling were identified and investigated. A state of the art of what was by then known as the theory of elastic stability was published by *S. Timoshenko* in 1936.[18] The studies were based on equilibrium considerations and just the buckling load was computed in this way. Interest in postbuckling studies may be found in 1937 in relation to aircraft design.[19] Major confusion in this field was experienced as a consequence of severe discrepancies between laboratory experiments and theoretical predictions of buckling loads for shells.[20]

A change in paradigm occurred with the work of a young Dutchman, *W. T. Koiter* (born in 1914), who employed bifurcation theory in continuum systems.[21] His thesis was ready years before its formal defense in 1945 but had to wait for the war to end to be published. In the new approach, the information given by critical loads was

[13] A. G. Bryan, *Proc. Cambridge Philosophical Society*, vol. 6, 1889.

[14] H. Poincare, Sur l'Equilibre d'Une Masse Fluide Animee d'Un Mouvement de Rotation, *Acta Math.*, vol. 7, 1885.

[15] A. Lyapunov, *Probleme General de la Stabilite du Mouvement*, Kharkov, 1892.

[16] I. Todhunter and K. Pearson, *History of the Theory of Elasticity*, Cambridge University Press, 1886 and 1893.

[17] A. E. H. Love, *A Treatise on the Mathematical Theory of Elasticity*, Fourth edition, 1927, also available from Dover Publications, New York, 1944.

[18] S. Timoshenko, *Theory of Elastic Stability*, London, 1936.

[19] A. Kromm and K. Marguerre, *Luftfahrt-Forsch*, vol. 14, 1937.

[20] T. von Karman and H. S. Tsien, The buckling of thin cylindrical shells under axial compression, *J. Aeronautical Science*, 8, 303–312, 1941.

[21] W. T. Koiter, *On the Stability of Elastic Equilibrium*, Ph.D. thesis, Delft Institute of Technology, Delft, Holland, 1945. There is an English translation by NASA, 1967.

seen as insufficient, and Koiter employed perturbation theory to develop an asymptotic analysis that allowed him to follow the postbuckling path in its early stages.

The work of Koiter was ignored by other researchers in Western countries until some 20 years later, when it was rediscovered at Harvard by *B. Budianski* and *J. Hutchinson*.[22] A most interesting review of postbuckling theory made in 1970 was made by Hutchinson and Koiter.[23] In England, *M. Thompson*, *M. Sewell*, and *A. H. Chilver* contributed to the study of stability of systems defined in terms of generalized coordinates since the early 1960s. The largest research effort at a single center started in 1963 at University College London, in England, and lasted for some 20 years. An account of the contributions in the first 10 years at *UCL* may be found in the monograph [37]. Russian scientists were also interested in the problem of nonlinear stability analysis, notably *V. Bolotin*, who greatly influenced the developments of stability theory.[24]

1.3 BASIC RECENT BIBLIOGRAPHY

A review of the literature in the form of primary sources (journal papers) in the last 30 years will not be attempted here. There are many journals in English in which problems of buckling and postbuckling of structures are reported. Some of the journals include the *Journal of Solids and Structures*, *Thin-Walled Structures*, *Journal of Engineering Mechanics* (ASCE), *Journal of Applied Mechanics* (ASME), *American Institute of Aeronautics and Astronautics* (AIAA) *Journal*, *Journal of Mechanical Sciences*, *Journal of the Mechanics of Structures and Machines*, *Journal of Nonlinear Mechanics*, *Journal of Structural Engineering* (ASCE), *Journal of Composite Materials*, *Journal of the Mechanics and Physics of Solids*, and *Applied Mechanics Reviews*. Stability papers with a computer flavor are published in the *International Journal of Numerical Methods in Engineering, Computers and Structures*, and *Computer Methods in Applied Mechanics and Engineering*. Other important journals on the subject are published in Russian, German, Japanese, and Spanish.

There are many books on this subject, and a list of some 40, mainly written in the last 20 years, is presented at the end of this chapter. A close look at those references shows that there are different perspectives from which this area of mechanics can be tackled. In broad terms, it is possible to see differences in approaches and differences in contents in the literature.

Let us first consider the approaches.[25] In a **theoretical approach**, the main aim is to produce a consistent mathematical formulation to model the nonlinear phenomena

[22] B. Budianski and J. Hutchinson, A survey of some buckling problems, *AIAA J.*, 4, 1505–1510, 1966.

[23] J. W. Hutchinson and W. T. Koiter, Post-buckling theory, *Appl. Mech. Rev.*, 23, 1970.

[24] V. V. Bolotin, Nonlinear theory of elasticity and stability in the case of large deflections, *Raschety na Prochnost*, No. 3, Mashgiz, Moscow, 1959.

[25] Consideration that a certain work can be classified in a given approach is sometimes a matter of argument. Often a book can contain more than one approach, and this is also true of the present book.

encountered in buckling and postbuckling. It should also provide a classification of the types of behavior that can occur. It works with theoretical models, finds equilibrium paths, and investigates stability. Critical states are classified in detail, and usually perturbation techniques are used as a theoretical and numerical tool (see, for example, [2, 8, 9, 17, 19, 25, 30, 37, 46, 16]).

Another approach is to grasp knowledge directly from a physical model, and this may be called an **experimental approach**. It is necessary to build an experimental model that could be in small, medium, or large scale, or have no scale relation with any real structure. The model is instrumented, tested, and conclusions are obtained from the response. This approach is present in, for example, [20, 45].

In a **computational approach** [12, 44] the emphasis is on the development of numerical techniques of analysis that allow computation of nonlinear paths. There are many topics of discussion here, such as the accurate evaluation of paths, how to change from one equilibrium path into another, problems of convergence, and modeling of nonlinear geometric and material properties.

The **behavioral approach** attempts to transfer experience through the study of certain problems. It develops a certain intuition about the expected behavior of a structural component. With this intuitive knowledge, the engineer should be in a better situation to understand that a certain design may be liable to buckling [12, 17].

In a **problem-solving approach**, the aim is to obtain solutions (mainly analytical) for problems that are of interest to the designer. Particular problems are considered and solved [41, 13, 10].

Finally, there is also a **code-of-practice approach**, which teaches how to design according to a given code of practice [24, 22, 15]. It helps the engineer in the design, even though he or she might not understand well the physics of the problem. It should contain the maximum number of rules, anticipate a large number of cases, and contain simple formulas and graphics with appropriate safety coefficients. It should not be ambiguous, so that there is a minimum of confusion in the practical application.

The kind of information in which the references are interested varies not only in the approach taken but also in the contents. There is a **buckling content**, in which the interest of the author concentrates on the computation of the first critical state [41, 29]; and there is a **postbuckling content**, which follows the Koiter tradition [17, 37, 8, 25, 2, 19]. Additionally, we find contents that go beyond elastic buckling: for example, **dynamic buckling content** [5, 6, 34] and **irreversible stability problems** [3, 7].

1.4 SCOPE OF THIS BOOK

The present study is restricted to structural systems with the following characteristic features:

(a) **Discrete** systems, in which the energy is written in terms of multiple generalized coordinates. The continuum counterpart has been developed by several authors, starting with Koiter himself, but it requires a more elaborate analysis. The discrete system, on the other hand, allows for an elementary treatment, so that it can be made accessible as an introduction to the field. "It facilitates a quicker and more explicit development than in the continuous theory, and it can also make the ideas seem more tangible because of the comparative ease with which simple complete solutions can be constructed."[26]

(b) **Elastic** constitutive relations, leading to reversible processes. Elastic instability occurs in many practical engineering structures, including those made of composite materials (such as fiber-reinforced plastics). Furthermore, there are many structures that behave elastically during the initial stages of buckling and only develop plasticity at advanced postbuckling states. In practical structures the validity of the constitutive relations at the equilibrium state considered should always be checked to make sure that the material remains elastic. The occurrence of plasticity or other irreversible effects requires the use of the principle of virtual work (instead of the total potential energy) and substantial modifications in the formulation.

(c) **Static** conditions. This means that the load is applied in such a way that inertia effects are not relevant.

(d) **Conservative** systems, such that an energy functional exists. In the developments presented in this book, use is made of the total potential energy criterion. This limits the analysis to structures for which such an energy functional exists. There are, however, structures in which the actual work done depends on the specific path taken to reach a given equilibrium state. Those are said to be nonconservative systems and should be investigated using a dynamic approach.

(e) **Deterministic** analysis, so that there are no uncertainties introduced at the level of the formulation.

(f) **One load parameter** is employed to increase all the loads acting on the structure.

(g) **Holonomic** systems, such that there are no constraints on the values that the displacements can take, and the boundary conditions do not change during the loading process.

In the present book we try to follow a **theoretical** approach, with emphasis in the classification of critical states and the asymptotic representation of the postcritical paths. Several chapters are **behavioral** and are dedicated to the study of cases from which some generalizations can be made.

Great emphasis is placed throughout the text on **postbuckling** analysis and behavior and on the **sensitivity** of buckling states to imperfections and to changes in design parameters.

A comment should be made regarding examples. We shall not teach by means of examples, nor is it intended to construct a catalogue of solutions for typical cases. In our presentation, the theoretical aspects are first introduced and then examples follow to illustrate the theory.

[26]M. J. Sewell, On the connection between stability and the shape of the equilibirum surface, *J. Mech. Physics Solids*, 14, 203, 1966.

1.5 PROBLEMS

Review questions. (*a*) Describe in your own words what is meant by buckling. (*b*) Why is postbuckling behavior important? (*c*) Give examples of structures for which you think that buckling is important. (*d*) What is the difference between buckling and collapse? (*e*) Draw the problem that was originally studied by Euler. (*f*) When did bifurcation theory develop? (*g*) Compare the contributions of Timoshenko and Koiter. (*h*) List different approaches that can be used to study the stability of structures. (*i*) List the properties of structural systems to be studied in this book. (*j*) Find sentences in newspapers or magazines in which the words equilibrium, stability, unstable are used. Are they used in a way similar to what is described in section 1.1? (*k*) Find what journals of those listed in section 1.3 are available in your library. (*l*) Find what books (of those listed in the references) are available in your library.

1.6 BIBLIOGRAPHY

[1] Allen, H. G., and Bulson, P. S., *Background to Buckling*, McGraw-Hill, London, 1980.

[2] Batista, R. C., *Estabilidade Elastica de Sistemas Mecanicos Estruturais* (in Portuguese), LNCC/CNPq, Rio de Janeiro, 1982.

[3] Bazant, Z. P., and Cedolin, L., *Stability of Structures: Elastic, Inelastic Fracture and Damage Theories*, Oxford University Press, New York, 1991.

[4] Bleich, F., *Buckling Strength of Metal Structures*, McGraw-Hill, New York, 1952.

[5] Bolotin, V. V., *The Dynamic Stability of Elastic Systems*, Holden-Day, San Francisco, 1964.

[6] Bolotin, V. V., *Non Conservative Problems in the Theory of Elastic Stability*, Pergamon Press, London, 1965.

[7] Bolotin, V. V., *Stability Problems in Fracture Mechanics*, Wiley, New York, 1996.

[8] Britvec, S. J., *The Stability of Elastic Systems*, Pergamon Unified Eng. Series, vol. 12, Pergamon, Elmsford, NY, 1973.

[9] Britvec, S. J., *Stability and Optimization of Flexible Space Structures*, Birkhauser, Cambridge, MA, 1995.

[10] Brush, D. O., and Almroth, B. O., *Buckling of Bars, Plates and Shells*, McGraw-Hill-Kogakusha, Tokyo, 1975.

[11] Bulson, P. S., *The Stability of Flat Plates*, Chatto and Windus, London, 1970.

[12] Bushnell, D., *Computerized Buckling Analysis of Shells*, Martinus Nijhoff, Dordrecht, 1985.

[13] Chajes, A., *Principles of Structural Stability Theory*, Prentice-Hall, Englewood Cliffs, NJ, 1974.

[14] Chen, W. F., and Lui, E. M., *Structural Stability: Theory and Implementation*, Elsevier, London, 1987.

[15] Chen, W. F., and Lui, E. M., *Stability Design of Steel Frames*, CRC Press, Boca Raton, FL, 1991.

[16] Como, M., and Grimaldi, A., *Theory of Stability of Continuous Elastic Structures*, CRC Press, Boca Raton, FL, 1995.

[17] Croll, J. G. A., and Walker, A. C., *Elements of Structural Stability*, Macmillan, London, 1972.

[18] Dym, C. L., *Stability Theory and Its Applications to Structural Mechanics*, Noordhoff Int., Leyden, 1974.

[19] El Naschie, M. S., *Stress, Stability and Chaos in Structural Engineering: An Energy Approach*, McGraw-Hill, London, 1990.

[20] Esslinger, M., and Geier, B., *Postbuckling Behavior of Structures*, Springer-Verlag, Berlin, 1975.

[21] Farshad, M., *Stability of Structures*, Series: Developments in Civil Engineering, vol. 43, Elsevier, Amsterdam, 1994.

[22] Fukumoto, Y. (Ed.), *Structural Stability Design*, Pergamon, Oxford, 1997.

[23] Gajewski, A., and Zyczkowski, M., *Optimal Structural Design under Stability Constraints*, Kluwer, Dordrecht, 1988.

[24] Galambos, T. V., *Guide to Stability Design Criteria for Metal Structures*, Fifth ed., John Wiley, New York, 1998.

[25] Huseyin, K., *Nonlinear Theory of Elastic Stability*, Noordhoff Int., Leyden, 1975.

[26] Huseyin, K., *Vibrations and Stability of Multiple Parameter Systems*, Ivoardhoff, Alphen, 1978.

[27] Huseyin, K., *Multiple Parameter Stability Theory and Its Applications*, Clarendon Press, Oxford, 1986.

[28] Iyengar, N. G. R., *Structural Stability of Columns and Plates*, Ellis Horwood, Chichester, 1988.

[29] Kumar, A., *Stability Theory of Structures*, Tata McGraw-Hill, New Delhi, 1985.

[30] Leipholtz, H., *Stability Theory*, Academic Press, New York, 1970.

[31] Pignataro, M., Rizzi N., and Luongo, A., *Stability, Bifurcation and Postcritical Behavior of Elastic Structures*, Series: Developments in Civil Engineering, vol. 39, Elsevier, Amsterdam, 1991.

[32] Roorda, J., *Buckling of Structures*, University of Waterloo Press, Waterloo, 1980.

[33] Simitses, G., *An Introduction to the Elastic Stability of Structures*, Krieger, Malabar, FL, 1986.

[34] Simitses, G., *Dynamic Stability of Suddenly Loaded Structures*, Springer-Verlag, New York, 1990.

[35] Supple, W. J. (Ed.), *Structural Instability*, IPC Science and Technology Press, Guildford, UK, 1973.

[36] Szabo, J., Gaspar, Z., and Tarnaic, T. (Ed.), *Post Buckling of Elastic Structures*, Series: Developments in Civil Engineering, vol. 17, Elsevier, Amsterdam, 1986.

[37] Thompson, J. M. T., and Hunt, G. W., *A General Theory of Elastic Stability*, John Wiley and Sons, London, 1973.

[38] Thompson, J. M. T., *Instabilities and Catastrophes in Science and Engineering*, John Wiley and Sons, London, 1982.

[39] Thompson, J. M. T., and Hunt, G. W. (Ed.), *Collapse: The Buckling of Structures in Theory and Practice*, Cambridge University Press, 1983.

[40] Thompson, J. M. T., and Hunt, G. W., *Elastic Instability Phenomena*, J. Wiley and Sons, New York, 1984.

[41] Timoshenko, S. P., and Gere, J. M., *Theory of Elastic Stability*, McGraw-Hill, New York, 1961.

[42] Trahair, N. S., *Flexural-Torsional Buckling of Structures*, CRC Press, Boca Raton, FL, 1993.

[43] Vardoulakis, I., and Sulem, J., *Bifurcation Analysis in Geomechanics*, Blackie, London, 1995.

[44] Waszczyszyn, Z., Cichon, C., and Radwanska, M., *Stability of Structures by Finite Element Methods*, Elsevier, Amsterdam, 1994.

[45] Yamaki, N., *Elastic Stability of Circular Cylindrical Shells*, North-Holland, Amsterdam, 1984.

[46] Ziegler, H., *Principles of Structural Stability*, Blaisdell Publishing Company, Waltham, MA, 1968.

CHAPTER
TWO

EQUILIBRIUM STATES IN NONLINEAR ELASTICITY

2.1 INTRODUCTION

In the last 200 years there have been impressive advances regarding the formulation of the nonlinear theory of elasticity. This progress was dominated by the contributions of French researchers such as Cauchy, Lagrange, and Navier, who developed the mathematical theory of elasticity between 1800–1850. The history of the theory of elasticity has been reviewed by Todhunter and Pearson at the end of the last century [23] and by other researchers in this century [2] and shows the painful process of developing concepts such as strains or stresses.

The stability approach that we follow in this book uses the energy as a key ingredient of the theory, and for this reason we introduce in this chapter the total potential energy formulation.

The total potential energy is presented in section 2.2 using a continuous formulation, with examples of a circular arch under a central point load, and a circular plate under in-plane compression. The two problems are continued in the following sections to show examples of the formulation as it progresses. The conditions of stationary energy and equilibrium are reviewed in section 2.3. Section 2.4 deals with the total potential energy formulation of discrete (or discretized) systems, defined in terms of multiple generalized coordinates or degrees of freedom. A reader who is familiar with energy formulations could skip this chapter but should be aware of the notation employed to write the energy in terms of generalized coordinates in section 2.4.

Problems in three-dimensional elasticity are defined in terms of displacements, strains, and stresses. A presentation of the nature of each one of these variables and their relations is outside the scope of this book, but a brief summary is presented in

Appendix A. For a more detailed account the reader is referred to [19, 9, 7, 16, 17, 3, 5, 8, 10, 15] and other books.

2.2 TOTAL POTENTIAL ENERGY

Definitions. The total potential energy is one of several functionals that can be used in theoretical elasticity. Other functionals derived from energy considerations are the complementary energy, functionals of two fields (notably the Hellinger–Reissner functional) and functionals of three fields (i.e., the Hu–Washizu functional). An account of such functionals may be found in [24].

In the total potential energy, only the displacement field is taken as unknown and susceptible to variations. Strains and stresses satisfy kinematic and elastic (either linear or nonlinear) constitutive relations. The usefulness of a functional in mechanics arises from the use of the calculus of variations [21]: Thus the first variation of a functional should lead to some fundamental equation of the problem, such as equilibrium (as in the total potential energy) or compatibility (as in the complementary energy formulation). In the context of elastic stability theory there are good reasons to choose the total potential energy (denoted here as V), mainly because we are concerned with stability of equilibrium states, so that a functional related to equilibrium has a starting advantage. Furthermore, higher-order variations of V are related to important properties of critical stability states and provide much light to the problem. This was initially discussed by Koiter in his doctoral thesis [12].

Let us define the **total potential energy** V of a stressed body under load as

$$V \equiv \int_{\mathcal{V}} \omega \, d\mathcal{V} + \int_{\mathcal{V}} \rho \, d\mathcal{V} + \int_{\mathcal{S}_f} \psi \, d\mathcal{S}, \tag{2.1}$$

where ω is the **strain energy density** per unit volume of the body, ρ is the **potential of the volume forces,** ψ is the **potential of the forces acting on the boundary**, $\mathcal{V}$ is the volume, and $\mathcal{S}$ is the surface. To obtain the energy density ω it is necessary to consider the internal energy produced by the stress tensor σ_{ij} acting on the strain tensor ε_{ij}. For a linearly elastic material this takes the form

$$\omega = \frac{1}{2}\sigma_{ij}\varepsilon_{ij}. \tag{2.2}$$

A summation is implied in (2.2) for $i = 1, 2, 3$ and $j = 1, 2, 3$. The two tensors σ_{ij} and ε_{ij} are generalized stress and strain measures, which should be consistent with each other. A brief discussion on stresses and strains is given in Appendix A, but for this topic the reader should consider texts that specialize on the subject, such as [16] or others listed at the end of this chapter.

Nonlinear elastic materials require the definition of the differential of ω as

$$d\omega = \sigma_{ij} d\varepsilon_{ij}, \tag{2.3}$$

which should be integrated between $\varepsilon_{ij} = 0$ and the strain level ε_{ij} reached at a certain point of the body for a given load. We assume that the differential should be computed as a function of strains

$$d\omega = \frac{\partial \omega}{\partial \varepsilon_{ij}} d\varepsilon_{ij}. \tag{2.4}$$

This means that the energy density ω is equivalent to a constitutive equation, in the sense that its partial derivatives with respect to the components of the strain tensor provide the components of the associated stress tensor.

Furthermore, it is assumed that the external loads F_i (force per unit volume) and f_i (forces per unit surface) may be obtained from a load potential. Simple forms of such potentials (valid for most cases considered in this book) are

$$\begin{aligned} \rho &= -F_i u_i \qquad \text{in } \mathcal{V}, \\ \psi &= -f_i u_i \qquad \text{in } \mathcal{S}_f, \end{aligned} \tag{2.5}$$

where u_i are the displacements and $\mathcal{S}_f$ is the part of the boundary where forces are specified.

The forces F_i and f_i of the system are increased from an initial value (zero) to a given value by means of control parameters. There may be several forces acting on the structure, but throughout this book we assume that all forces are increased at the same rate; i.e., there is only one parameter Λ that controls the increments of all load components. Problems with multiple control parameters have been treated by Huseyin [11] and other researchers.

In problems in which there is a single load parameter Λ it may be convenient to deal with potentials in the form $\psi(u_i, \Lambda)$ and $\rho(u_i, \Lambda)$, in which Λ and not F_i and f_i are specified. Thus, one writes

$$\begin{aligned} \rho &= -\Lambda\ F_i u_i \qquad \text{in } \mathcal{V}, \\ \psi &= -\Lambda\ f_i u_i \qquad \text{in } \mathcal{S}_f. \end{aligned} \tag{2.6}$$

A distinction is usually made depending on the form of the load potential. If the load system is such that the load potentials are linear in Λ (as in (2.6)), it is said that we are in the presence of a **specialized system**. In a specialized system the displacements do not have to be linear in the load potential. Systems in which ψ or ρ are nonlinear functions of Λ are known as **general systems**. This distinction is taken from [22].

Finally, in the specialized system, one has

$$V[\varepsilon_{ij}, u_i, \Lambda] = \int_{\mathcal{V}} \omega[\varepsilon_{ij}]\, d\mathcal{V} + \Lambda \int_{\mathcal{V}} \rho[u_i]\, d\mathcal{V} + \Lambda \int_{\mathcal{S}_f} \psi[u_i]\, d\mathcal{S}, \tag{2.7}$$

where ε_{ij} and u_i are linked by kinematic relations (see Appendix A).

Slender and thin-walled elements, which are prone to buckling, are not modeled using three-dimensional elasticity but employ what are known as technical theories. For beam columns, the theory of Bernoulli has been used for over 300 years; Kirchhoff theory is employed for thin-walled plates; and thin-walled shell structures are often studied by means of the Love–Kirchhoff hypothesis. Technical theories introduce assumptions regarding the deformations of the plane sections that are initially normal to the midsurface of the element and neglect changes in the thickness. An account of nonlinear theories for beams, plates, and shells may be found in the classical book by Donnell [4], and the contradictions introduced by the hypothesis in technical theories of shells are discussed by Novozhilov [20].

The strain energy density ω in a thin-walled element including membrane and bending action takes the form

$$\omega = \frac{1}{2}\left(M_{ij}\chi_{ij} + N_{ij}\varepsilon_{ij}\right) \qquad \text{for} \qquad i, j = 1, 3, \tag{2.8}$$

which is employed instead of (2.2). Here, M_{ij} are the components of moments; χ_{ij} are the changes in curvature of the midsurface; N_{ij} are the in-plane stress resultants; and ε_{ij} are the membrane strains of the midsurface. In thin-walled members, shear stress components are assumed to produce negligible strain energy.

For the specialized system one has

$$V[\varepsilon_{ij}, \chi_{ij}, u_i, \Lambda] = \int\int \omega[\varepsilon_{ij}, \chi_{ij}]\, dx_1 dx_2 + \Lambda \int\int \rho[u_i]\, dx_1 dx_2 + \Lambda \int_{S_f} \psi[u_i]\, dS, \tag{2.9}$$

where $\mathcal{S}$ is the boundary. The strains ε_{ij} and χ_{ij} are linked to u_i by kinematic relations.

Example 2.1 (Energy of a circular arch). *A circular arch with hinge supports at the ends, under a central point load, is shown in Figure* 2.1. *A convenient coordinate system for this circular arch is the angular coordinate θ, measured from the central section of the arch. The angle at the supports is $\theta = \pm\theta_0$.*

We consider longitudinal strains ε_θ and changes in curvature χ_θ, and the associated stress resultants are N_θ and M_θ. Now

$$\omega = \frac{1}{2}(N_\theta \varepsilon_\theta + M_\theta \chi_\theta),$$

$$\psi = -P\, w|_{\theta=0}.$$

The total potential energy results in

$$V = \frac{1}{2}\int_{\theta=-\theta_0}^{\theta_0} (N_\theta \varepsilon_\theta + M_\theta \chi_\theta)\, R\, d\theta - P\, w|_{\theta=0}.$$

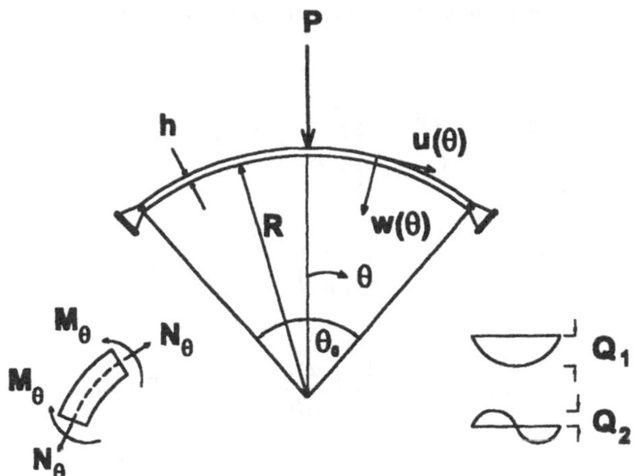

Figure 2.1 Circular arch under point load.

The geometric parameters are indicated in Figure 2.1. The first term $N_\theta \varepsilon_\theta$ leads to the membrane contribution, while the second term $M_\theta \chi_\theta$ represents the bending contribution to V.

The kinematic relations assumed are

$$\varepsilon_\theta = \varepsilon'_\theta + \varepsilon''_\theta, \qquad \chi_\theta = \frac{1}{R^2}\left(\frac{d^2 w}{d\theta^2}\right),$$

where the linear and nonlinear components of ε_θ are

$$\varepsilon'_\theta = \frac{1}{R}\frac{du}{d\theta} - \frac{w}{R}, \qquad \varepsilon''_\theta = \frac{1}{2R^2}\left(\frac{dw}{d\theta}\right)^2.$$

The in-plane component u is only present in the linear part of the membrane strain. The constitutive equations for a linearly elastic material result in

$$N_\theta = EA\varepsilon_\theta, \qquad M_\theta = EI\chi_\theta,$$

where E is the modulus of elasticity; A is the cross-sectional area; and I is the moment of inertia of the arch. Prismatic properties are assumed for this structure.

The energy may be written in terms of deformations as

$$V[\varepsilon_\theta, \chi_\theta, P] = \frac{R}{2}\int_{\theta=-\theta_0}^{\theta_0} \left\{EA\left[(\varepsilon'_\theta)^2 + 2\varepsilon'_\theta\varepsilon''_\theta + (\varepsilon''_\theta)^2\right] + EI\,(\chi_\theta)^2\right\} d\theta - Pw|_{\theta=0}.$$

Finally, V can be computed from the displacement components (u, v) and takes the form

$$V[u, w, P] = \frac{1}{2}EAR\int_{\theta=-\theta_0}^{\theta_0}\left\{\left[\frac{1}{R^2}\left(\frac{du}{d\theta}\right)^2 - \frac{2}{R^2}\frac{du}{d\theta}w + \frac{1}{R^2}w^2\right]\right.$$

$$\left.+\frac{1}{R^3}\left[\frac{du}{d\theta}\left(\frac{dw}{d\theta}\right)^2 - w\left(\frac{dw}{d\theta}\right)^2\right] + \frac{1}{R^4}\left[\frac{1}{4}\left(\frac{dw}{d\theta}\right)^4\right]\right\}d\theta$$

$$+\frac{1}{2}EIR\int_{\theta=-\theta_0}^{\theta_0}\frac{1}{R^4}\left(\frac{d^2w}{d\theta^2}\right)^2 d\theta - Pw\,|_{\theta=0}. \tag{2.10}$$

The energy thus derived has quartic terms in w and quadratic terms in u.

Example 2.2 (Energy of a circular arch with constant strain). *Let us rewrite the energy of the arch of* (2.10), *assuming that the central line has a constant strain* ε_θ. *This uniform strain is denoted as* $\varepsilon_\theta^{average}$ *and is computed as an average value of the actual variation of the strain along the arch, in the form*

$$\varepsilon_\theta^{average} \approx \frac{1}{2\theta_0}\int_{\theta=-\theta_0}^{\theta_0}\varepsilon_\theta(\theta)\,d\theta.$$

Next it is assumed that the in-plane displacements u are small in comparison with out-of-plane displacements w. This leads to

$$\varepsilon_\theta' \approx -\frac{w}{R},$$

and the average value of the hoop strain is given by

$$\varepsilon_\theta^{average} = \frac{1}{2\theta_0}\int_{\theta=-\theta_0}^{\theta_0}\left[-\frac{w}{R} + \frac{1}{2R^2}\left(\frac{dw}{d\theta}\right)^2\right]d\theta.$$

Then the energy V takes the form

$$V[w, P] = \frac{1}{2}(2\theta_0 R)\ AE\ (\varepsilon_\theta^{average})^2 + \frac{1}{2}EI\int_{-\theta_0}^{\theta_0}\left(\frac{d^2w}{d\theta^2}\right)^2 Rd\theta - P\ w|_{\theta=0}. \tag{2.11}$$

This new expansion for V differs from (2.10) *in several respects. First, u is no longer present in* (2.11). *Second, there are terms to the fourth power before integration in* (2.10), *while only square terms appear in* (2.11) *before integration.*

Example 2.3 (The energy of a circular plate). *A circular plate is shown schematically in Figure 2.2: It is assumed to be simply supported around the circumferential boundary, with radius R and uniform thickness h. The plate is loaded by a radial load p acting as a pressure perpendicular to the thickness. Convenient coordinates in this case are a polar system defined in terms of the radius r and an angular coordinate θ. The displacement components in the out-of-plane (transverse) and in-plane (radial) directions with respect to the middle surface are denoted by w and u, respectively. Since the problem is axisymmetric, it is not necessary to consider a third displacement component.*

The formulation is carried out in terms of radial and circumferential strains, ε_r and ε_θ; radial and circumferential curvatures, χ_r and χ_θ; and their associated stress and moment resultants, N_r, N_θ, M_r, and M_θ. The potential energy V may be written as

$$V = \frac{1}{2}\int_{r=0}^{R} (N_\theta \varepsilon_\theta + N_r \varepsilon_r + M_\theta \chi_\theta + M_r \chi_r) 2\pi r \, dr - 2\pi R \Lambda u|_{r=R},$$

where $\Lambda = ph$ is the resultant of the load, applied at the midsurface and on the edges.

The kinematic equations are

$$\varepsilon_r = \varepsilon_r' + \varepsilon_r'' = \frac{\partial u}{\partial r} + \frac{1}{2}\left(\frac{\partial w}{\partial r}\right)^2, \qquad \varepsilon_\theta = \frac{u}{r},$$

$$\chi_r = \frac{\partial^2 w}{\partial r^2}, \qquad \chi_\theta = \frac{1}{r}\frac{\partial w}{\partial r}. \tag{2.12}$$

Notice that because of the axisymmetric nature of the problem, all derivatives with respect to the circumferential direction are zero; i.e., $\partial(\#)/\partial\theta = 0$.

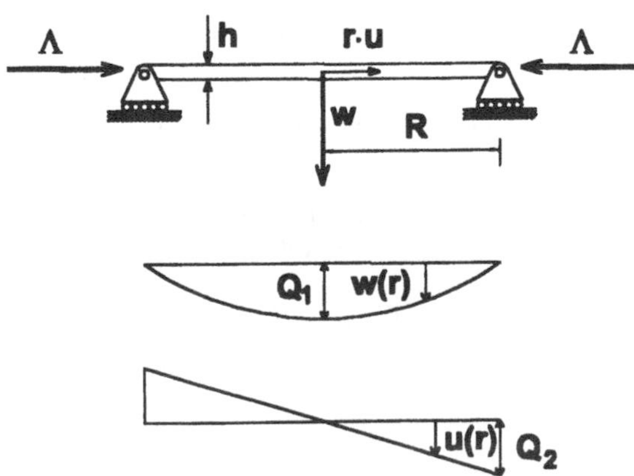

Figure 2.2 Circular plate under in-plane load.

The constitutive equations for a linear, homogeneous, and elastic material take the form

$$N_r = C(\varepsilon_r + \nu\varepsilon_\theta), \qquad N_\theta = C(\varepsilon_\theta + \nu\varepsilon_r),$$
$$M_r = D(\chi_r + \nu\chi_\theta), \qquad M_\theta = D(\chi_\theta + \nu\chi_r), \tag{2.13}$$

where the membrane and bending stiffness of the plate are given in terms of the modulus of elasticity E and Poisson's ratio ν as

$$C = \frac{Eh}{1-\nu^2}, \qquad D = \frac{Eh^3}{12(1-\nu^2)}. \tag{2.14}$$

The only sources of nonlinearity in this problem are the kinematic relations of ε_r, with a linear part in terms of u, given by

$$\varepsilon_r' = \frac{\partial u}{\partial r},$$

and a nonlinear (quadratic) part in terms of w, given by

$$\varepsilon_r'' = \frac{1}{2}\left(\frac{\partial w}{\partial r}\right)^2.$$

This has the consequence that both N_r and N_θ are nonlinear functions of the displacement w.

Next we write the energy in the form

$$V = \pi C \int_{r=0}^{R} \left\{\left[(\varepsilon_r')^2 + (\varepsilon_\theta)^2 + 2\nu\varepsilon_\theta\varepsilon_r'\right] + 2\left[\varepsilon_r'\varepsilon_r'' + 2\varepsilon_r''\varepsilon_\theta\right] + (\varepsilon_r'')^2\right\} r\,dr$$
$$+\pi D \int_{r=0}^{R} \left[(\chi_r)^2 + (\chi_\theta)^2 + 2\nu\chi_r\chi_\theta\right] r\,dr - 2\pi R\Lambda u|_{r=R}.$$

Finally, substituting the kinematic equations into V results in

$$V = \pi C \int_{r=0}^{R} \left\{\left[\left(\frac{\partial u}{\partial r}\right)^2 + \left(\frac{u}{r}\right)^2 + 2\nu\frac{u}{r}\left(\frac{\partial u}{\partial r}\right)\right] + 2\left[\frac{\partial u}{\partial r}\left(\frac{\partial w}{\partial r}\right)^2 + \frac{u}{r}\nu\left(\frac{\partial w}{\partial r}\right)^2\right]\right.$$
$$\left. + \frac{1}{4}\left(\frac{\partial w}{\partial r}\right)^4\right\} r\,dr + \pi D \int_{r=0}^{R} \left[\left(\frac{\partial^2 w}{\partial r^2}\right)^2 + \left(\frac{1}{r}\frac{\partial w}{\partial r}\right)^2 + 2\nu\frac{1}{r}\frac{\partial w}{\partial r}\frac{\partial^2 w}{\partial r^2}\right] r\,dr$$
$$-2\pi R\Lambda u|_{r=R}. \tag{2.15}$$

Equation (2.15) *has quartic terms in w and quadratic terms in u.*

2.3 STATIONARY TOTAL POTENTIAL ENERGY AND EQUILIBRIUM

Using the previous definition of V in terms of ε_{ij} and u_i, we can now perform the first variation of V with respect to the displacement field [21, 24] in the form

$$\delta V = \int_V \frac{\partial \omega}{\partial \varepsilon_{ij}} \delta \varepsilon_{ij}\, d\mathcal{V} + \Lambda \int_V \frac{\partial \rho}{\partial u_i} \delta u_i d\mathcal{V} + \Lambda \int_S \frac{\partial \psi}{\partial u_i} \delta u_i\, d\mathcal{S}, \qquad (2.16)$$

where $\delta\varepsilon_{ij}$ and δu_i are the variations of strains and displacements, which satisfy the kinematic relations

$$\delta \varepsilon_{ij} = \frac{1}{2}\left(\frac{\partial \delta u_i}{\partial x_j} + \frac{\partial \delta u_j}{\partial x_i} + \frac{\partial \delta u_m}{\partial x_i}\frac{\partial \delta u_m}{\partial x_j} \right). \qquad (2.17)$$

From section 2.3, the first variation δV now results in

$$\delta V = \int \sigma_{ij} \delta \varepsilon_{ij}\, d\mathcal{V} - \Lambda \int F_i \delta u_i\, d\mathcal{V} - \Lambda \int_{S_f} f_i \delta u_i\, d\mathcal{S}. \qquad (2.18)$$

This first variation is next interpreted in the context of virtual work [14, 21]. The principle of virtual work states that

A necessary and sufficient condition for equilibrium of a set of forces and stresses is that, for any virtual displacement field, the internal virtual work should be equal to the external virtual work.

Let us denote the virtual strains by $\delta\varepsilon_{ij}$, associated with the virtual displacements δu_i by means of kinematic relations. Furthermore, we impose the condition that δu_i should satisfy the kinematic boundary conditions $\delta u_i = 0$ at $\mathcal{S}_u$. Let σ_{ij} be real stresses, and F_i and f_i the real forces of the problem. Thus

$$\text{Internal virtual work } = \int \sigma_{ij} \delta \varepsilon_{ij}\, d\mathcal{V},$$

$$\text{External virtual work } = \Lambda \int F_i \delta u_i\, d\mathcal{V} + \Lambda \int_{S_f} f_i \delta u_i\, d\mathcal{S}.$$

But equality of internal and external virtual work leads to an equation similar to the first variation of V, provided variations in ε_{ij} and u_i satisfy the conditions of virtual strains and displacements (that is, variations of strains should satisfy the kinematic equations in terms of variations in displacements, and displacements should be zero at the boundaries $\mathcal{S}_u$). This means that when stresses are in equilibrium with the loads,

$$\boxed{\delta V = 0 \quad \Rightarrow \quad \text{equilibrium}} \qquad (2.19)$$

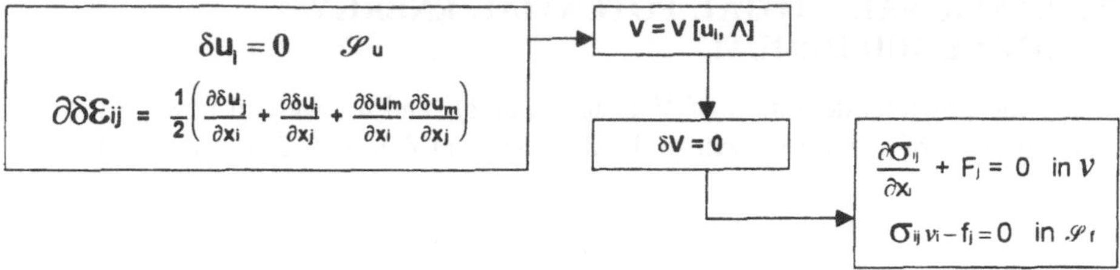

Figure 2.3 Total potential energy.

This is also the condition required so that V has a stationary value with respect to variations in the field δu_i. We may now state a new condition for equilibrium in the following form:

A necessary and sufficient condition for equilibrium of a set of stresses and forces is that the total potential energy of the system should be stationary with respect to kinematically admissible displacements.

Notice that equilibrium does not require that V should be a minimum; in fact, the information involved in $\delta V = 0$ is only related to a stationary condition. With reference to Figure 2.3, the energy V can be thought of as a "black box," constructed using kinematically admissible variables, and the condition of first variation of V equal to zero produces equilibrium.

The boundary conditions that should be satisfied before entering the functional are called **essential boundary conditions**, and in the case of V they are geometric constraints. Those boundary conditions that are satisfied as a consequence of the first variation of the functional are the **natural boundary conditions**, and they are force conditions in relation to V.

2.4 ENERGY FORMULATION IN TERMS OF GENERALIZED COORDINATES

2.4.1 Discrete Form of the Energy

The solution of problems in practical engineering situations leads to the use of numerical methods. Nowadays, the most commonly employed numerical methods are the finite element method and the boundary element method. Some years ago, the finite difference method would have been the first choice of an analyst, and perhaps new methods will be developed in the next decade. What numerical methods have in common is that they reduce the number of unknowns of a problem to a finite set, usually called the generalized coordinates, or the degrees of freedom of a problem.

If we denote the generalized coordinates by Q_i, then the displacements result in $u_j = u_j(Q_i)$. The total potential energy can be written in the form

$$V = V[Q_i, \Lambda], \tag{2.20}$$

where $i = 1, 2, \ldots, N$, and N is the number of generalized coordinates employed in the discretization.

The scalar V is a nonlinear functional in terms of the Q_i. We may group terms that are linear in Q_i, those that are quadratic in the generalized coordinates $(Q_i Q_j)$, cubic terms $(Q_i Q_j Q_k)$, etc. In the rest of the book we shall employ the notation

$$\boxed{V[Q_i, \Lambda] = A_i Q_i + \tfrac{1}{2!} A_{ij} Q_i Q_j + \tfrac{1}{3!} A_{ijk} Q_i Q_j Q_k + \tfrac{1}{4!} A_{ijkl} Q_i Q_j Q_k Q_l} \quad (2.21)$$

to write the total potential energy, where i, j, k, l range from 1 to the number of generalized coordinates employed in the analysis. The coefficients A are symmetric, so that

$$A_{ij} = A_{ji},$$
$$A_{ijk} = A_{ikj} = A_{kji} = A_{jik} = A_{jki} = A_{kji}, \quad (2.22)$$

etc. The load parameter Λ is found inside the coefficients.

For example, the energy for a two-degrees-of-freedom system takes the form

$$V[Q_1, Q_2, \Lambda] = A_1 Q_1 + A_2 Q_2 + \frac{1}{2}\left(A_{11} Q_1^2 + 2A_{12} Q_1 Q_2 + A_{22} Q_2^2\right)$$
$$+ \frac{1}{6}\left(A_{111} Q_1^3 + 3A_{112} Q_1^2 Q_2 + 3A_{122} Q_1 Q_2^2 + A_{222} Q_2^3\right)$$
$$+ \frac{1}{4!}(A_{1111} Q_1^4 + 4A_{1112} Q_1^3 Q_2 + 6A_{1122} Q_1^2 Q_2^2 + 4A_{1222} Q_1 Q_2^3 + A_{2222} Q_2^4). \quad (2.23)$$

Of course, many terms can be zero in (2.23), as will be shown for specific examples in this chapter.

The load parameter Λ is usually contained in the A_i terms in the energy. A convenient way to identify the load terms is to obtain the derivatives of the coefficients with respect to Λ in the form of a Taylor expansion

$$A_i = A_i \Big|_{\Lambda=0} + \Lambda \frac{\partial A_i}{\partial \Lambda}\Big|_{\Lambda=0}.$$

The same could be done if A_{ij} was a function of the load parameter; i.e.,

$$A_{ij} = A_{ij} \Big|_{\Lambda=0} + \Lambda \frac{\partial A_{ij}}{\partial \Lambda}\Big|_{\Lambda=0}.$$

In most structural problems the load potential is linear in the load and in the displacements.

The presence of the load in the total potential energy is very important in the context of the present analysis, and in all examples we shall identify the nonzero coefficients in the energy expansion (2.21) and also the nonzero derivatives with respect to the load.

Example 2.4 (Generalized coordinates of a circular plate). *Let us consider again the plate problem of Example 2.3. There are many ways to obtain a discretization of a circular plate problem. For example, Bessel functions are employed in Chapter 7 of [22]. An approximate solution to the nonlinear equilibrium equations is obtained in this section by assuming a linear variation of the displacement u with the coordinate r and a trigonometric variation of w; that is,*

$$u(r) = \frac{r}{R}Q_2, \qquad w(r) = Q_1 \cos\left(\frac{\pi r}{2R}\right). \tag{2.24}$$

Here Q_1 and Q_2 have dimensions of displacements.

Such displacements fields are plotted in Figure 2.2 to illustrate the meaning of the generalized coordinates Q_1 and Q_2.

The strains become

$$\varepsilon_r = \frac{1}{2}\left(\frac{\pi}{2R}\right)^2 Q_1^2 \left[\sin\left(\frac{\pi r}{2R}\right)\right]^2 + \frac{1}{R}Q_2,$$

$$\varepsilon_\theta = \frac{1}{R}Q_2,$$

$$\chi_r = -\left(\frac{\pi}{2R}\right)^2 Q_1 \cos\left(\frac{\pi r}{2R}\right),$$

$$\chi_\theta = -\frac{1}{r}\left(\frac{\pi}{2R}\right) Q_1 \sin\left(\frac{\pi r}{2R}\right).$$

Notice that ε_r is linear in Q_2, but it is nonlinear in Q_1.

The energy V can now be computed from Q_1 and Q_2 and yields

$$V[Q_1, Q_2, \Lambda] = -2R\Lambda Q_2 + \frac{1}{2}\left[\frac{1}{2}\pi^2 (1.191 + \nu)\frac{D}{R^2}Q_1^2 + 2C(1+\nu)Q_2^2\right]$$
$$+\frac{1}{6}\left[3\frac{\pi^2+4}{8R}C(1+\nu)\right]Q_1^2 Q_2 + \frac{1}{24}\left[\frac{3}{8}\left(\frac{\pi}{4}\right)^2\left(3\pi^2+16\right)\frac{C}{R^2}\right]Q_1^4. \tag{2.25}$$

A compact form of V can be obtained by definition of the following coefficients:

$$A_2' = -2R,$$

$$A_{11} = \frac{1}{2}\pi^2 (1.191 + \nu)\frac{D}{R^2}, \qquad A_{22} = 2C(1+\nu),$$

$$A_{112} = A_{121} = A_{211} = \frac{\pi^2+4}{8R}C(1+\nu) = 1.7337\frac{C}{R}(1+\nu),$$

$$A_{1111} = \frac{3}{8}\left(\frac{\pi}{4}\right)^2\left(3\pi^2+16\right)\frac{C}{R^2} = 10.55\frac{C}{R^2}.$$

Thus, the energy expression reduces to

$$V[Q_1, Q_2, \Lambda] = A_2'\Lambda Q_2 + \frac{1}{2}(A_{11}Q_1^2 + A_{22}Q_2^2) + \frac{1}{6}(3A_{112})Q_1^2Q_2 + \frac{1}{24}A_{1111}Q_1^4. \tag{2.26}$$

Example 2.5 (Generalized coordinates of a circular arch). *For the circular arch of Example 2.2 we assume a discretization based on two degrees of freedom*

$$w = Q_1 R \cos\frac{\pi\theta}{2\theta_0} + Q_2 R \sin\frac{\pi\theta}{\theta_0}$$

as illustrated in Figure 2.1. Thus, Q_1 and Q_2 are nondimensional quantities.

The approximate solution proposed satisfies the following conditions at the boundaries:

$$w(\theta = -\theta_0) = 0, \qquad w(\theta = \theta_0) = 0,$$

so it is kinematically admissible for simple supports.

Substitution of $w[Q_1, Q_2]$ into V in (2.15) and dividing all terms by a factor $(EAR\theta_0)$ leads to the following coefficients of the energy expansion:

$$A_1' = -1,$$

$$A_{11} = \left(\frac{\pi}{2\theta_0}\right)^4 \frac{IA}{R^2} + \frac{8}{\pi^2}, \qquad A_{22} = \left(\frac{\pi}{\theta_0}\right)^4 \frac{IA}{R^2},$$

$$A_{111} = -\frac{3}{2}\frac{\pi}{\theta_0^2}, \qquad A_{122} = -2\frac{\pi}{\theta_0^2},$$

$$A_{1111} = \frac{3}{2}\left(\frac{\pi}{2\theta_0}\right)^4, \qquad A_{2222} = \frac{3}{2}\left(\frac{\pi}{\theta_0}\right)^4, \qquad A_{1122} = \frac{1}{8}\left(\frac{\pi}{\theta_0}\right)^4,$$

where

$$\Lambda = \frac{P}{EA\theta_0}.$$

2.4.2 Equilibrium Conditions

It is not evident that the same principles governing the continuous version of V also apply to a discrete version. Thus, we employ a new axiom for discrete systems as follows:

Axiom I. *A necessary and sufficient condition for equilibrium is that the total potential energy of the system should be stationary with respect to the generalized coordinates.*

The condition of stationary V can be written as

$$\delta V = \frac{\partial V}{\partial Q_i}\delta Q_i = 0, \tag{2.27}$$

where δQ_i are variations in the generalized coordinates of the system. In this book we use the following notation for derivatives of V:

$$V_i = \frac{\partial V}{\partial Q_i},$$

$$V' = \frac{\partial V}{\partial \Lambda}.$$

The first variation is written as

$$\delta V = V_i \delta Q_i = 0$$

leading to

$$V_i\left[Q_j, \Lambda\right] = 0 \qquad \Rightarrow \qquad \text{equilibrium.}$$

For a good discussion of the continuous versus discrete versions of V, the reader is referred to [6].

If the total potential energy is written as in (2.21), then one can write the equilibrium conditions as

$$V_i[Q_i, \Lambda] = A_i + A_{ij}Q_j + \frac{1}{2}A_{ijk}Q_jQ_k + \frac{1}{6}A_{ijkl}Q_jQ_kQ_l. \tag{2.28}$$

The constant term is associated with the load, and in general it is found that the other terms do not contain Λ. The second term is linear in the displacement field.

It was seen that the energy is also a function of the load level Λ. For a fixed value $\Lambda = \Lambda^F$, an equilibrium configuration is given by a set of values of generalized coordinates $Q_i = Q_i^F$, which define the deflected shape of the structure. As Λ is modified, the equilibrium configuration also changes. The evolution of equilibrium states can be conveniently visualized in a space with Λ and Q_i as coordinates; such space has $N + 1$ axis, where $i = 1, \ldots, N$. The set of equilibrium states in such a space defines what is called an equilibrium path, that is, a sequence of values Λ and Q_i for which the system is in equilibrium. A path is a curve in the space, as shown in Figure 2.4.

A nonlinear structure may have more than one equilibrium path. The most important path starts from the unloaded state; i.e., it contains the state for which $\Lambda = 0$. Such path is called the **primary, fundamental**, or **precritical path**. There are equilibrium paths that do not contain the origin of the equilibrium space; i.e., they do not start from the unloaded configuration but cross the fundamental path at a finite value of Λ. Paths like those are known as **secondary** or **postcritical paths**. There may also

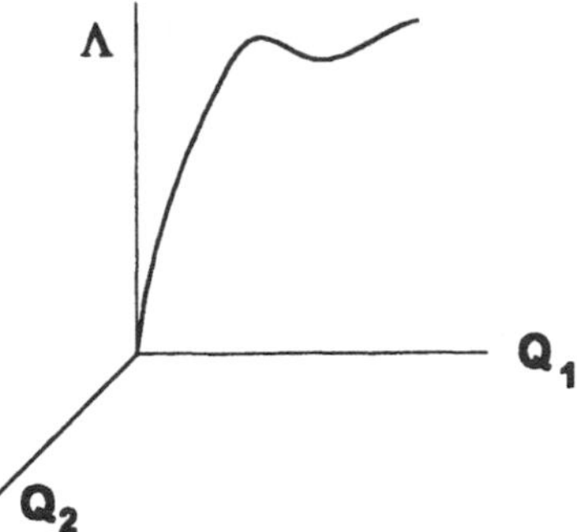

Figure 2.4 Equilibrium path for a two-degrees-of-freedom system.

be **tertiary paths**, which cross a secondary path. Finally, there are **complementary paths**, which do not contain the unloaded state or points in other paths.

An advantage of the energy formulation is that once V is obtained in terms of Q_i and Λ, then the equilibrium and stability analysis can be formulated and carried out in a completely general way. The specific features of the defining equations and discretization are necessary at the stage considered in this chapter, and in future chapters we shall often make use of available energies to carry out the examples.

2.4.3 Matrix Form

The coefficients A_{ij} can be grouped in a matrix

$$A_{ij} = \begin{bmatrix} A_{11} & A_{12} & \cdots & A_{1n} \\ A_{21} & A_{22} & \cdots & A_{2n} \\ \cdots & \cdots & \cdots & \cdots \\ A_{n1} & A_{n2} & \cdots & A_{nn} \end{bmatrix}.$$

This is a symmetric matrix and is the traditional stiffness matrix for linear problems, usually called K^0_{ij}.

The third- and fourth-order terms depend on higher-order contributions to the stiffness. They may be written in matrix form as

$$\left[A_{ijk}Q_j\right]\, Q_k = \begin{bmatrix} A_{111}Q_1 + A_{112}Q_2 + \cdots & \cdots & A_{1n1}Q_1 + A_{1n2}Q_2 + \cdots \\ \cdots & \cdots & \cdots \\ A_{n11}Q_1 + A_{n12}Q_2 + \cdots & \cdots & A_{nn1}Q_1 + A_{nn2}Q_2 + \cdots \end{bmatrix} \begin{Bmatrix} Q_1 \\ \cdots \\ Q_n \end{Bmatrix}.$$

Thus, to write the quadratic terms of the equilibrium equation in matrix form, we require the contraction of A_{ijk} by Q_j. This contracted matrix can be called K^1_{ij}, so that

$$K^1_{ij}\, Q_j = \left[A_{ijk}Q_k\right]\, Q_j.$$

To have a matrix representation of the cubic term in equilibrium we need a double contraction, i.e.,

$$K^2_{ij}\ Q_j = \left[A_{ijkl} Q_k Q_l\right]\ Q_j.$$

The contracted matrix can be identified as K^2_{ij}. Finally, we write the equilibrium of the system as

$$V_i[Q_i, \Lambda] = A_i + K^0_{ij} Q_j + \frac{1}{2} K^1_{ij} Q_j + \frac{1}{6} K^2_{ij} Q_j.$$

In the theory of elastic stability we shall make use of second-, third-, and fourth-order derivatives of V with respect to the Q_j. The second-order derivatives can be obtained as

$$V_{ij}[Q_i, \Lambda] = A_{ij} + A_{ijk} Q_k + \frac{1}{2} A_{ijkl} Q_k Q_l.$$

The right-hand side can also be interpreted as a matrix and is usually called the tangent stiffness matrix given by

$$V_{ij}[Q_i, \Lambda] = K^0_{ij} + K^1_{ij} + \frac{1}{2} K^2_{ij}.$$

Example 2.6 (Equilibrium of a circular plate). *For the circular plate under in-plane loading of Example 2.4, let us find and solve the equilibrium conditions. Equilibrium is given by two equations:*

$$V_1 = A_{11} Q_1 + A_{112} Q_1 Q_2 + \frac{1}{6} A_{1111} Q_1^3 = 0,$$

$$V_2 = \Lambda A_2' + A_{22} Q_2 + \frac{1}{2} A_{112} Q_1^2 = 0.$$

Both equations depend on Q_1 and Q_2, but the second equation can be solved as

$$Q_2 = -\frac{A_2'}{A_{22}} \Lambda - \frac{1}{2} \frac{A_{112}}{A_{22}} Q_1^2.$$

Next, Q_2 is substituted in the first equation to obtain Q_1:

$$A_{11} Q_1 - \frac{A_2'}{A_{22}} A_{112} \Lambda Q_1 - \frac{1}{2} \frac{(A_{112})^2}{A_{22}} Q_1^3 + \frac{1}{6} A_{1111} Q_1^3 = 0$$

or else

$$\left[(A_{11} A_{22} - A_2' A_{112} \Lambda) + \left(\frac{1}{6} A_{22} A_{1111} - \frac{1}{2} A_{112} A_{112}\right) Q_1^2\right] Q_1 = 0.$$

Fundamental equilibrium path: *One solution of the latter equation is*

$$Q_1^F = 0, \qquad Q_2^F = -\frac{A_2'}{A_{22}} \Lambda.$$

This is an equilibrium path emerging from the origin, since for $\Lambda = 0$ *the displacements are* $Q_1^F = Q_2^F = 0$. *Because of this reason, the path is called the fundamental or primary path. Notice that within the present model, linearity of the fundamental path is not an approximation, but it is the exact solution.*

Secondary path: *Let us investigate other solutions different than the fundamental path. If* $Q_1^F \neq 0$, *then the solution of the first equilibrium equation is*

$$(A_{11}A_{22} - \Lambda A_2' A_{112}) + \frac{1}{2}\left(\frac{1}{3}A_{22}A_{1111} - A_{112}A_{112}\right) Q_1^2 = 0.$$

We call $Q_1 = Q_1^s$ *the solution of this equation, to identify the new solution. For each value of* Λ, *there are two solutions with the same absolute value but different sign*

$$(Q_1^s)^2 = 2\frac{A_{11}A_{22} - \Lambda A_2' A_{112}}{A_{112}A_{112} - \frac{1}{3}A_{22}A_{1111}},$$

$$Q_2^s = Q_2^F - \frac{1}{2}\frac{A_{112}}{A_{22}}(Q_1^s)^2.$$

The new path contains the displacements in the primary path, to which it adds a new contribution. In the present two-degrees-of-freedom problem, it was easy to find the states that satisfy equilibrium by directly solving the nonlinear (cubic) equations. However, this is not always possible in general cases with multiple degrees of freedom, and we employ approximate numerical techniques for the solution of nonlinear equilibrium equations.

Bifurcation in the circular plate: *A plot of the equilibrium paths of the circular plate is shown in Figure 2.5. Notice that to have real solutions of* Q_1^s *it is necessary to assume that the load is negative, i.e., compressive.*

One can see that the fundamental path is linear in this case, and the nonlinearity of the kinematics of the problem is reflected in the new path that occurs at higher load levels. This is called the secondary path (also known as the postcritical path).

Let us follow the loading process in the circular plate: As the load increases from zero, the structure follows the fundamental path, in which only in-plane displacements occur. At the load level indicated by C, there is an intersection between two equilibrium paths: The linear fundamental path and the nonlinear secondary path. Notice that the secondary path involves both in-plane and out-of-plane displacement components, so that the shape of the plate significantly changes. The state at C in which there are two equilibrium paths is called a critical state, and we shall see that there are important changes in the stability of the system as it crosses a critical state. The stability of equilibrium states is considered in Chapter 4.

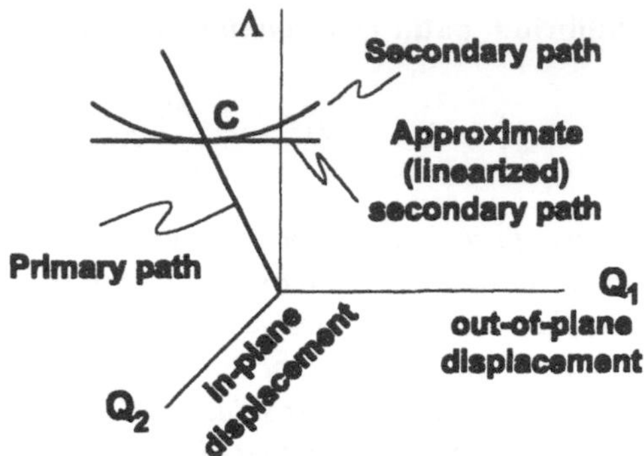

Figure 2.5 Equilibrium paths for a circular plate.

But for loads higher than the state given by C, there seems to be three possible equilibrium states. Which one will the structure follow in practice? An answer is that the structure will follow stable states, not unstable ones. This eliminates the part of the fundamental path that will be shown to be unstable. But one still has two possibilities, both along the secondary path. In theory both have the same probability because they are stable; however, the structure will follow one or another based on the presence of small effects, such as perturbations, imperfections, eccentricities in the loads, etc. This is considered in chapters dealing with imperfection sensitivity.

The overall behavior of the circular plate is known as **bifurcation behavior**. *A more specific name is given to it: stable symmetric bifurcation, as will be shown in Chapter 7. When the structure reaches the state C, then it changes to the secondary path and follows it. The secondary path has zero stiffness at the critical state and slowly increases the stiffness as deflections in the out-of-plane displacements occur. Because of the need to have large displacements in order to equilibrate a small additional load, the structure changes its shape; thus we say that buckling occurs at C.*

Of course, the model of buckling of a plate based on only two degrees of freedom is a crude approximation to the real behavior of the structure, and a better approximation would be required to improve the solution. But the present simplified approach shows the nature of the buckling behavior, which is governed by an almost linear fundamental path and a stable bifurcation.

Example 2.7 (Simplifications in the circular plate). *Next, we investigate the influence of neglecting cubic terms in the equilibrium equations of the circular plate of Example 2.6.*

Only equation $V_1 = 0$ is cubic in Q_1. A simplified form can be obtained neglecting cubic terms and yields

$$V_1 = A_{11}Q_1 + A_{112}Q_1Q_2 = 0,$$

$$V_2 = \Lambda A_2' + A_{22}Q_2 + \frac{1}{2}A_{112}Q_1^2 = 0.$$

Solving Q_2 from the second equation leads to the same expression as the one obtained in Example 2.6. Substitution into $V_1 = 0$ now yields

$$\left[(A_{11}A_{22} - A_2' A_{122}\Lambda) + \left(-\frac{1}{2}A_{112}A_{112}\right)Q_1^2\right]Q_1 = 0.$$

The fundamental path is the same as in the original solution. But since in $V_1 = 0$ we have already neglected cubic terms in Q_1, it is consistent to eliminate the new cubic term $\left(-\frac{1}{2}A_{112}A_{112}\right)Q_1^3$. Thus, for all values of Q_1, one has

$$A_{11}A_{22} - \Lambda A_2' A_{112} = 0$$

or else

$$\Lambda^s = \frac{A_{11}A_{22}}{A_2' A_{112}}, \qquad Q_2^s = -\frac{A_2'}{A_{22}}\Lambda^s.$$

This secondary path is plotted in Figure 2.5 and is horizontal, occurring for a fixed value of load $\Lambda = \Lambda^S$. Notice that while in the present model the exact solution of the fundamental path is linear, a linear secondary path is a consequence of an approximate analysis. The problem with assuming a linearized secondary path is that it does not provide information about the load reserve of the structure if it reaches this path. We do not employ such linearizations of the secondary path in this text.

Example 2.8 (Equilibrium of a circular arch). *Let us find the equilibrium equations for the arch problem under a central load discretized by two degrees of freedom (Example 2.5).*

The equilibrium conditions of the problem are

$$V_1 = A_1' + A_{11}Q_1 + \frac{1}{2}(A_{111}Q_1^2 + A_{122}Q_2^2) + \frac{1}{6}(A_{1111}Q_1^3 + 2A_{1122}Q_1Q_2^2) = 0,$$

$$V_2 = \left(A_{22} + A_{122}Q_1 + \frac{1}{2}A_{1122}Q_1^2 + \frac{1}{6}A_{2222}Q_2^2\right)Q_2 = 0.$$

Fundamental path: *One solution of equilibrium is given by*

$$Q_2^F = 0.$$

This value of Q_2^F is next replaced in the other equilibrium equation, leading to a cubic equation in terms of Q_1 and linear in Λ:

$$V_1 = \Lambda A_1' + A_{11}Q_1 + \frac{1}{2}A_{111}Q_1^2 + \frac{1}{6}A_{1111}Q_1^3 = 0.$$

In this case it is simpler to make explicit the values of Λ *as a function of* Q_1 *in the form*

$$\Lambda^F = \frac{1}{A_1'}\left(A_{11} + \frac{1}{2}A_{111}Q_1 + \frac{1}{6}A_{1111}Q_1^2\right)Q_1.$$

This is the equation governing the primary or fundamental path. Notice that at $\Lambda^F = 0$ *the displacements become* $Q_1^F = Q_2^F = 0$.

Secondary path: *Another solution may be obtained if* $Q_2 \neq 0$ *in* $V_2 = 0$. *The new conditions are*

$$\Lambda A_1' + A_{11}Q_1 + \frac{1}{2}A_{111}Q_1^2 + \frac{1}{6}A_{1111}Q_1^3 = 0,$$

$$A_{22} + A_{122}Q_1 + \frac{1}{2}A_{1122}Q_1^2 = -\frac{1}{6}A_{2222}Q_2.$$

From the second condition, one obtains

$$Q_2^S = -\frac{6}{A_{2222}}\left(A_{22} + A_{122}Q_1 + \frac{1}{2}A_{1122}Q_1^2\right).$$

This can be replaced in the equation $V_1 = 0$ *and leads to a polynomial in* Q_1. *Terms of order higher than three should be neglected in this new equation for consistency. Thus, a cubic secondary path in terms of both* Q_1 *and* Q_2 *will result for the present arch problem.*

Limit and bifurcation states in the arch: *The results of the primary and secondary paths are shown in Figure* 2.6.

Notice that, unlike the case of the circular plate, in the arch under central load the fundamental path is nonlinear. Furthermore, had we considered a one-degree-of-freedom model with Q_1 *as the only generalized coordinate, the primary path would have been the same: a cubic curve in the* $\Lambda - Q_1$ *space.*

At point L we find that the stiffness of the structure, as given by the tangent to the path, becomes zero (a horizontal tangent). The path descends following the

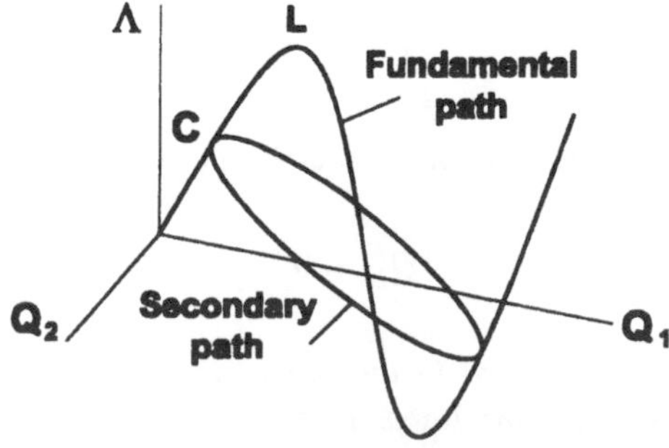

Figure 2.6 Equilibrium paths for a circular arch.

state at L, so that rather than increasing loads one can only increase displacements. Furthermore, a stability analysis would show that the fundamental path becomes unstable for displacements larger than L. Unable to find stable equilibrium states in the vicinity, the arch will experience dynamic displacements in search of a new shape with which stiffness can be regained. Again, the state at L is critical, and the process of deformation that initiates at L is called buckling. But the critical state is very different from what we identified in the circular plate, with not two but just one equilibrium path. The critical state at L is called **limit point***, and a detailed study is presented in Chapter* 6.

The arch also presents bifurcation state at point C, where a secondary path is initiated. This behavior is closer to what was observed in the circular plate; however, the branches are not upright but tend to grow displacements with a decrease in the load. This is called **unstable symmetric bifurcation***, and more details are given in Chapter 7.*

Which critical state (either L or C) occurs at the lower load level depends on the geometry and load of the arch. In some cases only a state L exists, and we say that buckling behavior is dominated by a limit point, or limit point behavior.

Example 2.9 (Simplifications in the circular arch). *We consider again the circular arch and investigate now the influence of neglecting terms in the nonlinear equilibrium conditions.*

When we neglect cubic terms in equilibrium, the two conditions with up to quadratic terms included become

$$V_2 = \left(A_{22} + A_{122} Q_1 + \frac{1}{6} A_{2222} Q_2 \right) Q_2 = 0,$$

$$V_1 = \Lambda \, A_1' + A_{11} Q_1 + \frac{1}{2} A_{111} Q_1^2 + \frac{1}{2} A_{122} Q_2^2 = 0.$$

The fundamental path is now given by

$$Q_2^F = 0,$$

$$\Lambda^F = -\frac{1}{A_1'} \left(A_{11} + \frac{1}{2} A_{111} Q_1 \right) Q_1^F.$$

This approximate formulation leads to a quadratic variation of the primary path, which is contained in the plane defined by Λ^F and Q_1^F.

The secondary path satisfies $Q_2^S \neq 0$, so that

$$A_{22} + A_{122} Q_1^S + \frac{1}{6} A_{2222} Q_2^S = 0$$

and

$$Q_2^S = -\frac{6}{A_{2222}} (A_{22} + A_{122} Q_1^S).$$

Substituting Q_2^S into $V_1 = 0$ allows us to compute Q_1^S by retaining up to quadratic terms for consistency.

Example 2.10 (Linearization of the arch problem). *If we neglect quadratic and cubic terms in the equilibrium of an arch, then only linear terms are kept in the equilibrium equations; i.e.,*

$$V_2 = A_{22} Q_2 = 0,$$

$$V_1 = \Lambda\, A_1' + A_{11} Q_1 = 0.$$

This set has only one solution, which is linear,

$$Q_2^F = 0,$$

$$\Lambda^F = -\frac{A_{11}}{A_1'} Q_1^F.$$

Clearly, a linearization of the equilibrium equations of an arch does not provide reliable information about its true behavior. The fundamental path becomes linear, even though from our previous studies we know of the importance of the large deflections. Furthermore, the information about the secondary path detected in Examples 2.8 and 2.9 is lost in the linear formulation.

We shall not admit this level of simplification in our studies, and, whenever possible, we shall retain up to fourth-order terms in the energy expansion V.

Example 2.11 (Energy of an inextensible bar). *A bar that can be part of a frame (Figure 2.7) has length L and is oriented at θ_x from the x-axis. It is assumed that the bar is inextensional; i.e., the length remains L following deformation of the structure. The strain energy is given by*

$$\Omega = \frac{1}{2} EI \int \chi^2 dx,$$

where the change in curvature can be obtained from the nonlinear relation

$$\chi = \frac{d^2 w}{dx^2}\left[1 - \left(\frac{dw}{dx}\right)^2\right]^{-1/2}.$$

Substitution of χ into the strain energy leads to the approximate result

$$\Omega = \frac{1}{2} EI \int \left[\left(\frac{d^2 w}{dx^2}\right)^2 + \left(\frac{dw}{dx}\right)^2 \left(\frac{d^2 w}{dx^2}\right)^2 + \cdots\right] dx.$$

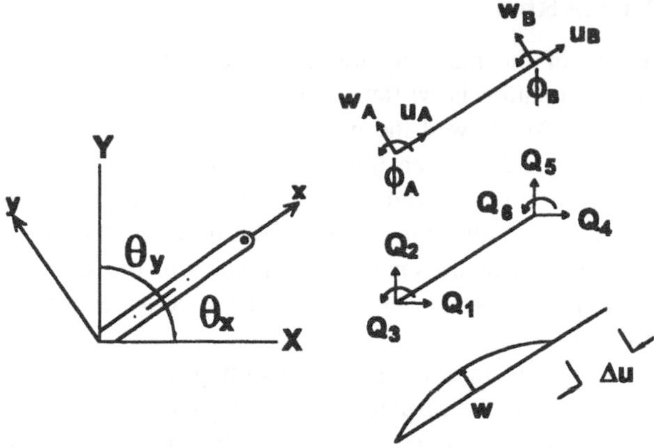

Figure 2.7 Bar of Example 2.11.

The end shortening of the member due to bending is

$$\Delta u = L - \int \left[1 - \left(\frac{dw}{dx}\right)^2\right]^{1/2} dx$$

$$= \int \left[\frac{1}{2}\left(\frac{dw}{dx}\right)^2 + \frac{1}{8}\left(\frac{dw}{dx}\right)^4\right] dx.$$

Notice that Δu does not contribute to the strain energy, but it contributes to the load potential.

The out-of-plane displacements can be approximated by cubic polynomials

$$w(x) = \left(1 - 3\,\overline{x}^2 + 2\,\overline{x}^3\right) w_A + \left(3\,\overline{x}^2 - 2\,\overline{x}^3\right) w_B$$

$$+ \left(\overline{x} - 2\,\overline{x}^2 + \overline{x}^3\right) L\phi_A + \left(-\,\overline{x}^2 + \overline{x}^3\right) L\phi_B,$$

where w_A and ϕ_B are the displacement and rotation at node A, and $\overline{x} = x/L$.

To have generalized coordinates in global coordinate axis, the nodal variables are rotated as indicated in Appendix B *for extensional bars, but now*

$$u_B = u_A + \Delta u.$$

It is left to the reader to carry out the substitutions to get an explicit form of the energy of an inextensional bar.

2.5 FINAL REMARKS

In this short review of the main equations of elasticity under nonlinear kinematic assumptions, the formulation is written in terms of the total potential energy of the structural component. We showed the process of obtaining the energy in terms of a continuous formulation and then substituted the displacements in terms of generalized coordinates. This led to a discretized version of the total potential energy, for which we employ an axiom for equilibrium of the system based on the first variation. A zero value of the first variation of the energy with respect to the generalized coordinates indicates equilibrium of the system. For nonlinear systems, equilibrium paths can be computed in this way.

Two main examples were developed in this chapter. In one of them (the circular plate under in-plane loads) the displacements along the fundamental path are linear, and nonlinearity is reflected along a new, secondary path that crosses the primary path. In the second example (the circular arch under a central concentrated load) the nonlinearity of the load-displacement relations is reflected from the beginning along the fundamental path, and also leads to a secondary path at higher load values. Simplifications were introduced in both cases to show their influence on the behavior of the model and to recognize that linearizations of the equilibrium conditions may not be acceptable in the present context.

If the total potential energy is restricted to quadratic terms, then the equilibrium conditions become linear and information about the postbuckling behavior of the structure is lost. To investigate stability of a structural system it is necessary to retain up to fourth-order terms in the total potential energy.

The problems posed in this chapter lead to nonlinear equilibrium conditions in terms of displacement degrees of freedom. The solution of such problems requires the use of numerical techniques of analysis to follow a nonlinear path. In the next chapter we look at a technique that is useful for the computation of nonlinear equilibrium paths and also allows us to obtain a classification of the possible solutions and detect the nature of singularities along a path.

2.6 PROBLEMS

In solving the problems of this book it is advisable to use an algebraic symbolic manipulator to carry out the computations. This will simplify the substitutions and computation of derivatives and integrals, so that the reader can concentrate on the mechanics of the problem.

Review questions (*Hint*: see Appendix A). (*a*) What is the difference between a specialized system and a general system? (*b*) Explain the differences between a theory for large strains, a theory for large rotations and small strains, and a theory for small rotations. (*c*) What are the definitions of specific strains and strain tensor in the theory of large rotations and small strains? (*d*) What is the difference between the Cauchy and the Piola–Kirchhoff stress tensors? (*e*) Give examples of load potentials other than those presented in (2.6). (*f*) According to the hypothesis accepted in this chapter, why should loads be applied at a slow rate? (*g*) Explain the difference between natural and essential boundary conditions. (*h*) Is a minimum of V required for equilibrium? (*i*) Explain the difference between a primary and a secondary path. (*j*) What is the maximum number of cubic terms in V that one may find in a three-degrees-of-freedom system? (*k*) Explain the differences between total potential energy and virtual work principles.

Problem 2.1 (Circular ring). Obtain the total potential energy for an elastic circular ring under radial pressure (see Figure 5.9). Consider the hoop membrane stress resultant N_θ and the bending resultant M_θ

$$N_\theta = AE\ \varepsilon_\theta, \qquad M_\theta = EI\ \chi_\theta.$$

Use the following kinematic relations:

$$\varepsilon_\theta = \frac{1}{R}\frac{du}{d\theta} + \frac{1}{2}\left(\frac{u}{R} - \frac{dw}{Rd\theta}\right)^2 + \frac{w}{R},$$

$$\chi_\theta = \frac{1}{R^2}\frac{du}{d\theta} - \frac{d^2w}{R^2 d\theta^2},$$

where w is the transverse displacement, and u is the in-plane displacement component of the ring. The pressure p acts in the direction of the transverse displacement w. Employ the following discretization for the displacements:

$$w = Q_1 R + Q_2 R \cos(n\theta), \qquad u = Q_3 R \sin(n\theta)$$

for $n = 0, 1, 2, 3, \ldots$.

Problem 2.2 (Circular ring). For the ring of Problem 2.1 obtain the equilibrium path that starts from the unloaded state. Consider $E = 70GPa$, $R = 200mm$, $A = 5 \times 30.5mm^2$, $I = 5^3 \times 30.5/12mm^4$. Consider $n = 2$ and $n = 3$.

Problem 2.3 (Cantilever beam). Consider the cantilever beam shown in Figure 2.8. Design a laboratory experiment to obtain the total potential energy from experimental measurements. Propose a way to implement the test.

Problem 2.4 (Circular plate). For the circular plate considered in Example 2.6, plot the strain energy, the load potential, and the total potential energy, as a function of the generalized coordinates Q_1 and Q_2 for a fixed value of the load $\Lambda = 0.003$. Discuss which term dominates as the values of generalized coordinates are increased. The radius of the plate is $500mm$, the thickness is $3.175mm$, and the plate is made of aluminum, with $E = 70GPa$ and $\nu = 0.3$.

Problem 2.5 (Column made of composite material) [1]. A column in Figure 2.9 is made with a laminated composite material. The constitutive relations are such that bending-extension coupling is allowed, leading to

$$N = A\ \varepsilon + B\ \chi, \qquad M = B\ \varepsilon + D\ \chi,$$

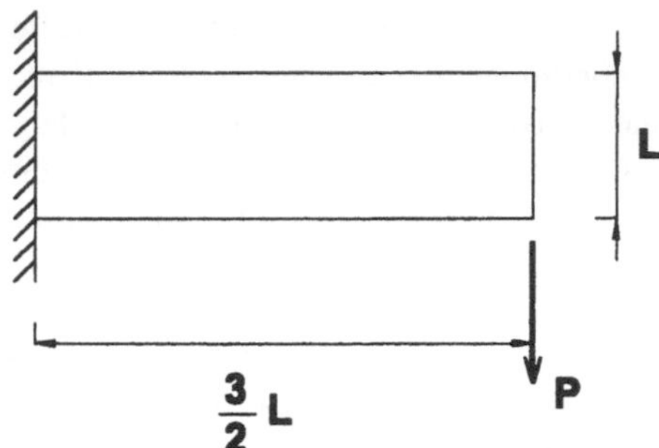

Figure 2.8 Cantilever beam of Problem 2.3.

where A is the membrane stiffness; D is the bending stiffness; and B is the bending-extension coupling coefficient. We employ the nonlinear kinematic relations

$$\varepsilon = \frac{du}{dx} + \frac{1}{2}\left(\frac{dw}{dx}\right)^2, \qquad \chi = -\frac{d^2w}{dx^2}.$$

The total potential energy is written as

$$V = \frac{1}{2}\int \left(N\,\varepsilon + M\,\chi + 2\Lambda P\,\frac{du}{dx}\right) dx.$$

For simple supported boundary conditions, we use an approximate solution with three degrees of freedom

$$\frac{du}{dx} = Q_1 + \Phi(x)\,Q_2, \qquad w = [\Phi(x) - 1]\,Q_3,$$

with

$$\Phi(x) = \sin\left(\frac{\pi x}{2L}\right) + \cos\left(\frac{\pi x}{2L}\right).$$

Find the total potential energy in the form $V = V\,[Q_1, Q_2, Q_3, \Lambda]$.

Problem 2.6 (Mechanical system). The structure in Figure 2.10 is formed by two rigid members and two flexible elastic springs with stiffness C (rotational) and

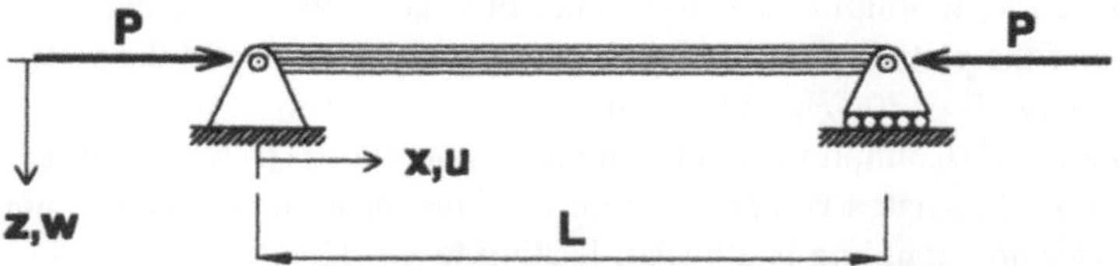

Figure 2.9 Problem 2.5.

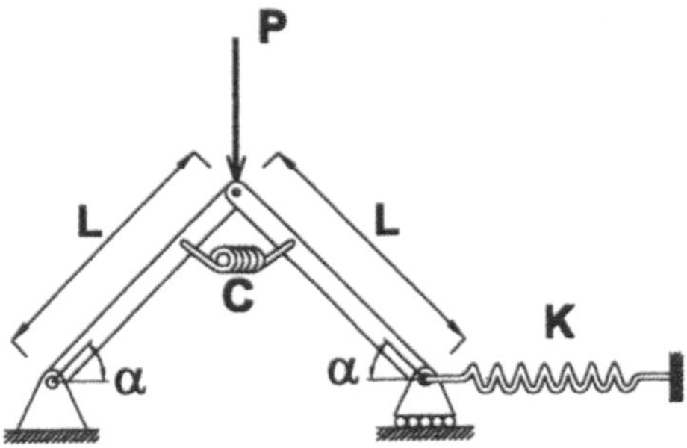

Figure 2.10 Problem 2.6.

K (linear). Write the total potential energy as a function of the rotation of the central pin including up to fourth-order terms. Find the equilibrium path of the system.

Problem 2.7 (Inextensional column). A vertical column is assumed inextensional and deflects as depicted in Figure 2.11. The deformation is symmetric in the sense that the rotations at the ends have the same values. Approximate the displacement field by a cubic polynomial. Write the energy in the deflected configuration. Evaluate equilibrium paths, and find for what load the paths intersect. Draw the equilibrium paths for $E = 200GPa$, $I = 8.4661mm^4$, $L = 580mm$, $A = 40.645mm^2$.

Problem 2.8 (Plane frame). A steel frame structure is shown in Figure 2.12, with EI constant and $\alpha = 127.5o$. Write the total potential energy using cubic interpolation for out-of-plane displacements and linear interpolation for in-plane displacements in each member, as indicated in Appendix B. Data: $E = 200GPa$, $L = 3000mm$, $I = 2.7055 \times 10^8 mm^4$, $A = 12903mm^2$.

Problem 2.9 (Plane frame). A symmetric frame, shown in Figure 2.13, is loaded at the joints. It is assumed that the frame has bending and stretching, with $L = 5m$, $E = 200GPa$, $A = 7742mm^2$, $I = 2.5 \times 10^8 mm^4$, $\alpha = 127.5$, $P_2 = P_1 = 0.5P_3$. Employ cubic functions (as in Appendix B) to approximate the displacement field. Write the total potential energy of the frame, up to quartic terms. Find the equilibrium conditions. Try to solve them with the help of a symbolic manipulator. If you cannot solve the equilibrium equations, simply trying is useful to introduce the need of numerical techniques.

Problem 2.10 (Rectangular plate). Write the total potential energy of a rectangular flat plate, including nonlinear kinematic relations. Assume simply supported boundary conditions and in-plane loads.

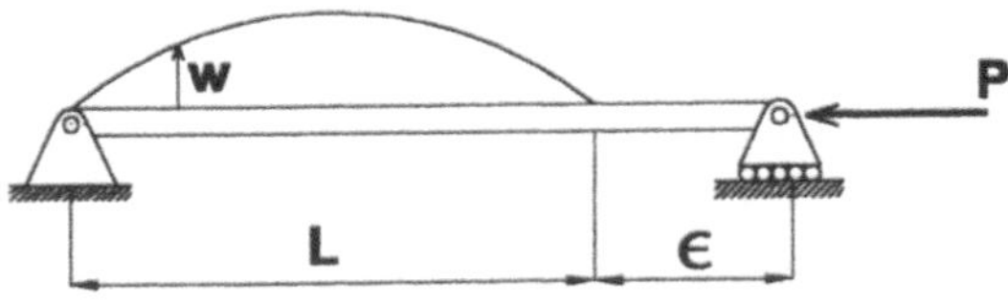

Figure 2.11 Problem 2.7.

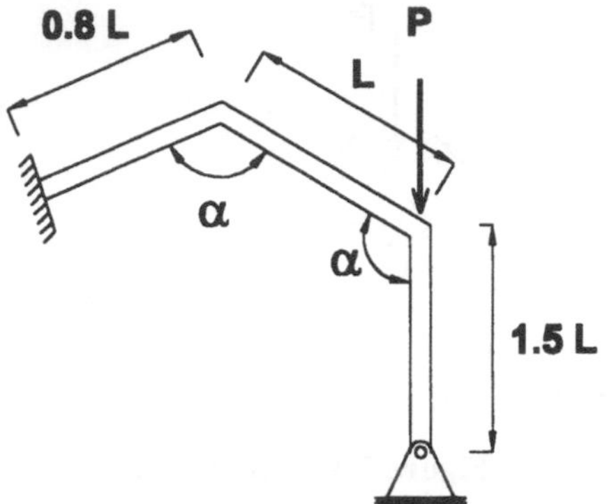

Figure 2.12 Problem 2.8.

Problem 2.11 (Circular arch). Find the discrete form of V for a circular arch with

$$w = Q_1 R \cos\left(\frac{\pi\theta}{2\theta_0}\right) + Q_2 R \cos\left(\frac{3\pi\theta}{2\theta_0}\right).$$

Write the equilibrium conditions using the following data: $E = 210GPa$; the length of the arch is $600mm$; the central angle is θ_0; and the cross section has width $30mm$ and thickness $5mm$.

Problem 2.12 (Truss structure). The truss of Figure 2.14 is formed by two in-extensional members. Member AB is stiffer than member BC, so that AB cannot buckle, but as the structure rotates θ, then member BC buckles. The maximum out-of-plane displacement w_0 occurs when the two members are on a line. Assume the out-of-plane displacement as in trusses (Appendix B). Write the total potential energy

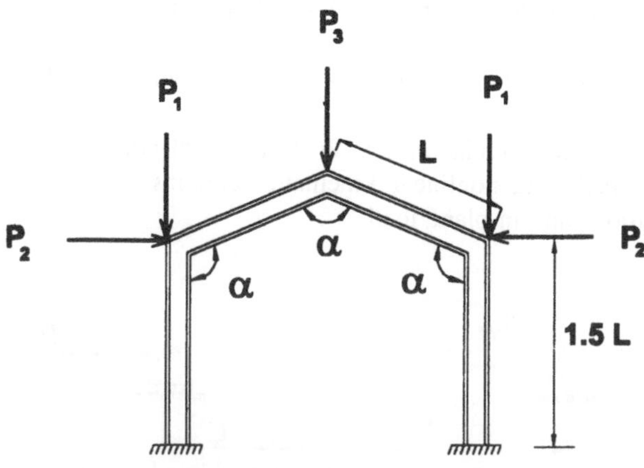

Figure 2.13 Problem 2.9.

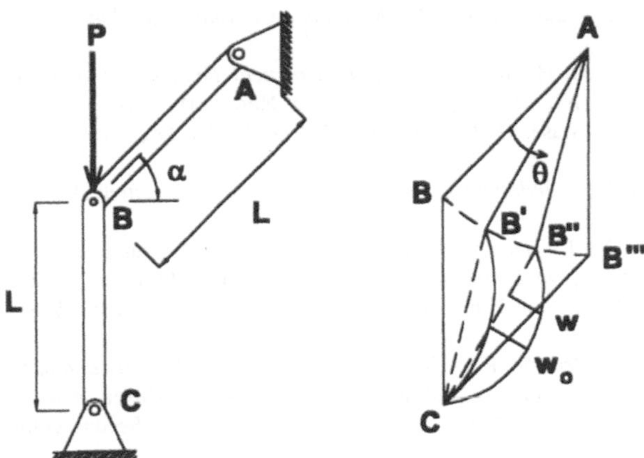

Figure 2.14 Problem 2.12.

of the structure. Find the equilibrium configurations for the buckled structure. Notice that this is a one-way system (nonholonomic) because the displacements can only occur with anticlockwise rotations.

2.7 BIBLIOGRAPHY

[1] Barbero, E. J., and Godoy, L. A., Influence of bending-extension coupling on buckling of composite column, *Mechanics of Composite Materials and Structures*, 4, 191–207, 1997.

[2] Benvenutto, E., *An Introduction to the History of Structural Mechanics, Part 1: Statics and Resistance of Solids; Part 2: Vaulted Structures and Elastic Systems*, Springer-Verlag, New York, 1991.

[3] Brush, D. O., and Almroth, B. O., *Buckling of Bars, Plates, and Shells*, McGraw-Hill-Kogakusha, Tokyo, 1975.

[4] Donnell, L., *Beams, Plates and Shells*, McGraw-Hill, New York, 1976.

[5] Dym, C., and Shames, I., *Solid Mechanics: A Variational Approach*, McGraw-Hill, New York, 1973.

[6] El Naschie, M. S., *Stress, Stability and Chaos in Structural Engineering: An Energy Approach*, McGraw-Hill, London, 1990.

[7] Fung, Y. C., *Foundations of Solid Mechanics*, Prentice-Hall, Englewood Cliffs, NJ, 1965.

[8] Fung, Y. C., *A First Course in Continuum Mechanics*, Prentice-Hall, Englewood Cliffs, NJ, 1977.

[9] Green, A. E., and Zerna, W., *Theoretical Elasticity*, Oxford University Press, London, 1954.

[10] Green, A. E., and Adkins, J. E., *Large Elastic Deformations and Non-Linear Continuum Mechanics*, Clarendon Press, Oxford, 1960.

[11] Huseyin, K., *Nonlinear Theory of Elastic Stability*, Noordhoff Int., Leyden, 1975.

[12] Koiter, W. T., *On the Stability of Elastic Equilibrium*, Ph.D. thesis, Delft Institute of Technology, Delft, The Netherlands, 1945.

[13] Koiter, W. T., The application of the initial post-buckling analysis to shells, In: *Buckling of Shells*, Ed. E. Ramm, Springer-Verlag, Berlin, 3–17, 1987.

[14] Langhaar, H. L., *Energy Methods in Applied Mechanics*, Wiley, New York, 1962; reprinted by Krieger, Malabar, FL, 1989.

[15] Love, A. E. H., *The Mathematical Theory of Elasticity*, Cambridge University Press, Cambridge, 1892.

[16] Malvern, L. E., *Introduction to the Mechanics of a Continuous Medium*, Prentice-Hall, Englewood Cliffs, NJ, 1969.
[17] Marsden, J. E., and Hughes, T. J. R., *Mathematical Foundations of Elasticity*, Prentice-Hall, Englewood Cliffs, NJ, 1983; reprinted by Dover, New York, 1994.
[18] Mase, G. E., and Mase, G. T., *Continuum Mechanics for Engineers*, CRC Press, Boca Raton, FL, 1992.
[19] Novozhilov, V. V., *Foundations of the Nonlinear Theory of Elasticity*, Graylock Press, Rochester, NY, 1953.
[20] Novozhilov, V. V., *The Theory of Thin Shells*, Noordhoff, Holland, 1959.
[21] Reddy, J. N., *Energy and Variational Methods in Applied Mechanics*, Wiley, New York, 1984.
[22] Thompson, J. M. T., and Hunt, G. W., *A General Theory of Elastic Stability*, Wiley, London, 1973.
[23] Todhunter, I., and Pearson, K., *A History of the Theory of Elasticity and of Strength of Materials*, Cambridge University Press, Cambridge, 1893; reprinted by Dover, New York, 1960.
[24] Washizu, K., *Variational Methods in Elasticity and Plasticity*, Second edition, Pergamon Press, Oxford, 1975.

CHAPTER

THREE

PERTURBATION TECHNIQUES

3.1 INTRODUCTION

We already found in Chapter 2 that equilibrium problems in elastic systems with large rotations yield a nonlinear problem in terms of displacements. Solutions of equilibrium paths, stability of states of equilibrium, identification of critical states, evaluation of postcritical paths, imperfection sensitivity of states, and design sensitivity analysis are all problems that require dealing with nonlinear systems. Thus, all formulations required to produce a meaningful analysis of stability of structures lead to nonlinear systems. For this reason, we explore some details of the solution of nonlinear problems in this chapter.

There are several families of techniques that can be effectively used to solve nonlinear problems.

- One such family includes **continuation methods**, also called **step-by-step methods**, in which the solution is obtained in incremental form. One of the variables is increased, and the response is computed for this small increment to update the solution. Then a new increment is given and a new solution is computed. There are many ways to carry out such analysis depending on what is increased, and in the context of solid mechanics the most popular continuation techniques are load increment, displacement increment, arc length, work control techniques, etc. [5, 2]. These are efficient methods of analysis and are included in many finite element general-purpose programs.
- Another family of solutions uses **static perturbation techniques**. Here there is a state at which the solution is known, and an approximate analytical solution is constructed by means of a power series representation of the solution in terms of a perturbation parameter. The literature on perturbation techniques is extensive, and only a few works are indicated at the end of this chapter [3, 4, 8, 11, 9, 13, 10].

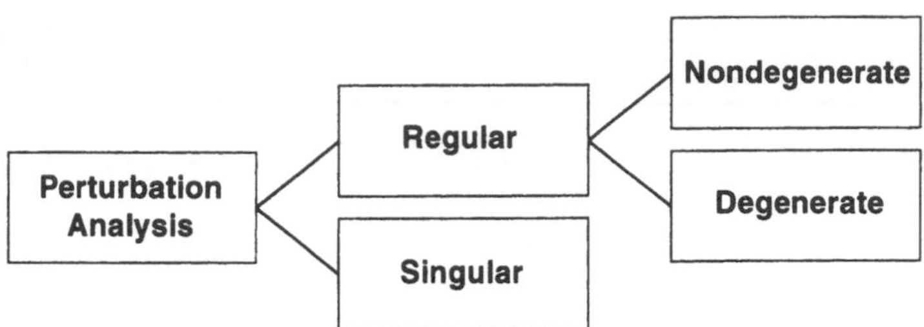

Figure 3.1 A summary of perturbation techniques.

The static perturbation technique for the buckling of systems that can be specified by a number of generalized coordinates was discussed by Sewell in 1965 [12]. Sewell stressed the importance of the perturbation technique in the understanding of buckling problems and pointed out that early buckling studies limited to the computation of a critical load were in fact using lower-order perturbation analysis. The importance of higher-order perturbation analysis in the context of buckling was presented, for example, in [12, 14] in the 1960s.

In this chapter we review perturbation techniques because they are very convenient in the development of a general theory of elastic stability. Not only is it possible to solve the nonlinear problems involved, but also to provide an important classification of the solutions and thus identify different features of the behavior. Section 3.2 deals with Taylor expansion for discrete systems, in which the solution of a problem is expanded in terms of a perturbation parameter. The technique of perturbation via implicit differentiation is developed in section 3.3, and it is shown how it applies to equilibrium problems. There are two perturbation parameters considered: the load parameter and one of the components of the displacement vector. The explicit substitution technique is presented in section 3.4. This is an alternative way to develop the perturbation equations and leads to the same results as those obtained via implicit differentiation. Problems of degenerate perturbations are studied in section 3.5, and the contraction mechanism is presented as a standard way to introduce a new condition when it is required. Singular perturbations are the subject of section 3.6. We present a technique that uses the least degenerate solution, in which a polynomial with unknown exponents, rather than a Taylor expansion, is employed. Finally, section 3.7 contains an application of the above ideas to compute the nonlinear path of an arch under a central point load. This is a case of regular, nondegenerate perturbations.

A summary of the perturbation techniques discussed in this chapter may be seen in Figure 3.1.

3.2 TAYLOR SERIES EXPANSION

Taylor series expansions are approximations of an analytical function by means of power series. A Taylor series provides a means to predict the value of a function at one point in terms of the value of the function and its derivatives at another point.

Example 3.1. *For example, a quartic approximation to the cosine function of θ is*

$$\cos\theta = 1 - \frac{1}{2!}\theta^2 + \frac{1}{4!}\theta^4 - \cdots.$$

There is always an error involved when we truncate the expansion, and the error is a function of what terms are not included in the expansion. The error in this example is proportional to θ^6. Plots of the Taylor expansion for different orders of approximation are shown in Figure 3.2.

Instead of expanding a known function in a power series, it is also possible to expand the solution of an equation (or set of equations) in a power series. For example, consider a nonlinear equation of the form

$$F(Q_j, \Lambda) = 0, \tag{3.1}$$

where Q_j are the components of the displacements of the system, in vector form, and Λ is a scalar control parameter. Equation (3.1) can be a scalar equation, but in general it is a vector equation, and it will be used as a vector in the applications to nonlinear equilibrium paths.

The solution of (3.1) is given by values of Q_j for different values of Λ. We seek to represent the solution of this problem as a function of another parameter s, called the **perturbation parameter**

$$Q_j = Q_j(s), \tag{3.2}$$

$$\Lambda = \Lambda(s). \tag{3.3}$$

Thus, we want to obtain a parametric solution in terms of s. The perturbation parameter acts as an increment along the solution under study.

To carry out the analysis, a perturbation parameter has to be chosen. This parameter should be adequate to follow the nonlinear response of the system as it evolves

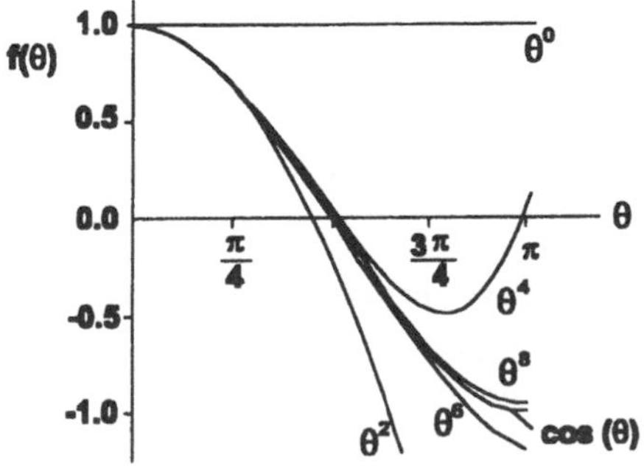

Figure 3.2 Taylor expansions of a cosine function with different levels of approximation.

in the $(Q_j - \Lambda)$ space. In general, we choose a parameter that is already present in the definition of the problem, and it could be Λ, or one of the components of Q_j, or another scalar parameter.

Next we assume that for a given value of s, say, $s = s_0$, the response of the system is known, so that it satisfies (3.1):

$$\Lambda(s_0) = \Lambda^E, \qquad Q_j(s_0) = Q_j^E, \qquad \text{and} \qquad F\left[Q_j^E, \Lambda^E\right] = 0. \tag{3.4}$$

This is called the **reference solution**, and usually it is readily available or easy to compute. The properties of the solution at $s = s_0$ are known, so that we can assure continuity and differentiability.

Then we can write an expansion for the solution as a Taylor series

$$\boxed{Q_j(s) = Q_j^E + (s - s_0)\,\dot{Q}_j + \tfrac{1}{2}(s - s_0)^2\,\ddot{Q}_j + \tfrac{1}{6}(s - s_0)^3\,\dddot{Q}_j + \cdots} \tag{3.5}$$

$$\boxed{\Lambda(s) = \Lambda^E + (s - s_0)\,\dot{\Lambda} + \tfrac{1}{2}(s - s_0)^2\,\ddot{\Lambda} + \tfrac{1}{6}(s - s_0)^3\,\dddot{\Lambda} + \cdots} \tag{3.6}$$

where the coefficients in the above series are the following derivatives:

$$\dot{Q}_j \equiv \frac{dQ}{ds}\Big|_{s=s_0}, \qquad \dot{\Lambda} \equiv \frac{d\Lambda}{ds}\Big|_{s=s_0},$$
$$\ddot{Q}_j \equiv \frac{d^2Q}{ds^2}\Big|_{s=s_0}, \qquad \ddot{\Lambda} \equiv \frac{d^2\Lambda}{ds^2}\Big|_{s=s_0}. \tag{3.7}$$

Because the expansions (3.5) and (3.6) are written in a power series, the scalar $(s - s_0)$ should be a small parameter; i.e., $(s - s_0) \ll 1$. The approximations become more accurate as the number of terms is increased. A solution with terms up to $(s - s_0)$ is called a linear approximation; if the solution includes $(s - s_0)^2$, then it is a quadratic approximation, etc.

Not all functions can be represented in this simple way, and special techniques have to be devised in some cases. This is discussed with reference to singular perturbations in section 3.6.

3.3 REGULAR PERTURBATIONS VIA IMPLICIT DIFFERENTIATION

Perturbation techniques are a systematic procedure to identify the coefficients in the series expansion of a nonlinear problem.

There are several ways to formalize the procedure of perturbations: The first one, considered in this section, is by means of implicit differentiation. In the second method, presented in section 3.4, we employ explicit substitution. Both techniques lead to the same results and only differ in the formalisms involved. Implicit differentiation is employed, for example, in [15].

3.3.1 Basic Procedure

Consider a problem governed by (3.1) and the solution expanded in Taylor series as in (3.5) and (3.6).

Substitution of (3.5)–(3.6) into (3.1) leads to

$$F\left[Q_j(s), \Lambda(s), s\right] = 0. \tag{3.8}$$

Notice that the parameter s can also be present in F independent of the control Λ and response Q_j parameters.

Let us compute the values of the unknown derivatives in (3.5) and (3.6). First, we differentiate F with respect to s and obtain the **first-order perturbation equation**

$$\frac{dF}{ds} = \frac{\partial F}{\partial Q_j}\dot{Q}_j + \frac{\partial F}{\partial \Lambda}\dot{\Lambda} + \frac{\partial F}{\partial s} = 0. \tag{3.9}$$

In brief notation, (3.9) can be written as

$$G\left(Q_j, \Lambda, s\right) = 0.$$

The term $\left[\frac{\partial F}{\partial s}\right]$ is known, while the values of $\dot{Q}_j$ and $\dot{\Lambda}$ are unknown at this stage, so we write

$$\boxed{\frac{\partial F}{\partial Q_j}\dot{Q}_j + \frac{\partial F}{\partial \Lambda}\dot{\Lambda} = \bar{G}} \tag{3.10}$$

with

$$\bar{G} = -\frac{\partial F}{\partial s}.$$

The **second-order perturbation equation** is

$$\frac{d^2F}{ds^2} = \frac{\partial G}{\partial Q_k}\dot{Q}_k + \frac{\partial G}{\partial \Lambda}\dot{\Lambda} + \frac{\partial G}{\partial s} = 0 = H\left(Q_k, \Lambda, s\right). \tag{3.11}$$

Alternatively, this may be written in terms of F as follows:

$$\left[\frac{\partial^2 F}{\partial Q_j \partial Q_k}\dot{Q}_j + \frac{\partial^2 F}{\partial Q_k \partial \Lambda}\dot{\Lambda} + 2\frac{\partial^2 F}{\partial Q_k \partial s}\right]\dot{Q}_k$$

$$+\left[\frac{\partial^2 F}{\partial Q_j \partial \Lambda}\dot{Q}_j + \frac{\partial^2 F}{\partial \Lambda^2}\dot{\Lambda} + 2\frac{\partial^2 F}{\partial \Lambda \partial s}\right]\dot{\Lambda}$$

$$+\left[\frac{\partial F}{\partial Q_j}\ddot{Q}_j + \frac{\partial F}{\partial \Lambda}\ddot{\Lambda} + \frac{\partial^2 F}{\partial s^2}\right] = 0,$$

where we made use of the following relations:

$$\frac{\partial}{\partial s}\dot{Q}_j = \ddot{Q}_j, \qquad \frac{\partial}{\partial s}\dot{\Lambda} = \ddot{\Lambda}.$$

Equation (3.11) can also be written as

$$\frac{\partial F}{\partial Q_j}\ddot{Q}_j + \frac{\partial F}{\partial \Lambda}\ddot{\Lambda} = -\left(2\frac{\partial^2 F}{\partial Q_k \partial s}\dot{Q}_k + \frac{\partial^2 F}{\partial \Lambda \partial s}\dot{\Lambda} + \frac{\partial^2 F}{\partial s^2}\right)$$
$$-\left(\frac{\partial^2 F}{\partial Q_j \partial Q_k}\dot{Q}_j\dot{Q}_k + 2\frac{\partial^2 F}{\partial Q_k \partial \Lambda}\dot{\Lambda}\dot{Q}_k + \frac{\partial^2 F}{\partial \Lambda^2}\dot{\Lambda}\dot{\Lambda}\right) \tag{3.12}$$

or, in compact form, as

$$\boxed{\left(\frac{\partial F}{\partial Q_j}\right)\ddot{Q}_j + \left(\frac{\partial F}{\partial \Lambda}\right)\ddot{\Lambda} = -\bar{H}} \tag{3.13}$$

where $\bar{H}$ groups all the terms on the right-hand side of (3.12); namely,

$$\bar{H} = \left(2\frac{\partial^2 F}{\partial Q_k \partial s}\dot{Q}_k + \frac{\partial^2 F}{\partial \Lambda \partial s}\dot{\Lambda} + \frac{\partial^2 F}{\partial s^2}\right)$$
$$+\left(\frac{\partial^2 F}{\partial Q_j \partial Q_k}\dot{Q}_j\dot{Q}_k + 2\frac{\partial^2 F}{\partial Q_k \partial \Lambda}\dot{\Lambda}\dot{Q}_k + \frac{\partial^2 F}{\partial \Lambda^2}\dot{\Lambda}\dot{\Lambda}\right). \tag{3.14}$$

The **third-order perturbation equation** is obtained as

$$\frac{d^3 F}{ds^3} = \frac{\partial H}{\partial Q_l}\dot{Q}_l + \frac{\partial H}{\partial \Lambda}\dot{\Lambda} + \frac{\partial H}{\partial s} = 0 = I\left(Q_l, \Lambda, s\right)$$

leading to

$$\boxed{\frac{\partial F}{\partial Q_j}\dddot{Q}_j + \frac{\partial F}{\partial \Lambda}\dddot{\Lambda} = -\bar{I}} \tag{3.15}$$

in which

$$\bar{I} = 3\left[\left(\frac{\partial^2 F}{\partial Q_j \partial Q_k}\dot{Q}_j + \frac{\partial^2 F}{\partial Q_k \partial \Lambda}\dot{\Lambda} + \frac{\partial^2 F}{\partial Q_k \partial s}\right)\ddot{Q}_k\right.$$
$$+\left(\frac{\partial^2 F}{\partial Q_j \partial \Lambda}\dot{Q}_j + \frac{\partial^2 F}{\partial \Lambda^2}\dot{\Lambda} + \frac{\partial^2 F}{\partial s \partial \Lambda}\right)\ddot{\Lambda}$$
$$+\left(\frac{1}{3}\frac{\partial^3 F}{\partial Q_j \partial Q_k \partial Q_l}\dot{Q}_l + \frac{\partial^3 F}{\partial Q_j \partial Q_k \partial \Lambda}\dot{\Lambda} + \frac{\partial^3 F}{\partial Q_j \partial Q_k \partial s}\right)\dot{Q}_j\dot{Q}_k$$

$$+\left(\frac{\partial^3 F}{\partial Q_j \partial \Lambda^2}\dot{Q}_j + \frac{\partial^3 F}{\partial s \partial \Lambda^2} + \frac{\partial^3 F}{\partial \Lambda^3}\dot{\Lambda}\right)\dot{\Lambda}\dot{\Lambda}$$

$$+\left(\frac{\partial^3 F}{\partial Q_j \partial s^2}\dot{Q}_j + 2\frac{\partial^3 F}{\partial Q_j \partial \Lambda \partial s}\dot{Q}_j\dot{\Lambda} + \frac{\partial^3 F}{\partial s^2 \partial \Lambda}\dot{\Lambda}\right) + \frac{1}{3}\frac{\partial^3 F}{\partial s^3}\Bigg]. \tag{3.16}$$

At this stage, it is important to remember that we seek solution to the coefficients in (3.5) and (3.6), that is, the values of $\dot{Q}_j$, $\dot{\Lambda}$, $\ddot{Q}_j$, $\ddot{\Lambda}$, etc. The coefficients are the same derivatives that now appear on the left-hand side of the perturbation equations but are evaluated at $s = s_0$. To obtain them we next evaluate our perturbation equations at $s = s_0$, and this yields

$$\begin{aligned}
F\left[Q_j(s_0), \Lambda(s_0), s_0\right] &= 0,\\
G\left[Q_j(s_0), \Lambda(s_0), s_0\right] &= 0,\\
H\left[Q_j(s_0), \Lambda(s_0), s_0\right] &= 0,\\
I\left[Q_j(s_0), \Lambda(s_0), s_0\right] &= 0,
\end{aligned} \tag{3.17}$$

or else

$$\begin{aligned}
\left(\frac{\partial F}{\partial Q_j}\dot{Q}_j + \frac{\partial F}{\partial \Lambda}\dot{\Lambda} + \bar{G}\right)\Bigg|_{s=s_0} &= 0,\\
\left(\frac{\partial F}{\partial Q_j}\ddot{Q}_j + \frac{\partial F}{\partial \Lambda}\ddot{\Lambda} + \bar{H}\right)\Bigg|_{s=s_0} &= 0,\\
\left(\frac{\partial F}{\partial Q_j}\dddot{Q}_j + \frac{\partial F}{\partial \Lambda}\dddot{\Lambda} + \bar{I}\right)\Bigg|_{s=s_0} &= 0.
\end{aligned} \tag{3.18}$$

The set of (3.18) represents linear conditions in terms of the unknown vectors $\dot{Q}_j$, $\ddot{Q}_j$, and unknown scalars $\dot{\Lambda}$, $\ddot{\Lambda}$, etc. If such unknowns are solved from (3.18), then the complete solution of this problem can now be written as in (3.5) and (3.6).

Some features of the perturbation equations include the following:

- The evaluated perturbation equations can be solved sequentially. This is true because in the first equation of (3.18) the only unknowns are $\dot{Q}_j$ and $\dot{\Lambda}$, and higher-order coefficients such as $\ddot{Q}_j$ and $\ddot{\Lambda}$ do not appear. In the second equation of (3.18) we can assume that $\dot{Q}_j$ and $\dot{\Lambda}$ are available from the first set, so that the only unknowns are $\ddot{Q}_j$ and $\ddot{\Lambda}$. Notice that the second set is independent of $\dddot{Q}_j$ and $\dddot{\Lambda}$.
- Every system has the same operators $(\partial F/\partial Q_j)$ and $(\partial F/\partial \Lambda)$ associated with the unknowns.
- Every system has $(N + 1)$ unknowns. If we include the choice of perturbation parameter as a new condition, then one of the unknown derivatives in each system becomes known.

- For example, if $s \equiv \Lambda$, then the first coefficient becomes $\dot{\Lambda} = d\Lambda/d\Lambda = 1$, and the rest of the derivatives of Λ are all zero $\ddot{\Lambda} = 0 = \dddot{\Lambda}$.
- On the other hand, if we choose $s \equiv Q_1$, then we get $\dot{Q}_1 = 1$, and $\ddot{Q}_1 = \dddot{Q}_1 = 0$. For $j \neq 1$, then $\dot{Q}_j \neq 0$, $\ddot{Q}_j \neq 0$.
- Each new (nonzero) term that is included in the perturbation expansion represents a small correction to the solution, because $(s - s_0)$ is a small parameter.
- If F is truncated due to physical assumptions, then a perturbation approximation cannot include an arbitrary number of terms. For consistency of the solution, if we neglected terms of order higher than three in F, then we should only consider a perturbation expansion up to third order.
- If the solution is not consistent with the degree of approximation adopted for F, then the accuracy decreases as new terms are included in the perturbation expansion.
- We assume that it is possible to solve each one of the systems of simultaneous (perturbation) equations. This means that they are nondegenerate perturbation systems.

3.3.2 Application to Equilibrium Problems

In this section we show that the information of the total potential energy functional and its derivatives at an equilibrium state are sufficient to trace an equilibrium path in the neighborhood of the state considered.

In this case the function F takes the form of N conditions

$$V_i = 0, \qquad i = 1, \dots, N. \tag{3.19}$$

Throughout this book, as in most references in this field, the derivatives of V are identified as follows:

$$V_i = \frac{\partial V}{\partial Q_i}, \qquad V' = \frac{\partial V}{\partial \Lambda}. \tag{3.20}$$

Let us expand the equilibrium path in the form

$$Q_j(s) = Q_j^E + s\dot{Q}_j^E + \frac{1}{2}s^2\ddot{Q}_j^E + \frac{1}{6}s^3\, \dddot{Q}^E{}_j \cdots, \tag{3.21}$$

$$\Lambda(s) = \Lambda^E + s\dot{\Lambda}^E + \frac{1}{2}s^2\ddot{\Lambda}^E + \frac{1}{6}s^3\, \dddot{\Lambda}^E \cdots, \tag{3.22}$$

where the state given by $\left[Q_j^E, \Lambda^E\right]$ is a known reference state, from which the expansion is obtained. Thus, the static perturbation technique allows to obtain information about neighboring states provided the energy and its derivatives are known at one equilibrium state.

If the energy V is truncated to quartic terms, the equilibrium condition includes up to cubic terms in the displacements, so that the perturbation expansion can only be carried out up to cubic terms.

Equilibrium now becomes

$$V_i\left[Q_j(s), \Lambda(s)\right] = 0, \tag{3.23}$$

where we notice that V_i does not depend directly on s.

In this problem we have identified F with V_i. The derivatives that are necessary for the perturbation equations are also identified as

$$\frac{\partial F}{\partial \Lambda} \rightarrow \frac{\partial}{\partial \Lambda}\left(\frac{\partial V}{\partial Q_i}\right) = \frac{\partial^2 F}{\partial \Lambda \partial Q_i} = V_i', \qquad \frac{\partial^2 F}{\partial \Lambda^2} \rightarrow V_i'',$$

$$\frac{\partial F}{\partial Q_j} \rightarrow V_{ij}, \qquad \frac{\partial^2 F}{\partial Q_j \partial Q_k} \rightarrow V_{ijk},$$

$$\frac{\partial^2 F}{\partial Q_k \partial \Lambda} \rightarrow V_{ij}'.$$

The first-order perturbation expansions, (3.9), are now

$$\boxed{V_{ij}\dot{Q}_j + V_i'\dot{\Lambda} = 0} \tag{3.24}$$

The second-order perturbation equations derived from (3.12) are

$$\boxed{V_{ij}\ddot{Q}_j + V_i'\ddot{\Lambda} = -\left(V_{ijk}\dot{Q}_j\dot{Q}_k + 2V_{ij}'\dot{\Lambda}\dot{Q}_j + V_i''\dot{\Lambda}\dot{\Lambda}\right)} \tag{3.25}$$

Such equations should be evaluated at the equilibrium state E, from which we have expanded the solution.

The load as perturbation parameter. A variety of perturbation parameters can be used to follow a nonlinear equilibrium path. The actual values of the derivatives in (3.21) and (3.22) change depending on the choice of s and usually lead to different levels of accuracy. Typical choices for s in nonlinear elasticity are a load parameter, or a displacement parameter, although other choices are also possible.

If we choose the load as the perturbation parameter, then

$$s = \Lambda, \qquad \dot{\Lambda} = 1, \qquad \text{and} \qquad \ddot{\Lambda} = 0.$$

The derivatives $(\dot{\#})$ with respect to the perturbation parameter became now derivatives with respect to the load parameter, and are denoted as $(\#)'$. The evaluated perturbation equations are

$$V_{ij}Q_j' \,|^E = -V_i' \,|^E,$$

$$V_{ij}Q_j'' \,|^E = -\left(V_{ijk}Q_j'Q_k' + 2V_{ij}'Q_j' + V_i''\right) |^E .$$

These equations can be solved sequentially to compute $Q_j'^E$, $Q_j'^E$, etc. The perturbation solution of this problem can be written in the form

$$Q_j(s) = Q_j^E + sQ_j'^E + \frac{1}{2}s^2 Q_j''^E + \frac{1}{6}s^3 Q_j'''^E + \cdots$$

$$\Lambda(s) = \Lambda^E + s.$$

A displacement as perturbation parameter. An alternative approach to solve this problem is to choose a displacement component as perturbation parameter. For example,

$$s = Q_1.$$

Without loss of generality, we assume that Q_1 is an adequate parameter to follow the path, but if Q_1 is zero along the path, or has a small component in it, then a better choice would be to take another component of the displacement vector. Now $\dot{(\#)} = d\,(\#)\,/dQ_1$, and

$$\dot{Q}_1 = 1, \qquad \ddot{Q}_1 = 0.$$

The perturbation equations are

$$\boxed{V_{im}\dot{Q}_m + V_i'\dot{\Lambda} = -V_{i1}}$$

$$\boxed{V_{im}\ddot{Q}_m + V_i'\ddot{\Lambda} = -\left(V_{ijk}\dot{Q}_j\dot{Q}_k + 2V_{ij}'\dot{\Lambda}\dot{Q}_j + V_i''\dot{\Lambda}\dot{\Lambda}\right)}$$

where $m = 2, 3, \ldots, N$. There are $(N-1)$ displacement unknowns and one load unknown in each perturbation equation.

The solution leads to the perturbation expansions

$$Q_1 = s,$$

$$Q_j(s) = Q_j^E + s\dot{Q}_j^E + \frac{1}{2}s^2\ddot{Q}_j^E + \cdots \qquad \text{for} \qquad j \neq 1,$$

$$\Lambda(s) = \Lambda^E + s\dot{\Lambda}^E + \frac{1}{2}s^2\ddot{\Lambda}^E + \cdots.$$

Important:

- It is convenient to adopt s as the largest component of the displacement vector Q_j into the path that we want to follow.

3.4 PERTURBATIONS VIA EXPLICIT SUBSTITUTION

Rather than presenting the technique in symbolic form, in which the derivatives still have to be computed, it is also possible to directly substitute the perturbation expansion in the equation to be solved, and carry out the algebraic operations. This is called an explicit substitution technique for perturbation analysis. This procedure is employed, for example, in [6, 11].

3.4.1 Basic Procedure

Again, we start from (3.8)

$$F\left[Q_j(s), \Lambda(s), s\right] = 0.$$

Now we substitute the expansions (3.2)–(3.3) into (3.8) to obtain

$$F\left\{\left[Q_j + (s - s_0)\dot{Q}_j + \frac{1}{2}(s - s_0)^2 \ddot{Q}_j + \cdots\right],\right.$$

$$\left.\left[\Lambda + (s - s_0)\dot{\Lambda} + \frac{1}{2}(s - s_0)^2 \ddot{\Lambda} + \cdots\right], s\right\} = 0. \tag{3.26}$$

Expanding and collecting terms of like powers in s we get a polynomial expression

$$P(s) = \alpha_0 + (s - s_0)\alpha_1 + (s - s_0)^2 \alpha_2 + \cdots = 0, \tag{3.27}$$

where α_0 is associated with constant terms; α_1 represents the terms associated with $(s - s_0)$; α_2 is associated with quadratic terms in $(s - s_0)$, etc.

- We can now use the **fundamental theorem of perturbations** [13]. This theorem states that if the coefficients α are independent of s, then each one of the coefficients has to be zero in order to satisfy the original equation $F = 0$. Thus, the following relations apply in order to satisfy (3.27)

$$\alpha_0 = 0, \qquad \alpha_1 = 0,$$

$$\alpha_2 = 0, \qquad \alpha_3 = 0. \tag{3.28}$$

These equations are equivalent to the evaluated perturbation equations of the implicit method.

- An alternative procedure is to differentiate $P(s)$ successively with respect to s and evaluate at $s = s_0$. This yields

$$P(s) = \alpha_0 + (s - s_0)\alpha_1 + (s - s_0)^2\alpha_2 + \cdots = 0,$$

$$\frac{dP(s)}{ds} = \alpha_1 + 2(s - s_0)\alpha_2 + 3(s - s_0)^2\alpha_3 + 4(s - s_0)^3\alpha_4 = 0,$$

$$\frac{d^2P(s)}{ds^2} = 2\left[\alpha_2 + 3(s - s_0)\alpha_3 + 6(s - s_0)^2\alpha_4 + \cdots\right] = 0,$$

$$\frac{d^3P(s)}{ds^3} = 3\left[\alpha_3 + 4(s - s_0)\alpha_4 + \cdots\right] = 0.$$

Next we evaluate the above perturbation equations at $s = s_0$ and obtain the following conditions:

$$\alpha_0 = 0, \qquad \alpha_1 = 0,$$

$$\alpha_2 = 0, \qquad \alpha_3 = 0.$$

The above equations are identical to those obtained by means of the fundamental theorem of perturbations in (3.28).

3.4.2 Application to Equilibrium

As an example relevant for the studies in this book, we consider an equilibrium path in terms of the energy V:

$$V = A_i' Q_i \Lambda + \frac{1}{2} A_{ij} Q_i Q_j + \frac{1}{3!} A_{ijk} Q_i Q_j Q_k + \frac{1}{4!} A_{ijkl} Q_i Q_j Q_k Q_l. \tag{3.29}$$

Equilibrium is given by

$$\frac{\partial V}{\partial Q_i} = V_i = 0. \tag{3.30}$$

Let us expand the equilibrium path in the form

$$Q_j(s) = Q_j^E + s\dot{Q}_j^E + \frac{1}{2}s^2\ddot{Q}_j^E + \frac{1}{6}s^3\dddot{Q}_j^E \cdots, \tag{3.31}$$

$$\Lambda(s) = \Lambda^E + s\dot{\Lambda}^E + \frac{1}{2}s^2\ddot{\Lambda}^E + \frac{1}{6}s^3\dddot{\Lambda}^E \cdots, \tag{3.32}$$

where the state given by $\left[Q_j^E, \Lambda^E\right]$ is a known reference state, from which the expansion is obtained.

Next we substitute the above expansions into the equilibrium condition and obtain

$$V_i[Q_j(s), \Lambda(s)] = A_i' \left(\Lambda^E + s\dot{\Lambda}^E + \frac{1}{2}s^2\ddot{\Lambda}^E + \cdots \right)$$

$$+A_{ij} \left(Q_j^E + s\dot{Q}_j^E + \frac{1}{2}s^2\ddot{Q}_j^E + \cdots \right)$$

$$+\frac{1}{2}A_{ijk} \left(Q_j^E + s\dot{Q}_j^E + \frac{1}{2}s^2\ddot{Q}_j^E + \cdots \right) \left(Q_k^E + s\dot{Q}_k^E + \frac{1}{2}s^2\ddot{Q}_k^E + \cdots \right)$$

$$+\frac{1}{6}A_{ijkl} \left(Q_j^E + s\dot{Q}_j^E + \frac{1}{2}s^2\ddot{Q}_j^E + \cdots \right) \left(Q_k^E + s\dot{Q}_k^E + \frac{1}{2}s^2\ddot{Q}_k^E + \cdots \right)$$

$$\times \left(Q_l^E + s\dot{Q}_l^E + \frac{1}{2}s^2\ddot{Q}_l^E + \cdots \right) + \cdots = 0. \tag{3.33}$$

Equation (3.33) is rearranged by putting together terms of like powers of s:

$$V_i(s) = [A_i'\Lambda^E]$$

$$+s \left[A_i'\dot{\Lambda}^E + A_{ij}\dot{Q}_j^E + A_{ijk}Q_j^E\dot{Q}_k^E + \frac{1}{2}A_{ijkl}Q_j^E Q_k^E \dot{Q}_l^E \right]$$

$$+\frac{1}{2}s^2 \left[A_i'\ddot{\Lambda}^E + A_{ij}\ddot{Q}_j^E + A_{ijk}(Q_j^E\ddot{Q}_k^E + \dot{Q}_j^E\dot{Q}_k^E) \right.$$

$$\left. +A_{ijkl} \left(Q_j^E \dot{Q}_k^E \dot{Q}_l^E + \frac{1}{2}Q_j^E Q_k^E \ddot{Q}_l^E \right) \right] + \cdots = 0. \tag{3.34}$$

This equation should be valid for small values of s, including $s = 0$. At $s = 0$, the system leads to the following set of conditions in the form

$$A_i'\dot{\Lambda}^E + A_{ij}\dot{Q}_j^E + A_{ijk}Q_j^E\dot{Q}_k^E + \frac{1}{2}A_{ijkl}Q_j^E Q_k^E \dot{Q}_l^E = 0, \tag{3.35}$$

$$A_i'\ddot{\Lambda}^E + A_{ij}\ddot{Q}_j^E + A_{ijk}(Q_j^E\ddot{Q}_k^E + \dot{Q}_j^E\dot{Q}_k^E)$$

$$+A_{ijkl} \left(Q_j^E \dot{Q}_k^E \dot{Q}_l^E + \frac{1}{2}Q_j^E Q_k^E \ddot{Q}_l^E \right) = 0. \tag{3.36}$$

The perturbation equation of order one, (3.35), can be also written as

$$\dot{\Lambda}^E A_i' + \left(A_{ij} + A_{ijk}Q_k^E + \frac{1}{2}A_{ijkl}Q_l^E Q_k^E \right) \dot{Q}_j^E = 0, \tag{3.37}$$

which should be solved in terms of $\dot{\Lambda}^E$ and $\dot{Q}_j^E$.

The second-order perturbation equation results in

$$A_i'\ddot{\Lambda}^E + \left(A_{ij} + A_{ijk}Q_k^E + \frac{1}{2}A_{ijkl}Q_j^E Q_k^E\right)\ddot{Q}_j^E$$

$$= -(A_{ijk} + A_{ijkl}Q_l^E)\,\dot{Q}_k^E \dot{Q}_j^E, \tag{3.38}$$

and they should be solved for $\ddot{\Lambda}^E$ and $\ddot{Q}_j^E$.

There are many problems for which the starting equilibrium point E is the unloaded and undeformed state (i.e., for $s = 0$, then $Q_j^E = 0$ and $\Lambda^E = 0$) and for them we get the simpler conditions

$$A_i'\dot{\Lambda}^E + A_{ij}\,\dot{Q}_j^E = 0, \tag{3.39}$$

$$A_i'\ddot{\Lambda}^E + A_{ij}\,\ddot{Q}_j^E = -A_{ijk}\dot{Q}_k^E \dot{Q}_j^E. \tag{3.40}$$

Load as a perturbation parameter. If the analysis is carried out using the load as a perturbation parameter, then we get the conditions

$$\left(A_{ij} + A_{ijk}Q_k^E + \frac{1}{2}A_{ijkl}Q_l^E Q_k^E\right)\dot{Q}_j^E = -A_i', \tag{3.41}$$

$$\left(A_{ij} + A_{ijk}Q_k^E + \frac{1}{2}A_{ijkl}Q_j^E Q_k^E\right)\ddot{Q}_j^E = -(A_{ijk} + A_{ijkl}Q_l^E)\,\dot{Q}_k^E \dot{Q}_j^E. \tag{3.42}$$

The solution of the above set leads to $\dot{Q}_j^E$, $\ddot{Q}_j^E$, etc., which are required in the perturbation expansion.

The case of a displacement as perturbation parameter is left as Problem 3.1 at the end of this chapter.

3.5 DEGENERATE PERTURBATION PROBLEMS

In sections 3.3 and 3.4 the analysis led to a set of perturbation equations in terms of the unknown coefficients required by the parametric solution of the problem. Until now we have assumed that it is possible to solve the algebraic system of equations arrived at by perturbations. This means that operators such as $\left[\frac{\partial F}{\partial Q_j}\right]$ have an inverse, and such cases are called **nondegenerate perturbation problems**.

There are, however, **degenerate perturbation problems** for which the symmetric operator $\left[\frac{\partial F}{\partial Q_j}\right]$ is singular.

The first-order perturbation equation, which was found to be

$$\frac{\partial F}{\partial Q_i}\dot{Q}_i + \frac{\partial F}{\partial \Lambda}\dot{\Lambda} + \bar{G} = 0,$$

has $(N+1)$ unknowns (the vector $\dot{Q}_i$ and one scalar unknown $\dot{\Lambda}$). Because we now consider the new situation where $\left[\frac{\partial F}{\partial Q_j}\right]$ is singular, then the independent equations are reduced to $(N-1)$. The usual situation is to have one extra equation derived from the choice of the perturbation parameter, so that the system still needs an extra condition to be solved.

This extra condition can be constructed using the **contraction mechanism**, which provides another scalar condition. The contraction mechanism is a systematic way to obtain a new scalar equation to eliminate the lack of determinacy of the degenerate system.

If $\left[\frac{\partial F}{\partial Q_j}\right]$ is singular, then there must exist a nonzero vector x_j that satisfies the condition

$$\left[\frac{\partial F}{\partial Q_j}\right] x_j = 0. \tag{3.43}$$

The vector x_j is an eigenvector of the symmetric matrix $\left[\frac{\partial F}{\partial Q_j}\right]$.

We said before that the equation of first-order perturbation cannot be solved directly. Let us premultiply this equation by the eigenvector x_j to obtain

$$x_j \frac{\partial F}{\partial Q_i} \dot{Q}_i + x_j \frac{\partial F}{\partial \Lambda} \dot{\Lambda} + x_j \bar{G} = 0. \tag{3.44}$$

Equation (3.44) is called the contracted equation of first order. If F is a vectorial equation, then the contracted form is a scalar equation. If $\left[\frac{\partial F}{\partial Q_i}\right]$ is symmetric, then the first term in (3.44) is zero because of (3.43):

$$x_j \frac{\partial F}{\partial Q_i} \dot{Q}_i = \dot{Q}_i \left(\frac{\partial F}{\partial Q_i} x_j \right) = 0.$$

We can compute the first-order coefficient from (3.44) in the form

$$\dot{\Lambda} = - \left. \frac{x_j \bar{G}}{x_j \left(\frac{\partial F}{\partial \Lambda}\right)} \right|_{s=s_0}. \tag{3.45}$$

With this value, we substitute $\dot{\Lambda}$ in the first-order perturbation condition

$$\frac{\partial F}{\partial Q_i} \dot{Q}_i = - \frac{\partial F}{\partial \Lambda} \dot{\Lambda} - \bar{G}. \tag{3.46}$$

Now we can solve (3.46) and compute $\dot{Q}_i$. Thus, the solution of a degenerate perturbation problem is carried out using the perturbation equations and the contracted form of the perturbation equations. A similar procedure is employed for higher-order perturbation equations. Problems similar to those arise whenever the starting point of the perturbation analysis is a critical state, such as initial postbuckling equilibrium paths.

3.6 SINGULAR PERTURBATION PROBLEMS

In sections 3.3 and 3.5, the existence of all derivatives $\dot{Q}_i$, $\ddot{Q}_i$, required for the perturbation expansions was assumed. For this reason, we could use perturbations based on Taylor series, and these are called **regular** perturbations.

However, if one or several derivatives in a Taylor expansion do not exist, or are infinite, then the perturbations are called **singular**, and new techniques have to be employed instead of the Taylor series approach. The causes of singularity in the perturbation analysis may be due to the following:

- A singularity in the domain, in which there are some features of the domain that present singularities, such as crack tip problems, or infinite domains in mechanics.
- A singularity in the model, and these are called **boundary layer problems**. Whenever there is a singularity in the model, this is only seen in the response. In boundary layer problems there is a narrow zone in the domain for which no matter how small the perturbation parameter is, the perturbation in the response does not disappear.

A case in which perturbations are singular at $s = s_0$ is shown in Figure 3.3.

A way to obtain an expansion of the solution starting at $s = s_0$ is to employ power series with fractional exponents. We do not know what exponents will be obtained in a given expansion, so that not only the coefficients but also the exponents are now treated as unknowns.

Rather than using the Taylor expansion, we can use a polynomial expansion of the form

$$Q_j(s) = Q_j^E + (s - s_0)^M Q_{j1} + (s - s_0)^O Q_{j2} + \cdots, \tag{3.47}$$

$$\Lambda(s) = \Lambda^E + (s - s_0)^N \Lambda_1 + (s - s_0)^P \Lambda_2 + \cdots, \tag{3.48}$$

where M, N, O, P are unknown exponents; and Q_{j1}, Q_{j2}, Λ_1, Λ_2 are not derivatives but coefficients to be evaluated as part of the perturbation analysis.

By explicit substitution we obtain

$$F\left[\left(Q_j^E + (s - s_0)^M Q_{j1}\right), \left(\Lambda^E + (s - s_0)^N \Lambda_1\right), s\right] = 0. \tag{3.49}$$

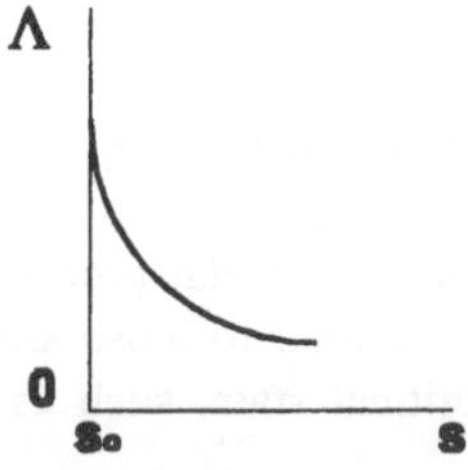

Figure 3.3 An example of singular perturbations at $s = s_0$.

To solve this equation, we can use trial and error: Try values of M, N and obtain a polynomial expression

$$(s - s_0)^R B_R + (s - s_0)^S B_S + \cdots 0 \tag{3.50}$$

with $R < S$. The first term must be zero to satisfy the above condition, i.e.,

$$B_R[Q_{j1}, \Lambda_1] = 0. \tag{3.51}$$

This equation can be solved for Q_{j1}, and Λ_1.

The solution for the true exponents M, N is the only consistent solution that can be obtained for a given problem; other exponents simply lead to conditions that cannot be satisfied. For this reason it is called the **least degenerate solution**. This technique was first employed in the context of stability theory for one-degree-of-freedom systems in [7], and a full account can be found in [1]. It will be used in Chapter 9 to evaluate the influence of imperfections in bifurcation analysis.

3.7 EQUILIBRIUM PATH OF A SHALLOW ARCH

In Chapter 2 we obtained the total potential energy V of a circular arch under a point load in order to investigate the exact equilibrium path and several approximations. In this section we employ the same energy to compute an approximate path using perturbation techniques and to illustrate different orders of approximation depending on the number of terms included in the expansion.

The energy V results in

$$V = A_1' + \frac{1}{2}(A_{11} + A_{22}) + \frac{1}{3!}(A_{111} + 3A_{122}) + \frac{1}{4!}(A_{1111} + A_{2222} + 6A_{1122}).$$

The energy coefficients are

$$A_1' = -1,$$

$$A_{11} = \left(\frac{\pi}{2\theta_0}\right)^4 \frac{I}{AR^2} + \frac{8}{\pi^2}, \qquad A_{22} = \left(\frac{\pi}{\theta_0}\right)^4 \frac{I}{AR^2},$$

$$A_{111} = -\frac{3}{2}\frac{\pi}{\theta_0^2}, \qquad A_{122} = -\frac{2\pi}{\theta_0^2},$$

$$A_{1111} = \frac{3}{2}\left(\frac{\pi}{2\theta_0}\right)^4, \qquad A_{2222} = \frac{3}{2}\left(\frac{\pi}{\theta_0}\right)^4,$$

$$A_{1122} = \frac{3}{4}\left(\frac{\pi}{\theta_0}\right)^4,$$

where $\Lambda = \frac{P}{EA\theta_0}$.

Let us now compute the equilibrium path starting at the unloaded state by means of perturbation techniques. We explore two possibilities: the load and one displacement component as perturbation parameters.

Example 3.2 (Load as perturbation parameter). *In this case we have*

$$s = \Lambda.$$

The first-order perturbation coefficients result in

$$\dot{Q}_1 = \frac{A_1'}{A_{11}} = \frac{1}{\frac{(\pi/\theta_0)^4}{16}\frac{I}{AR^2} + 8/\pi^2}, \qquad \dot{Q}_2 = 0, \qquad \dot{\Lambda} = 1.$$

The second-order perturbation coefficients are

$$\ddot{Q}_1 = \frac{3/2(\pi/\theta_0^2)\dot{Q}_1\dot{Q}_1}{\frac{(\pi/\theta_0)^4}{16}\frac{I}{AR^2} + 8/\pi^2}, \qquad \ddot{Q}_2 = 0, \qquad \ddot{\Lambda} = 0.$$

The third-order coefficients take the form

$$\dddot{Q}_1 = -\frac{3/32\,(\pi/\theta_0)^4\,\dot{Q}_1^3 + 9/2(\pi/\theta_0^2)\dot{Q}_1\ddot{Q}_1}{\frac{(\pi/\theta_0)^4}{16}\frac{I}{AR^2} + 8/\pi^2}, \qquad \dddot{Q}_2 = 0, \qquad \dddot{\Lambda} = 0.$$

The path is represented in terms of the load parameter Λ *as*

$$Q_1(\Lambda) = \Lambda\dot{Q}_1^E + \frac{1}{2}\Lambda^2\ddot{Q}_1^E + \frac{1}{6}\Lambda^3\,\dddot{Q}_1^E + \cdots,$$

$$Q_2(\Lambda) = 0.$$

These results indicate that for the present choice of generalized coordinates, the only active coordinate along the fundamental path is Q_1. *Figure* 3.4 *shows the path computed in this section for different levels of approximation. There are significant differences between the linear, quadratic, and cubic approximations in terms of the load parameter. Notice that we cannot improve the accuracy of the perturbation expansion by adding terms higher than three, because the equilibrium condition was truncated at the third order.*

Example 3.3 (A displacement component as a perturbation parameter). *Let us now consider the first displacement component as a perturbation parameter and follow the equilibrium path of the arch*

$$s = Q_1.$$

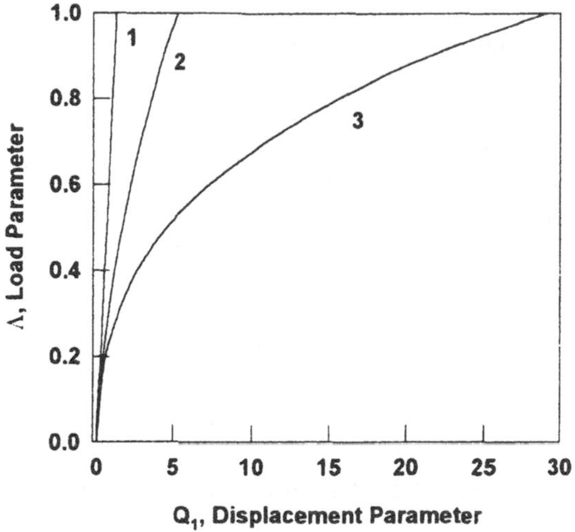

Figure 3.4 Equilibrium paths for a circular arch under a central point load. Load as a perturbation parameter. $r/h = 50$, $A = 1$, $I = 1/12$, and the central angle of the arch is $\theta_0 = 60$ degrees. (1) First-order, (2) second-order, (3) third-order perturbation parameter.

Again, the values of Q_2 are zero in all perturbation equations, and the coefficients of the load result in

$$\dot{\Lambda} = \frac{(\pi/\theta_0)^4}{16}\frac{I}{AR^2} + \frac{8}{\pi^2},$$

$$\ddot{\Lambda} = -\frac{3}{2}\left(\frac{\pi}{\theta_0^2}\right),$$

$$\dddot{\Lambda} = \frac{3}{32}\left(\frac{\pi}{\theta_0}\right)^4.$$

This leads to the following equation for the path:

$$\Lambda(Q_1) = Q_1\dot{\Lambda}^E + \frac{1}{2}Q_1^2\ddot{\Lambda}^E + \frac{1}{6}Q_1^3\dddot{\Lambda}^E.$$

Plots of the different approximations are shown in Figure 3.5. The accuracy improves with the number of terms included in the perturbation expansion, and the errors depend on how far the solution is computed from the original equilibrium state. For $Q_1 = 0.2$, the quadratic and cubic approximations are very similar; however, for $Q_1 = 0.6$ there are significant errors in the quadratic solution.

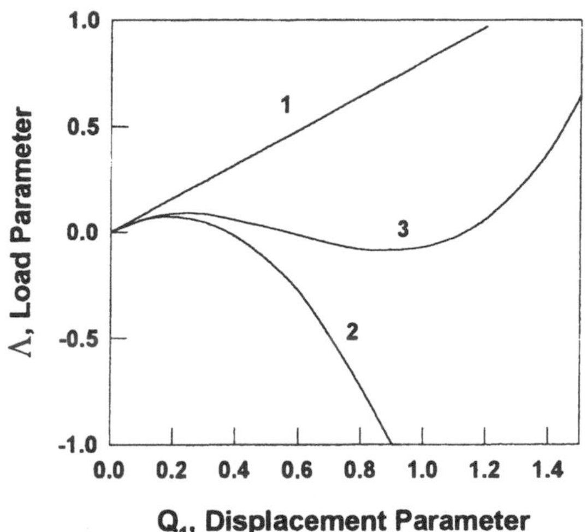

Figure 3.5 Displacement as a perturbation parameter in the arch problem of Example 3.3. (1) First-order, (2) second-order, (3) third-order perturbation parameter.

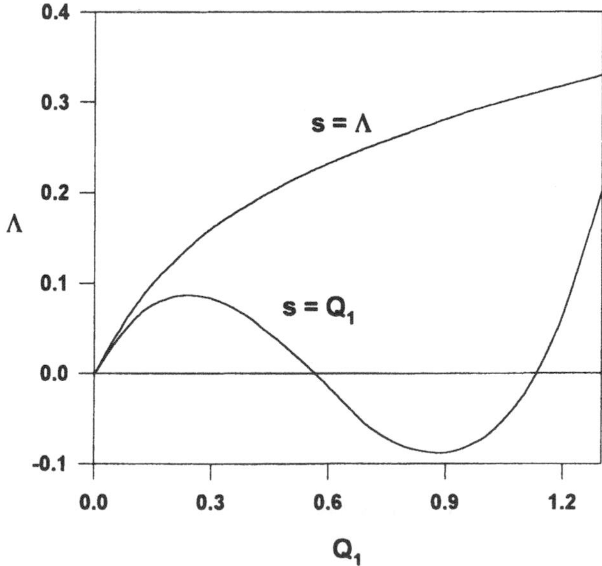

Figure 3.6 A comparison of third-order solutions for load versus displacement as perturbation parameter in the arch problem.

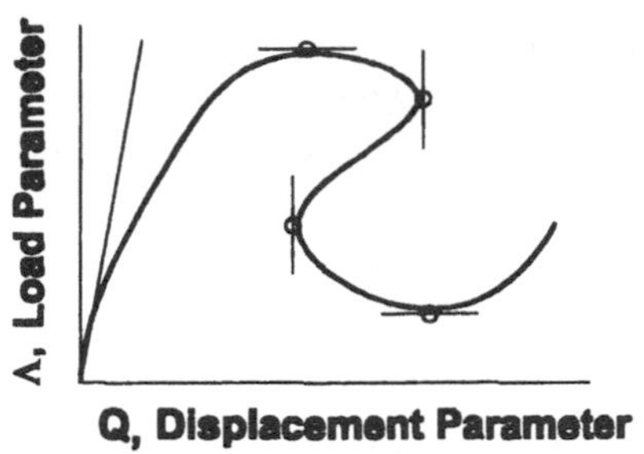

Figure 3.7 More complex equilibrium paths.

The fundamental equilibrium path of an arch under a central load has been computed using perturbations with the load and a displacement as perturbation parameters. For the same order of approximation, the results are very different. For example, we have plotted the third-order perturbation solution in Figure 3.6 *for both choices of the perturbation parameter, and they differ by large values.*

For the present problem, the choice of the load as perturbation leads to a path that raises monotonically with the load. This approach only approximates the exact solution for low values of the load and diverts from it for higher loads. Clearly, with this choice one cannot represent any path that decreases in the load with increasing displacements.

The choice of a displacement as perturbation parameter is more accurate in this case and represents well the exact solution of the problem.

The general conclusion is that whenever we expect a path to have a descending part with increasing displacements, then the load cannot be chosen as perturbation parameter, and a displacement component is a good choice.

Finally, there are cases that display decreasing load and displacement for some range of the variables, as indicated in Figure 3.7. In cases like this, there are two choices: either we update the origin of the analysis, denoted by E in this chapter, or else we choose a different perturbation parameter such as a scalar quantity associated with the work involved in the increment or another parameter.

3.8 PROBLEMS

Review questions. (a) Explain the differences between perturbation and continuation methods. (b) In what way are regular perturbations different from singular perturbations? (c) What are implicit differentiation and explicit substitution techniques in perturbation analysis? (d) Is the solution of a problem independent of the choice of perturbation parameter? (e) Distinguish between degenerate and nondegenerate perturbation problems.

Problem 3.1. Use the technique of explicit substitution with a displacement component as perturbation parameter to obtain the general perturbation equations for the equilibrium problem defined by $V_i = 0$. Use a symbolic manipulator to carry out the substitutions.

Problem 3.2 (Two-degrees-of-freedom model). For a given structure, the following energy coefficients have been computed:

$$A_1 = A_2 = -\frac{\Gamma}{2}\Lambda, \qquad A_{11} = A_{22} = 90.16\frac{\Gamma}{2}, \qquad A_{12} = 71.92\frac{\Gamma}{2},$$

$$A_{112} = A_{221} = 0.4\frac{\Gamma}{2}, \qquad A_{111} = A_{222} = -2402.4\frac{\Gamma}{2},$$

$$A_{1111} = A_{2222} = 24040\frac{\Gamma}{2}, \qquad A_{1112} = A_{2221} = -12026\frac{\Gamma}{2}, \qquad A_{1122} = 12024\frac{\Gamma}{2},$$

where Γ is a constant. Use the load as perturbation parameter and compute the equilibrium path.

Problem 3.3 (Two-degrees-of-freedom model). For the structure of Problem 3.2, use a displacement component as a perturbation parameter.

Problem 3.4 (Augusti column). The equilibrium condition of the Augusti column (more will be discussed about this in Chapters 13 and 14) is given by

$$V_1 = \left(A_{11} + \frac{1}{6}A_{1111}Q_1^2 + \frac{1}{2}A_{1122}Q_2^2\right)Q_1 + \Lambda K_1\bar{\xi} = 0,$$

$$V_2 = \left(A_{22} + \frac{1}{6}A_{2222}Q_2^2 + \frac{1}{2}A_{1122}Q_1^2\right)Q_2 + \Lambda\tau K_1\bar{\xi} = 0,$$

where

$$A_{11} = K_1(1-\Lambda), \qquad A_{22} = K_1(\tau-\Lambda), \qquad A_{1111} = A_{2222} = K_1\Lambda = -A_{1122}.$$

Obtain the equilibrium path using a displacement as perturbation parameter, with cubic perturbation terms. Plot the equilibrium path, with $\tau = 1$, $K_1 = 1$, and $\bar{\xi} = 0.06$. The solution is independent of the value of K_1, since it enters into all the energy coefficients.

Problem 3.5 (Augusti column). Solve Problem 3.4 using $\tau = 1.05$.

Problem 3.6. Write a program using Maple V or other symbolic manipulator to evaluate the derivatives of V up to fourth order using loops. Consider a four-degrees-of-freedom system.

Problem 3.7 (Plane frame). The frame considered in Problem 2.9 is loaded by horizontal and vertical loads. Compute the equilibrium path starting from the unloaded configuration, using perturbation techniques. Use data as in Problem 2.9.

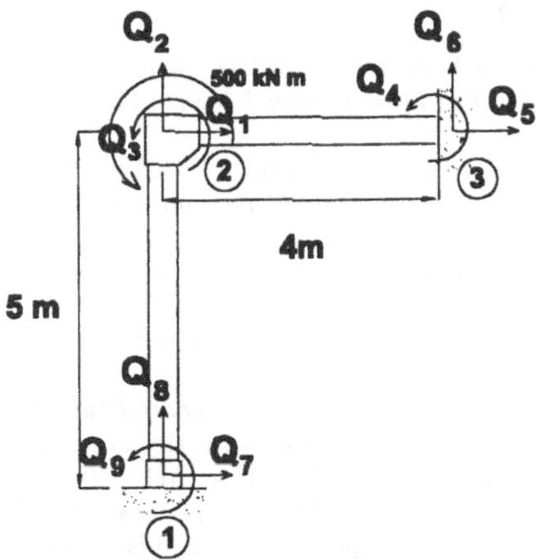

Figure 3.8 Frame of Problem 3.9.

Problem 3.8 (Column). For Problem 2.5 of a column with bending-extension coupling, use regular perturbations and $s = Q_3$ to evaluate the fundamental equilibrium path. Consider $A = 1.15MN$, $D = 3.47Nm^2$, $B = 1.3$, $E = 38.58GPa$, cross section $10 \times 3mm^2$.

Problem 3.9 (Frame). A wooden frame is shown in Figure 3.8. Calculate the equilibrium path using perturbation analysis, up to three times the load indicated. Assume $E = 10.5GPa$, $I = 15 \times 10^6 mm^4$, $A = 7500mm^2$. Joints 1 and 2 are fixed, and 3 is pinned. Choose Q_3 (rotation at the common joint) as perturbation parameter. Use the formulation of Appendix B to obtain the total potential energy.

Problem 3.10 (Frame). Solve Problem 3.9 with the load as perturbation parameter.

Problem 3.11 (Frame). The structure of Figure 3.9.a is idealized as shown in Figure 3.9.b, so that there is only one load parameter. Take $L = 6.1m$, $E = 200GPa$, $I = 2.4974 \times 10^8 mm^4$, $A = 7741.8mm^2$. Carry out the perturbation analysis using the central rotation as perturbation parameter. The two end nodes are fixed.

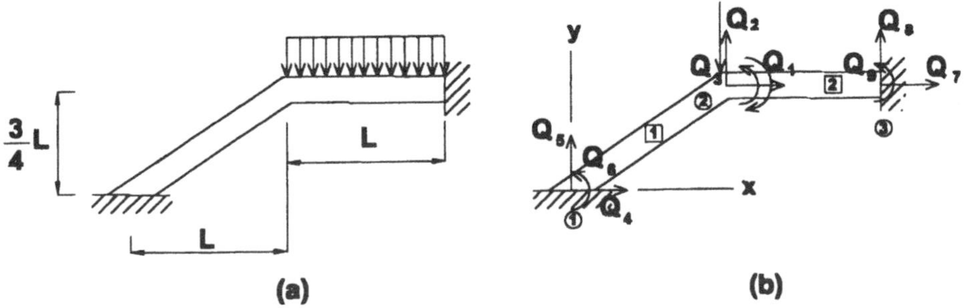

Figure 3.9 Frame of Problem 3.11 (a) actual load system; (b) idealization.

3.9 BIBLIOGRAPHY

[1] Banchio, E. G., and Godoy, L. A., Sensibilidad de estados criticos en inestabilidad estructural via perturbaciones singulares, *Rev. Internac. Metod. Numér. Calc. Diseñ. Ingr.*, 13(4), 1997, 467–486 (in Spanish).

[2] Bathe, K. J., *Finite Element Procedures*, Prentice-Hall, Englewood Cliffs, NJ, 1996.

[3] Bellman, R. E., *Perturbation Techniques in Mathematics, Physics and Engineering*, Holt Rinehart & Winston, New York, 1964.

[4] Bush, A. W., *Perturbation Methods for Engineers and Scientists*, CRC Press, Boca Raton, FL, 1994.

[5] Crisfield, M. A., *Nonlinear Finite Element Analysis of Solids and Structures, Vol. 1: Essentials*, Wiley, Chichester, UK, 1991.

[6] Croll, J. G. A., and Walker, A. C., *Elements of Structural Stability*, Macmillan, London, 1972.

[7] Godoy, L. A., and Mook, D. T., Higher order sensitivity to imperfections in bifurcation buckling analysis, *Internat. J. Solids Structures*, 33(4), 1996, 511–520.

[8] Godoy, L., Flores, F., Raichman, S., and Mirasso, A., *Tecnicas de Perturbacion en el Analisis No Linear Mediante Elementos Finitos*, Asociacion Argentina de Mecanica Computacional, Cordoba, Argentina, 1990.

[9] Keller, L. B., Perturbation Theory, in *Teoria de Bifurcacoes e suas Aplicacoes*, vol. 2, Laboratorio Nacional de Computacao Cientifica, LNCC/CNPq, Rio de Janeiro, 1985, 83–152.

[10] Kevorkian, J., and Cole, J. D., *Perturbation Methods in Applied Mathematics*, Applied Mathematical Sciences Series, vol. 34, Springer-Verlag, Berlin.

[11] Nayfeh, A., *Perturbation Methods*, Wiley, New York, 1973.

[12] Sewell, M. J., The static perturbation technique in buckling problems, *J. Mech. Phys. Solids*, 13, 1965, 247–265.

[13] Simmonds, J. G., and Mann, J. E., *A First Look at Perturbation Theory*, R. E. Krieger, Malabar, FL, 1986.

[14] Thompson, J. M. T., Discrete branching points in the general theory of elastic stability, *J. Mech. Phys. Solids*, 13, 1965, 295.

[15] Thompson, J. M. T., and Hunt, G. W., *A General Theory of Elastic Stability*, Wiley, London, 1973.

CHAPTER

FOUR

STABILITY OF AN EQUILIBRIUM STATE

4.1 SOME BASIC CONCEPTS ABOUT STABILITY CRITERIA

In Chapter 2 we considered equilibrium states in nonlinear elasticity, and in Chapter 3 we looked at one technique (perturbations) for the evaluation of equilibrium states. It is clear to any engineer that a structural design should satisfy the conditions of equilibrium. The same applies to situations in which the assessment of the safety of an existing structure is required. But equilibrium in a design office is not enough. The designer also needs to know if those equilibrium states calculated on paper will be observable when the structure is built and bears the design loads.

It could happen that a certain state of equilibrium of a structure computed by an engineer at a design stage is an idealization of conditions that cannot be achieved in practice. Furthermore, how will the structure in equilibrium behave if there are perturbations? Will the structure evolve into a different state in the presence of small environmental actions? These are very important matters in the context of engineering design. The concept of stability of equilibrium states deals with these questions.

As stated in [5], there is no universal answer to the question, "What is stability?" To identify if a system is at a stable or an unstable equilibrium configuration, it is necessary to employ some **criterion** of stability. In the context of mechanics, the stability of an equilibrium state is defined by introducing a physical **perturbation** into the system and investigating the subsequent response. An intuitive criteria is as follows:

- If the system returns to the original equilibrium state when the perturbation is eliminated, we say that the equilibrium state is **stable**.

- If the system does not return to the original state, but moves to a very different one, we say that the equilibrium state is **unstable**.
- **Neutral** equilibrium occurs if the system stays in the perturbed configuration and neither returns to the original equilibrium state nor goes to a different state.

For static problems, similar to those studied in this book, we may introduce a dynamic perturbation (a small motion of the system), and the study should be carried out within the realm of dynamics. Other possibilities exist for static systems: The perturbation can also be considered as static, and in such cases a dynamic analysis is not required.

In this chapter we investigate the **total potential energy criterion of stability**, which is employed in the rest of the book. But for the sake of completeness, the static and dynamic criteria are also presented.

The most general criterion of stability, valid for dynamic as well as static problems, is due to Lyapunov [10]. This criterion is described in section 4.2: It is dynamic in the sense that it deals with motions of the system and can be visualized in the so-called phase space, where we plot the displacements versus velocities of the system in motion. However, to assess the stability of a state of static equilibrium, there are simpler tests that can be employed, and they all differ on the nature of the perturbation introduced and on how they evaluate the response.

The central part of the chapter is section 4.3, in which the total potential energy criterion is introduced. This criterion is presented in an axiomatic way, and several forms to implement the stability test are shown. Stability in terms of a diagonal quadratic form is discussed in section 4.4. The concept of stability of an equilibrium path is introduced in section 4.5. In section 4.6 we present the dynamic criterion based on small vibrations of the system; this study falls within the field of linear vibrations. The static criterion is summarized in section 4.7. Finally, other stability criteria are mentioned in section 4.8.

In this book we investigate the stability of states that already satisfy equilibrium. However, one can also look at the stability of nonequilibrium states. This is a very useful concept in nonlinear thermodynamics, and the reader interested in this topic may find a good reference in the work of Prigogine [11].

4.2 STABILITY CRITERIA OF LYAPUNOV

4.2.1 Lyapunov's First Method

Let us consider a general nonlinear system that does not explicitly depend on the time t. Systems like that are called **autonomous** systems.

A state is said to be stable if any trajectory of motion that starts near to that state remains near to it during its motion.

To have a more formal definition, we need to specify what should be understood by "starts near" and "remains near."

For example, consider a dynamic system governed by the two conditions

$$\frac{dx}{dt} = f(x, y), \qquad \frac{dy}{dt} = g(x, y). \tag{4.1}$$

The actual meaning of the variables in the specific context of structural mechanics is not relevant at the moment. The two equations are independent of the parameter t; thus the system is autonomous.

A solution of the two equations is an equilibrium state, given by $x = x_0$ and $y = y_0$, so that

$$f(x_0, y_0) = 0, \qquad g(x_0, y_0) = 0. \tag{4.2}$$

Such a state is illustrated by point O in Figure 4.1.

Consider a solution of this problem as a function of t, given by

$$x = \varphi(t), \qquad y = \psi(t). \tag{4.3}$$

This is a possible response of the system, which does not satisfy the homogeneous conditions (4.2). The starting point of the motion ($t = 0$) defined by (4.3) is a state A in Figure 4.1. The distance between A and O is

$$\overline{OA} = \sqrt{[\varphi(0) - x_0]^2 + [\psi(0) - y_0]^2}. \tag{4.4}$$

During the motion with t, the distance between a generic point B along the trajectory and O is

$$\overline{OB} = \sqrt{[\varphi(t) - x_0]^2 + [\psi(t) - y_0]^2}. \tag{4.5}$$

Let us consider that

$$\overline{OB} < \varepsilon, \tag{4.6}$$

$$\overline{OA} < \delta. \tag{4.7}$$

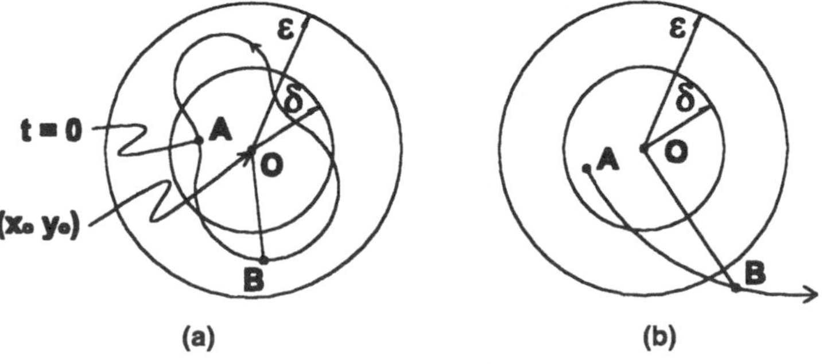

Figure 4.1 (a) Stable state, although not asymptotically stable; (b) unstable state.

We say that

A state (x_0, y_0) *of an autonomous system is stable if given a scalar value* $\varepsilon > 0$, *another* $\delta > 0$ *exists so that for every solution* $x = x(t)$, $y = y(t)$ *that falls inside a sphere of radius* δ *at* $t = 0$, *then the motion also falls inside a sphere of radius* ε *for all* $t \geq 0$.

The above condition is illustrated in Figure 4.1. The state A is close to the state O at the initial time, $t = 0$, let us say, at a distance that is smaller than δ. During the motion of the system, the state at A evolves and occupies different positions, but the distance to the state O is always smaller than a value ε.

But it may so happen that the trajectory goes away from point O as t increases. This is shown in Figure 4.1.b, where the motion starts close to the state (x_0, y_0), so that $\overline{OA}$ is smaller than δ. However, as time increases the distance $\overline{OB}$ increases. The original state (x_0, y_0) is said to be unstable. One cannot satisfy (4.6) and (4.7).

There is also a different condition of stability, known as **asymptotic stability**. This criterion considers what the distance $\overline{OB}$ is as $t \to \infty$. If $\overline{OB} \to 0$, i.e., the motion approaches the original state, then the state (x_0, y_0) is said to be asymptotically stable. Such a state is illustrated in Figure 4.2, and the condition may be written as follows:

A state (x_0, y_0) *is asymptotically stable if, given a scalar value* $\delta > 0$, *for every solution* $x = x(t)$, $y = y(t)$ *that falls inside a sphere of radius* δ *at* $t = 0$, *then it also approaches the state* (x_0, y_0) *as* $t \to \infty$.

Notice that in some cases a state may satisfy the condition to be stable but not asymptotically stable. This case is shown in Figure 4.1, where the trajectory as t increases forms a curve that encloses but does not contain the original state (x_0, y_0). In other words, if we introduce a perturbation in a state for which we investigate instability, then it remains close to the state investigated within a specified level of proximity but does not return to it.

On the other hand, a state can satisfy the above condition of asymptotic stability and yet be unstable. This is illustrated in Figure 4.3, where the state $\overline{OB}$ crosses through (x_0, y_0) but does not remain within a sphere of radius ε all the time. In this case, we say that the system is not stable.

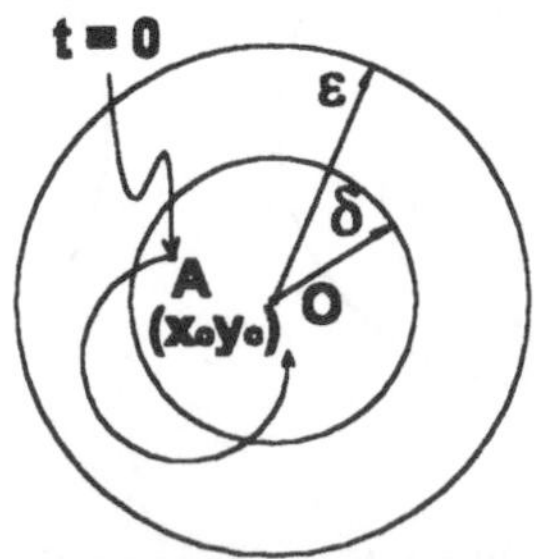

Figure 4.2 Asymptotically stable state.

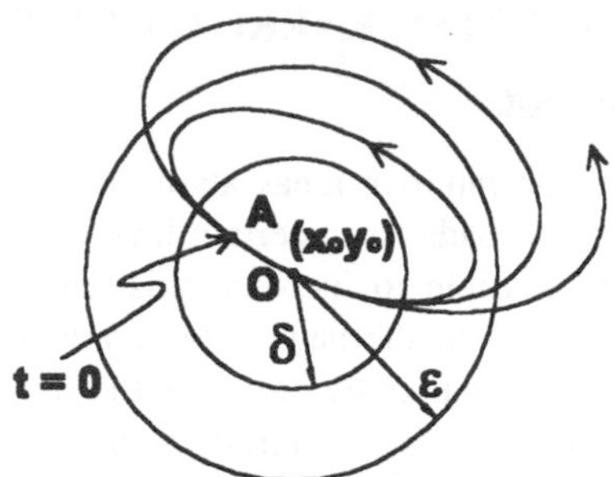

Figure 4.3 Unstable state.

Lyapunov's first method gives rise to the dynamic criterion of stability in structures, which is briefly described in section 4.6.

4.2.2 Lyapunov's Second Method

The evaluation of stability according to the criteria of Lyapunov's first method requires the actual computation of trajectories starting close to the original state. In Lyapunov's direct method (also called Lyapunov's second method), we are not required to solve the equations of motion to compute the trajectories, but to consider the behavior of a so-called Lyapunov function.

Lyapunov's function is a function $V(x, y)$ such that it defines a potential surface, with level curves that encircle the equilibrium state, i.e., the origin (x_0, y_0). The values of V should decrease along the trajectories of the system as they approach (x_0, y_0). We shall not discuss further properties of such functions in general systems, because in structural mechanics the total potential energy can be used as Lyapunov's function.

According to Lyapunov's stability theorem,

If V is a positive definite Lyapunov function in the domain studied containing (x_0, y_0), then a state (x_0, y_0) is stable if the derivative of V with respect to the system x, y is negative semidefinite. If the derivative is negative definite, then the state is asymptotically stable.

The derivatives of V with respect to the system are

$$\dot{V} = \frac{\partial V}{\partial x}\frac{dx}{dt} + \frac{\partial V}{\partial y}\frac{dy}{dt}. \tag{4.8}$$

For a detailed discussion of the method of stability due to Lyapunov, the reader is referred to [9, 12]. In the rest of the chapter we shall consider the total potential energy criterion, which is derived from Lyapunov's second method for elastic structures. Rather than defining a new Lyapunov function V, in structural mechanics we can employ the energy because it satisfies the requirements of defining a potential surface. This is explored in section 4.3.

4.3 THE TOTAL POTENTIAL ENERGY CRITERION

4.3.1 Preliminary Concepts

We start by considering some intuitive ideas about available energy and stability. If an elastic body under external loads has a certain level of energy in a given deformed configuration, and if it is possible to increase this energy level, then the changes in the system from the original to the new state should be done with some energy expense. However, the change to a configuration with a lower level of energy is done by releasing some of the elastic energy stored in the body. Changes in the energy levels as those mentioned above could be produced by some small perturbations. In a thought experiment one can imagine changes in the generalized coordinates of the system, but in real life these perturbations may arise from some environmental actions.

First, consider that a configuration in the deformed body is such that the energy cannot decrease because it has already reached a minimum. If a small perturbation is introduced, then one would say that this state will not change and the energy level will remain the same even in the presence of small ambient actions.

On the contrary, if the energy cannot increase because it has already reached a maximum, the system is not capable of absorbing an energy input. But since the energy may decrease due to environmental actions, then the system may release energy spontaneously, and in doing so it may change to another configuration without any energy input.

The first case (the energy does not decrease with perturbations) may be called stable, since the system spends the minimum possible energy to stay in that position, and it spontaneously prefers to be there rather than in another configuration. In the second case (the energy cannot increase with perturbations) the system spends the maximum possible energy to stay there, and it prefers to be at another configuration, less expensive in energy terms. Given a chance, the system tries to escape from this energy-expensive configuration and to find another, less demanding one. This second case may be called unstable: The system would not be there spontaneously.

4.3.2 Axiomatic Definition of Stability

In Chapter 2 we introduced an axiom regarding equilibrium via total potential energy (Axiom I in section 2.4). We shall state the energy criterion of stability also in an axiomatic way, as follows [14]:

Axiom II. *A necessary and sufficient condition for the stability of an equilibrium state is that the total potential energy should be a complete minimum with respect to changes in the generalized coordinates.*

Axiom III. *A necessary and sufficient condition for instability of an equilibrium state is that the total potential energy should be a complete maximum with respect to changes in the generalized coordinates.*

A special case occurs if the energy is locally at a stationary condition (so that equilibrium is satisfied), but it is neither a minimum nor a maximum. Such states are said to have **neutral** stability.

Notice that the two axioms make reference to a minimum (or maximum) with respect to other neighboring configurations obtained by changes in the response parameters of the system (the generalized coordinates) but for a fixed value of the control parameter (the load).

The energy criterion is not self-evident, and the need for an axiomatic definition may be found in the works of different authors in this field:

> The energy criterion "is essentially a statement of faith and can be confirmed only by correspondence between experimental results and analytical predictions made using the assumption" [4, p. 13].
>
> It seems clear, however, that a complete proof will eventually be found, and in the meantime we follow Koiter by expressing our complete faith in the axiom. [14, p. 87]

But the results of the energy criterion of stability can be successfully correlated with results from the dynamic criterion of stability (section 4.6). It is found that whenever the frequencies of vibration of a system are real, the total potential energy is at a minimum; and when at least one is imaginary, then V is at a maximum. A formal general proof of the relation between energy and stability is a difficult task, but the use of the above ideas in mechanics has been of great value to improve our understanding of the buckling and postbuckling of structures. Thus, validation of the energy criterion is based on the experience that it leads to the same conclusions as the dynamic criterion. For continuous systems, this idea was introduced by Koiter in his doctoral thesis [7].

Remarks

- The energy criterion of stability is based on a global measure (the energy of the complete body), not on local quantities such as stresses.
- It is similar to the dynamic criterion, in which the global quantities are the frequencies of vibration.
- One of the limitations of the energy criterion is that it only applies to elastic behavior, so that the stability of states involving plastic deformations cannot be assessed in this way.

There are several ways to implement the evaluation of stability of an equilibrium state via the total potential energy using Axiom I. Such ways lead to forms of the stability criterion in terms of

- the energy itself,
- a Taylor expansion of the energy,
- the matrix associated with the quadratic form (Hessian),
- the determinant of the quadratic form,
- the diagonal quadratic form.

Let us consider each one of them and illustrate the application to a simple two-degrees-of-freedom problem.

4.3.3 Stability Criterion in Terms of the Energy Itself

Consider an equilibrium state E, identified by a load level $\Lambda = \Lambda^E$ and generalized coordinates $Q_i = Q_i^E$. The energy in this case is given by $V^E = V[Q_i^E, \Lambda^E]$.

Next, consider a perturbation q_i in the generalized coordinates at the same load level, given by

$$\boxed{Q_i = Q_i^E + q_i, \qquad \Lambda = \Lambda^E} \tag{4.9}$$

Most probably the new state identified by $[Q_i, \Lambda^E]$ will not satisfy equilibrium, but it is still possible to compute its energy as $V = V[Q_i^E + q_i, \Lambda^E]$. According to Axiom II, if the equilibrium state $\left(Q_i^E, \Lambda^E\right)$ is stable, then its energy should be at a local minimum, so that any other state (Q_i, Λ) different from E should lead to a larger value of V. This condition may be written as

$$\boxed{V[Q_i^E + q_i, \ \Lambda^E] - V[Q_i^E, \Lambda^E] > 0 \Longrightarrow \text{stable}} \tag{4.10}$$

for any $q_i \neq 0$.

Unstable equilibrium, on the other hand, occurs if

$$V[Q_i^E + q_i, \ \Lambda^E] - V[Q_i^E, \Lambda^E] < 0 \Longrightarrow \text{unstable.} \tag{4.11}$$

The case of

$$V[Q_i^E + q_i, \ \Lambda^E] - V[Q_i^E, \Lambda^E] = 0 \Longrightarrow \text{neutral} \tag{4.12}$$

is known as neutral equilibrium.

This form of the energy criterion makes no distinction between regular and critical states, and for that we have to explore the criterion in more detail.

Example 4.1 (Stability using the energy). *Let us consider a two-degrees-of-freedom system, in which the energy is written in the form*

$$V = A_i' \Lambda Q_i + \frac{1}{2!} A_{ij} Q_i Q_j + \frac{1}{3!} A_{ijk} Q_i Q_j Q_k + \frac{1}{4!} A_{ijkl} Q_i Q_j Q_k Q_l.$$

Assume that the coefficients of V for this specific example are given by

$$A_1' = A_2' = -1, \qquad A_{11} = A_{22} = 90.16, \qquad A_{12} = 71.92,$$

$$A_{111} = A_{222} = -2402.4, \qquad A_{112} = A_{122} = 0.4,$$

$$A_{1111} = A_{2222} = 24040, \qquad A_{1112} = A_{2221} = -12026, \qquad A_{1122} = 12024.$$

We want to evaluate the stability of the equilibrium state $Q_1^E = Q_2^E = 0.005$ and $\Lambda^E = 0.78$ using the energy itself.

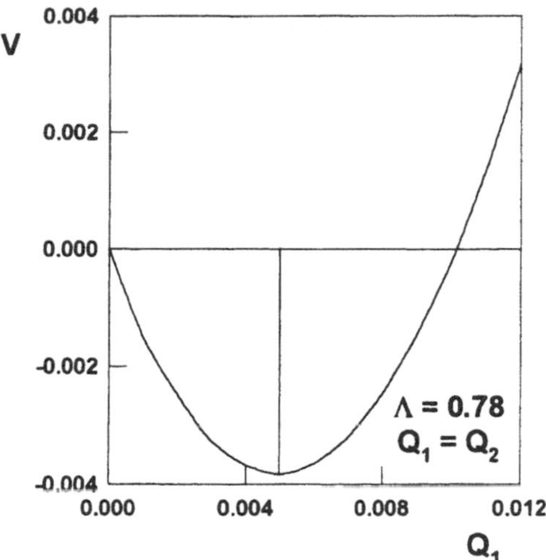

Figure 4.4 The energy at a stable (minimum) equilibrium state.

First, we compute the energy at the given state, leading to

$$V[Q_1^E, Q_2^E, \Lambda^E] = -0.00385.$$

We may plot values of V as a function of Q_1 (it would be the same for Q_2) as in Figure 4.4. The plot of the energy starts from

$$V[0, 0, 0.78] = 0$$

and takes negative values as Q_1 increases. The energy V is clearly a minimum at the equilibrium state E considered, and it increases until it becomes zero at a value close to $Q_1 = 0.01$. After that, V becomes positive and increases. Thus the original state considered, E, is stable.

4.3.4 Stability Criterion in Terms of the Energy Expansion

Since we are dealing with the study of the energy in the neighborhood of an equilibrium state, it is convenient to carry out a Taylor series expansion of V in terms of the increment in the generalized coordinates, q_i. Such expansion is

$$V[Q_i^E + q_i, \Lambda^E] = V[Q_i^E, \Lambda^E] + \frac{\partial V[Q_i^E, \Lambda^E]}{\partial Q_i} q_i$$

$$+ \frac{1}{2!} \frac{\partial^2 V[Q_i^E, \Lambda^E]}{\partial Q_i \partial Q_j} q_i q_j + \frac{1}{3!} \frac{\partial^3 V[Q_i^E, \Lambda^E]}{\partial Q_i \partial Q_j \partial Q_k} q_i q_j q_k + \cdots . \tag{4.13}$$

To simplify the notation, one may write

$$V^E = V[Q_i^E, \Lambda^E], \tag{4.14}$$

$$V_i^E = \frac{\partial V[Q_i^E, \Lambda^E]}{\partial Q_i}, \tag{4.15}$$

$$V_{ij}^E = \frac{\partial^2 V[Q_i^E, \Lambda^E]}{\partial Q_i \partial Q_j}, \tag{4.16}$$

so that (4.13) becomes

$$V = V^E + V_i^E q_i + \frac{1}{2!} V_{ij}^E q_i q_j + \frac{1}{3!} V_{ijk}^E q_i q_j q_k + \frac{1}{4!} V_{ijkl}^E q_i q_j q_k q_l + \cdots. \tag{4.17}$$

Remember that all derivatives are symmetric in the sense that

$$V_{ij} = V_{ji},$$

$$V_{ijk} = V_{ikj} = V_{jik} = V_{jki} = V_{kij} = V_{kji},$$

etc. In the above expansion near the equilibrium state we notice that

$$V_i^E = 0,$$

because E satisfies equilibrium. Thus, (4.17) reduces to

$$V - V^E = \frac{1}{2!} V_{ij}^E q_i q_j + \frac{1}{3!} V_{ijk}^E q_i q_j q_k + \frac{1}{4!} V_{ijkl}^E q_i q_j q_k q_l + \cdots. \tag{4.18}$$

At this stage we are only interested in the sign of $(V - V^E)$. If attention is restricted to small increments in displacements q_i, then quadratic terms dominate over cubic terms; and cubic terms dominate over quartic terms. Thus, stability is given by the sign of the first nonzero term in the energy expansion at fixed load. If the quadratic term $\frac{1}{2!} V_{ij}^E q_i q_j$ is not identically zero for all values of q_i, we say that the equilibrium state is normal or regular, and its stability is given by

$$\boxed{V_{ij}^E q_i q_j > 0 \Longrightarrow \text{stable}} \tag{4.19}$$

and the normal state is said to be stable.

However, if

$$V_{ij}^E q_i q_j < 0 \Longrightarrow \text{unstable}, \qquad (4.20)$$

the normal state is unstable.

If the quadratic form vanishes, i.e.,

$$V_{ij}^E q_i q_j = 0 \Longrightarrow \text{critical}, \qquad (4.21)$$

the state is said to be critical. States that do not satisfy (4.21) are called regular.

A critical state may be stable or unstable, and its stability should be investigated by returning to $V - V^E$ and looking at the next nonzero contribution in the energy expansion. Such analysis is carried out in the next chapter.

Example 4.2 (Stability using the quadratic form). *Example* 4.1 *is considered here using the quadratic form as a test of stability. The state considered is given by* $Q_1^E = Q_2^E = 0.022$ *and* $\Lambda^E = 3$.

Let us only obtain the first two terms of the quadratic form, i.e.,

$$V_{11}^E = V_{22}^E = A_{11} + A_{111}Q_1^E + A_{112}Q_2$$

$$= 90.16 + (0.4 - 2402.4)0.022 = 37.316,$$

$$V_{21}^E = A_{21} + A_{211}Q_1^E + A_{212}Q_2^E = 71.938.$$

With these results, it is possible to write

$$V_{ij}^E q_i q_j = V_{11}^E (q_1)^2 + V_{22}^E (q_2)^2 + 2V_{12}^E (q_1 q_2).$$

The above expression is plotted in Figure 4.5.a *as a function of the increments* q_1 *and* q_2. *In general terms, the quadratic form seems to be positive. However, a closer look at the surface in the direction given by* $q_1 = -q_2$ *(Figure* 4.5.b) *shows that the quadratic form there becomes negative. Thus, the equilibrium state considered is unstable.*

Notice that if the complete expression including quartic terms in the energy was used to compute V_{ij}^E, *then* $V_{11}^E = 40.22$ *and* $V_{21}^E = 71.93$, *and the results of stability would be the same. This illustrates that including higher-order terms does not modify the conclusions about the stability of the system.*

4.3.5 Stability Criterion in Terms with the Matrix Associated with the Quadratic Form

The term $V_{ij}^E q_i q_j$ is a quadratic form in terms of arbitrary increments q_i. Thus, what we should be testing there is not the q_i (since they simply represent all possible

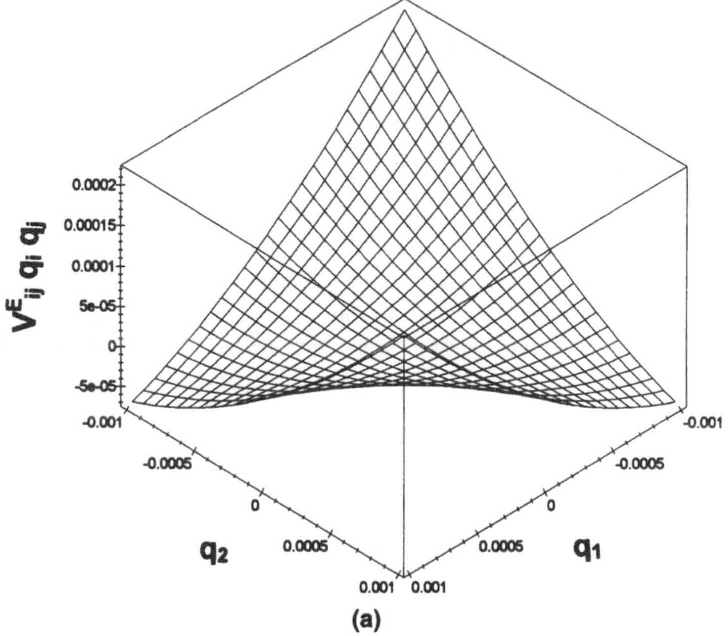

Figure 4.5 (a) Stability in terms of the quadratic form. *Continued.*

directions in the vicinity of E) but V^E_{ij} itself. The coefficients of the V^E_{ij} may be stored in matrix form leading to a symmetric matrix

$$V^E_{ij} = \begin{bmatrix} V_{11} & V_{12} & V_{13} & \cdots & V_{1n} \\ V_{21} & V_{22} & V_{23} & \cdots & V_{2n} \\ V_{31} & V_{32} & V_{33} & \cdots & V_{3n} \\ \cdots & \cdots & \cdots & \cdots & \cdots \\ V_{n1} & V_{n2} & V_{n3} & \cdots & V_{nn} \end{bmatrix}. \tag{4.22}$$

These coefficients are the second derivatives of the energy V with respect to the generalized coordinates of the problem, Q_i, and as such, V^E_{ij} is known in a mathematical context as the Hessian matrix. In general, this matrix does not have a diagonal form (that is, some $V^E_{ij} \neq 0$ for $i \neq j$).

There is a relation between the sign of the quadratic form and a property of the matrix V^E_{ij}, so that we can state the following:

A necessary and sufficient condition for stability of a normal equilibrium state is that the matrix of the Hessian of the total potential energy should be positive definite.

If the matrix of the Hessian is negative definite, then the normal equilibrium state is unstable. Cases of positive semidefinite or negative semidefinite matrices are associated with critical states.

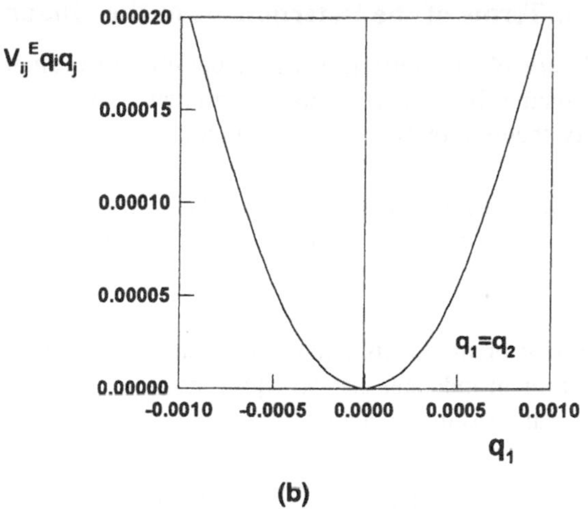

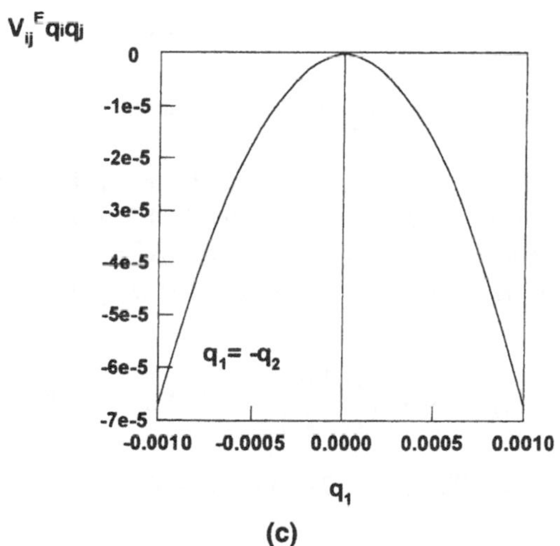

Figure 4.5 (Cont.) (b) Stability in the direction $q_1 = q_2$; (c) instability in the direction $q_2 = -q_1$.

The stability criteria can be summarized as follows:

$$\boxed{V_{ij}^E \text{ is positive definite} \Longrightarrow \text{stable}} \tag{4.23}$$

$$V_{ij}^E \text{ is negative definite} \Longrightarrow \text{unstable,} \tag{4.24}$$

$$V_{ij}^E \text{ is positive semidefinite or negative semidefinite} \Longrightarrow \text{critical.} \tag{4.25}$$

4.3.6 Stability in Terms of the Determinant of the Quadratic Form

Next, from the theory of quadratic forms it is possible to evaluate positive definite properties of the matrix by inspecting its determinant. This leads to the statement (called the stability theorem of Dirichlet–Lagrange):

A necessary and sufficient condition for stability of a normal state of equilibrium is that the determinant of the matrix of the quadratic form and all its minors should be positive.

If the determinant or one of its minors is negative, the system is at an unstable state. A zero determinant indicates critical equilibrium.

This criterion can be summarized as follows:

$$\boxed{\det[V_{ij}^E] > 0 \Longrightarrow \text{stable}} \tag{4.26}$$

$$\det[V_{ij}^E] < 0 \Longrightarrow \text{unstable,} \tag{4.27}$$

$$\det[V_{ij}^E] = 0 \Longrightarrow \text{critical.} \tag{4.28}$$

Example 4.3 (Stability using the determinant of the quadratic form). *We now use the determinant of the quadratic form in Example 4.2 to investigate stability of an equilibrium state at which $Q_1^E = Q_2^E = 0.022$ and $\Lambda^E = 0.3$.*

The complete determinant of V_{ij}^E is employed in the form

$$\det(V_{ij}^E) = \det \begin{vmatrix} V_{11} & V_{12} \\ V_{21} & V_{22} \end{vmatrix}^E = V_{11}V_{22} - V_{12}V_{21}\ |^E$$

$$= (37.316)^2 - (71.938)^2 = -3782.6 < 0.$$

We also verify that the minors are positive, and we obtain

$$V_{11} = 37.316 > 0, \qquad V_{22} = 37.316 > 0.$$

Thus, we conclude that the system is unstable at the equilibrium state considered.

4.4 STABILITY IN TERMS OF A DIAGONAL QUADRATIC FORM

4.4.1 Definition

Another way to carry out the test of stability on the matrix associated with the quadratic form is by transforming it into a diagonal form. This can be done by means of a rotation involving the eigenvectors of the matrix, and the resulting matrix may be called D_{ij}, with the form

$$D_{ij} = \begin{bmatrix} D_{11} & 0 & 0 & \cdots & 0 \\ 0 & D_{22} & 0 & \cdots & 0 \\ 0 & 0 & D_{33} & \cdots & 0 \\ \cdots & \cdots & \cdots & \cdots & \cdots \\ 0 & 0 & 0 & \cdots & D_{nn} \end{bmatrix}. \tag{4.29}$$

The elements in the diagonal of this matrix, D_{ii}, are called the stability coefficients. We state the stability criteria using D_{ij} as follows:

An equilibrium state is stable if all the stability coefficients in the diagonal form of the second variation of the total potential energy are positive.

An equilibrium state is unstable if at least one of the stability coefficients is negative.

An equilibrium state is critical if one of the stability coefficients is zero.

A proof of this criterion may be done with reference to the determinant of the quadratic form.

For two-degrees-of-freedom problems it is simple to obtain a diagonal matrix for the quadratic form. However, the transformation required to obtain the diagonal form D_{ij} in multiple-degrees-of-freedom systems is computationally expensive, and the stability approach based on D_{ij} will not be further explored here. This kind of analysis may be found in the books [14, 5]. Some hints on how to obtain the diagonal version of the quadratic form are given below, but the reader could skip this section without losing the main line of thought of this chapter.

Example 4.4 (Diagonal form). *For a five-degrees-of-freedom problem, the diagonal form of the second variation of the energy was found to be*

$$D_{ij} = \begin{bmatrix} -34 & 0 & 0 & 0 & 0 \\ 0 & 75 & 0 & 0 & 0 \\ 0 & 0 & 99 & 0 & 0 \\ 0 & 0 & 0 & 104 & 0 \\ 0 & 0 & 0 & 0 & 198 \end{bmatrix}.$$

At such equilibrium state, it is found that the system is unstable because $D_{11} = -34 < 0$.

4.4.2 Diagonal Form of the Hessian

A simple way to obtain a diagonal form for V_{ij} is to consider it as a matrix and use properties of equivalence and similarity between matrices.

A matrix B_{kl} is equivalent to another matrix V_{ij} if nonsingular matrices β_{ki} and α_{jl} exist, so that

$$B_{kl} = \beta_{ki} V_{ij} \alpha_{jl}. \tag{4.30}$$

For square matrices, then $\beta_{ij} = [\alpha_{ij}]^{-1}$ and

$$B_{kl} = \alpha_{ki}^{-1} V_{ij} \alpha_{jl}, \tag{4.31}$$

where α_{ki}^{-1} indicates the kth row and ith column of the inverse of matrix α. This is a transformation of similarity. A transformation of orthogonality can be made if $\alpha^{-1} = \alpha^T$ leading to

$$B_{kl} = \alpha_{ki}^T V_{ij} \alpha_{jl}. \tag{4.32}$$

Next, we may construct α in such a way that its columns are the eigenvectors of matrix V_{ij}; then

$$\alpha = [\mathbf{x}^1, \mathbf{x}^2, \ldots, \mathbf{x}^n], \tag{4.33}$$

and B becomes a diagonal matrix.

A condition to obtain a diagonal matrix D is that the eigenvectors $\mathbf{x}^n$ should be linearly independent vectors. A sufficient condition (but not necessary) to obtain D is that the eigenvectors $\mathbf{x}^n$ should be different. Furthermore, if the eigenvectors are linearly independent and orthogonal, then

$$D_{kl} = \alpha_{ki}^T V_{ij} \alpha_{jl}. \tag{4.34}$$

All symmetric matrices may be set to a diagonal form in this way.

Returning to our discussion on the quadratic form of V, since $V_{ij}^E = V_{ji}^E$ it is possible to construct a matrix α_{ij} with the eigenvectors and obtain a diagonal form.

In the new form, the diagonal matrix D_{ij} is associated with a new set of generalized coordinates. If Q_i are the original coordinates employed to define V_{ij}, the coordinates used to define D_{ij} are called u_j, and their relation is given by

$$Q_i = \alpha_{ij} u_j, \tag{4.35}$$

where

$$\det |\alpha_{ij}| \neq 0. \tag{4.36}$$

The inverse of this relation exists and is

$$u_i = \beta_{ij} Q_j \tag{4.37}$$

with

$$\det |\beta_{ij}| \neq 0. \tag{4.38}$$

An important feature to notice is that if the coordinates u_i are plotted in the Q_j space, they are not necessarily orthogonal.

Finally, we write

$$V_{ij} Q_i Q_j = D_{ij} u_i u_j.$$

The first member may be written as

$$V_{ij} Q_i Q_j = V_{ij} \alpha_{im} \alpha_{jn} u_m u_n,$$

so that

$$\boxed{D_{mn} = V_{ij}\alpha_{im}\alpha_{jn}} \tag{4.39}$$

where

$$\alpha_{im} = x_i^m. \tag{4.40}$$

The matrices D_{mn} and V_{ij} are equivalent and are related by a transformation of orthogonality.

Example 4.5. *As an example of diagonalization of a quadratic form, let us consider the two-degrees-of-freedom problem of Example* 4.1. *We limit the energy to cubic terms, so that*

$$V_{ij} = A_{ij} + A_{ijk}Q_k.$$

Let us study a state given by $Q_1^E = Q_2^E = 0.005$ *and* $\Lambda = 0.78$, *which satisfies equilibrium. Thus, one has*

$$V_{11}^E = V_{22}^E = 78.146, \qquad V_{12}^E = 71.924.$$

The eigenvalues ω *and eigenvectors of* V_{ij}^E *are*

$$\omega = 6.222, \qquad {}^1\mathbf{x} = 0.707 \begin{Bmatrix} 1 \\ -1 \end{Bmatrix},$$

$$\omega = 150.07, \qquad {}^2\mathbf{x} = 0.707 \begin{Bmatrix} -1 \\ -1 \end{Bmatrix},$$

$$\alpha = 0.707 \begin{bmatrix} 1 & -1 \\ -1 & -1 \end{bmatrix}.$$

We multiply the matrices as indicated in (4.39) *and get*

$$D_{ij} = \begin{bmatrix} 6.222 & 0 \\ 0 & 150.069 \end{bmatrix},$$

which is a diagonal form.

4.5 STABILITY OF AN EQUILIBRIUM PATH

Until now we have been concerned with stability of individual equilibrium states for a fixed load level. We can also be interested in the way the stability evolves along an equilibrium path, and to do this one has to reintroduce the load as a variable in the problem.

Example 4.6. *To show that the stability of equilibrium states changes with the load level, let us consider again the two-degrees-of-freedom problem of Example* 4.1. *The nonlinear equilibrium path is shown in Figure* 4.6.a, *where solid lines are used to depict stable states, whereas dotted lines indicate unstable paths.*

The equilibrium state at zero load and displacement is defined as a stable state. One way to show that is to compute the second variation of the energy there for increments q_i *at* $\Lambda = 0$, *and one would find a surface of second variation* $\delta^2 V = \frac{1}{2} V_{ij} q_i q_j$ *that is concave (Problem* 4.1).

Next, consider the state E_1 *for which* $\Lambda = 0.78$ *and* $Q_1 = Q_2 = 0.005$. *In Figure* 4.6.b *we show the surface of V as a function of* Q_1 *and* Q_2*; clearly such a surface represents a stable state. At another state, the load is increased to* $\Lambda = 3$, *with displacements at equilibrium* E_2 *given by* $Q_1 = Q_2 = 0.022$. *The energy is at a maximum in one direction and at a minimum in other direction (as shown in Figure* 4.6.c*); the state is now unstable. For the same load level* $\Lambda = 3$, *we explore now an equilibrium state* E_3 *on the descending path, at which* $Q_1 = Q_2 = 0.112$. *The state is again unstable, but now there are no directions in which a minimum is present (Figure* 4.6.d).

At this stage we do not know how stability was lost between states E_1 and E_2, and to find at what load level such a loss of stability occurred, one should investigate a continuous variation of the stability test along the path.

Thus, rather than looking at the stability of individual states, it may be more convenient to obtain a parametric representation of the stability of the path. To do that, we introduce a perturbation parameter s that follows the progress of equilibrium states along this path; this parameter could be the load, a displacement component, or any other parameter suitable to follow the path. Then we write

$$Q_i = Q_i(s), \qquad \Lambda = \Lambda(s),$$

and

$$V(s) = V[Q_i(s), \Lambda(s)].$$

All stability tests based on the total potential energy that have been described for individual states can now be extended to investigate stability of the path. A convenient way to implement this is by means of the determinant of V_{ij}, now in the form

$$\det|V_{ij}^E(s)| = \det|V_{ij}[Q_i(s), \Lambda(s)]|.$$

The kind of knowledge gained by investigation of the stability of a complete path is very important and provides rich information about the behavior of the structure.

Example 4.7. *Adopting* $s = Q_1$, *the stability determinant may be computed as in Figure* 4.7. *A short, initial part is stable, with the value of the determinant decreasing from nearly* 3000 *to* 0. *Negative values are obtained between approximately* $s = 0.01$ *and* $s = 0.09$. *A new change in the sign of the complete determinant occurs at*

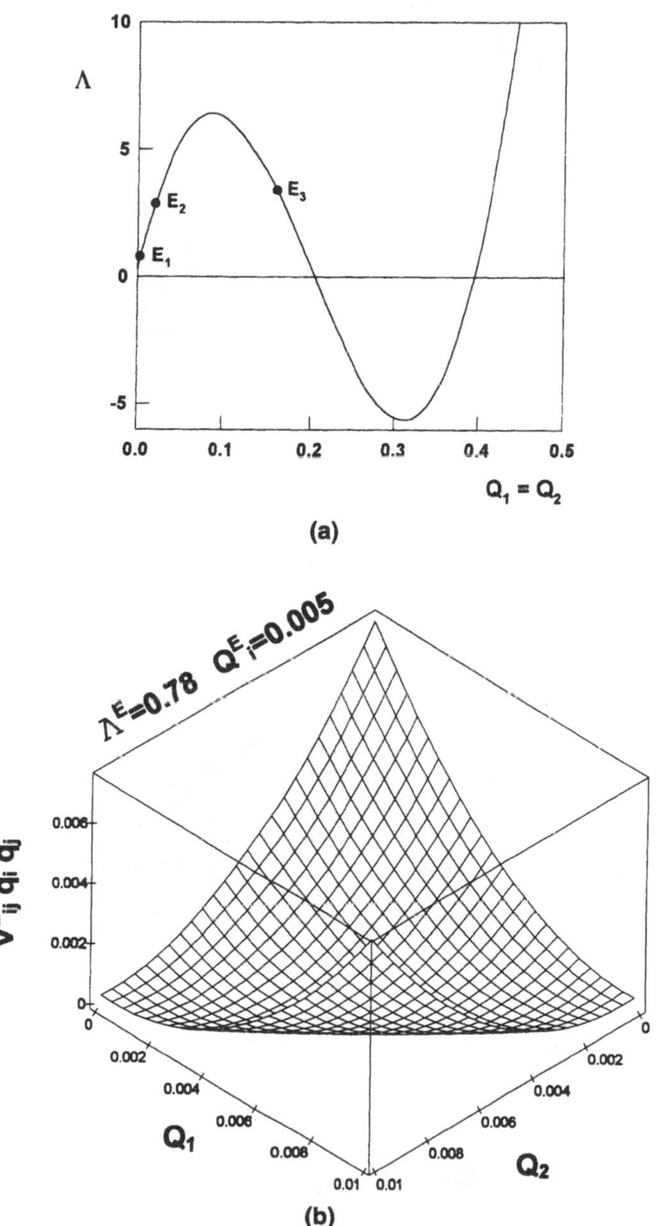

Figure 4.6 (a) Nonlinear equilibrium path; (b) a stable state E_1. *Continued.*

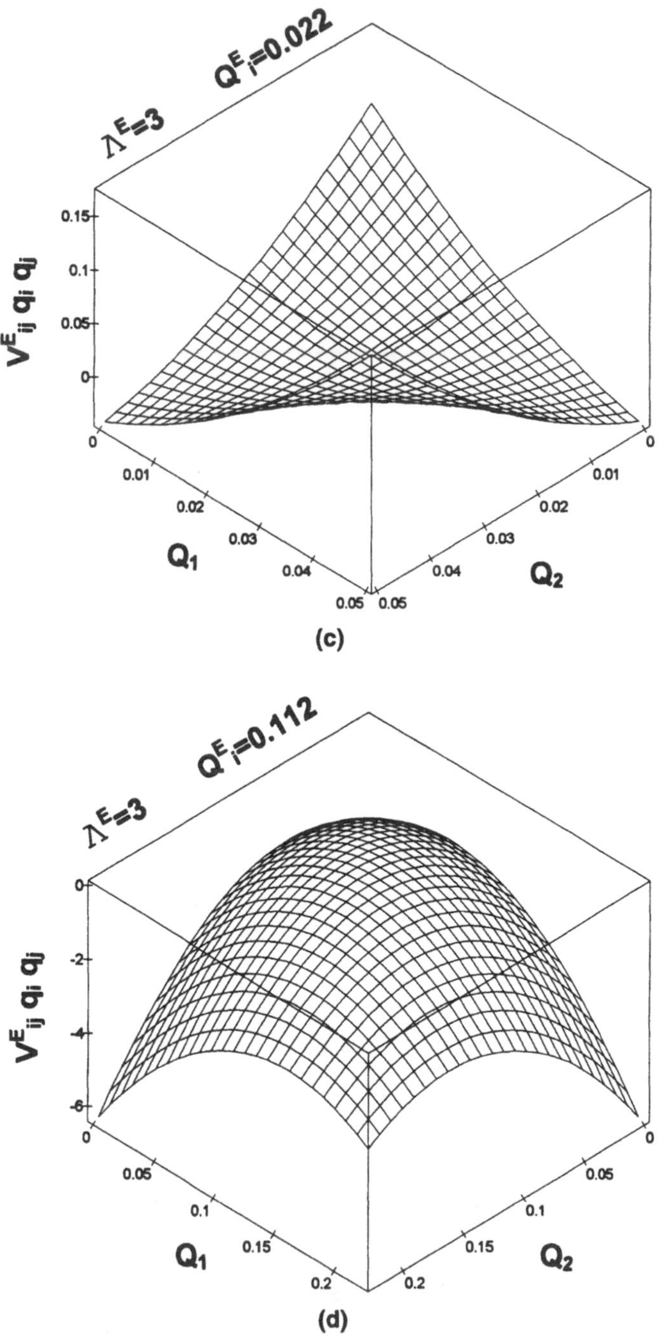

Figure 4.6 (Cont.) (c) Unstable state E_2, with a minimum in one direction and a maximum in another direction; (d) unstable state E_3, in which there is no minimum in any direction.

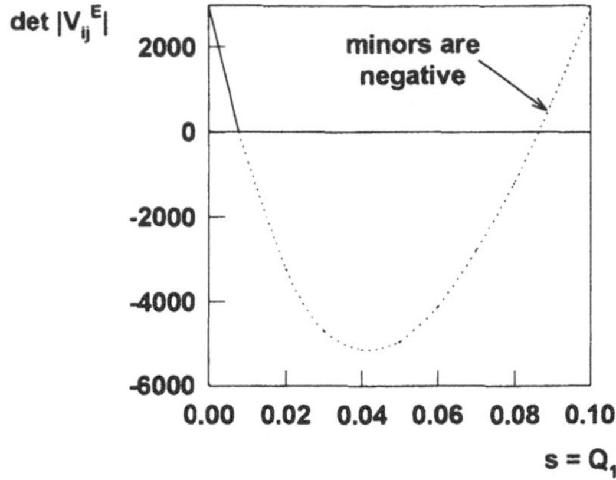

Figure 4.7 Stability determinant along the path of Figure 4.6.a.

values slightly lower than $s = 0.09$. However, this new change in the sign of the complete determinant (which now becomes positive) is accompanied by a negative minor, meaning that the states are unstable.

4.6 DYNAMIC CRITERION OF STABILITY

We consider here the stability of a static equilibrium state: A perturbation is introduced in the form of a vibration about the static equilibrium state, and we investigate the nature of the dynamic response of the system. In general, the study of a nonlinear dynamic response is required, but there are tests developed using small dynamic displacements and the initial linear dynamic response as indicators of the possible nonlinear dynamic behavior of the perturbed system.

This leads to stability in the local sense, and the dynamic criterion may be stated as follows:

An equilibrium state is stable if, for small vibrations about such a state, all the frequencies of vibration are real.

Thus, the measure of the dynamics of the perturbed system is carried out by considering the frequencies of vibration.

If at least one frequency is imaginary, we say that the equilibrium state is unstable.

If at least one frequency of vibration is zero, we say that the equilibrium state is critical. In this case stability can only be evaluated using nonlinear vibrations.

An excellent account of the dynamic criterion of stability is given in the book by Huseyin [6]. The criterion is next illustrated by a one-degree-of-freedom problem.

Example 4.8 (The dynamic criterion in a one-degree-of-freedom system). *To illustrate the use of the dynamic criterion of stability, let us consider a one-degree-of-freedom system, for which static equilibrium takes the form*

$$[K_0 + K_1(Q^E)]Q^E - \Lambda F = 0, \tag{4.41}$$

where K_0 represents the linear part of the stiffness of the system; $K_1(Q^E)$ is the non-linear part of stiffness and depends on the displacements at the equilibrium state Q^E; F is the load vector; and Λ is the load parameter. Next, we introduce a perturbation in the form of a small vibration q, which depends on time t, with initial amplitude $q = q_0$ and velocity $dq/dt = \dot{q}_0$ at $t = 0$. The total displacements are given by the sum of the static displacement plus the vibratory perturbation in the form

$$Q = Q^E + q(t). \tag{4.42}$$

The condition for small vibrations of the system involves the inertia forces, and the mass M has to be taken into account. The equation of motion is

$$M\frac{d^2q}{dt^2} + [K_0 + K_1(Q^E)]q = 0. \tag{4.43}$$

Equation (4.43) *may be written in the more convenient form*

$$\frac{d^2q}{dt^2} + \omega^2 q = 0, \tag{4.44}$$

where ω is the natural frequency of vibration of the system.

From the above equation it is possible to calculate the response of the system $q(t)$, and this is plotted in Figure 4.8 *for a general case. For real frequencies, $\omega^2 > 0$, the system oscillates with finite amplitude, and since damping was not included in the analysis, the amplitude of vibrations remains constant in time. However, if damping is included, the amplitude of vibrations decreases with time, until it becomes zero.*

For imaginary frequencies of vibration, $\omega^2 < 0$, the system is unstable. The motion is divergent, and the amplitude of vibration tends to infinity. Finally, if the frequency

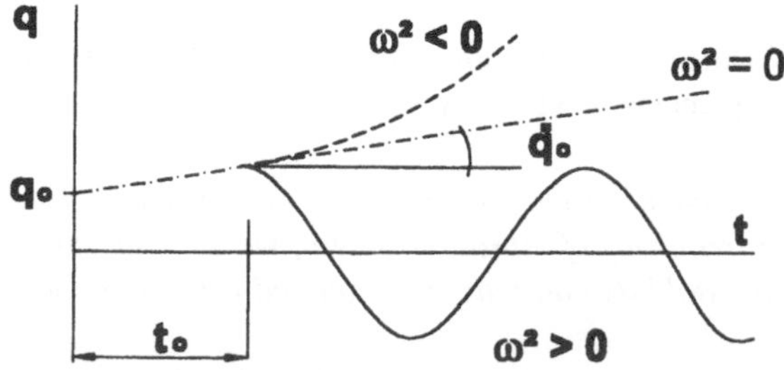

Figure 4.8 Dynamic criterion of stability in a one-degree-of-freedom system.

of vibration is zero, $\omega^2 = 0$, the system is said to be at a critical state: It continues to move with the initial velocity.

Remarks. We shall not employ the dynamic criterion of stability in this book. However, we can make some remarks about the dynamic criterion of stability applied to static systems:

- This criterion can be generalized to multiple degrees of freedom and to continuum systems.
- It is necessary to introduce the nonlinearity of the original state; otherwise, all states appear to be stable. Thus, the term $K_1(Q^E)$ cannot be omitted in the evaluation of stability.
- The evaluation of frequencies of vibrations is a linear eigenproblem.
- The criterion is based on a global measure (the frequencies of vibration of the body) rather than on local variables.
- No assumptions have been made on the constitutive law of the material of the system.
- This is the most general criterion of stability for static problems.

Finally, we should comment on the dynamic versus static analysis in stability. A system may be static or dynamic, according to the nature of the external loads applied. In general, the perturbation applied to test stability is dynamic, in the sense that it implies a motion of the system from the perturbed state. For dynamic motions of the system, the stability test is also dynamic, and we should employ nonlinear dynamics to investigate stability. Different studies are required depending on the nature of the external loads, their relation with the natural frequencies of the system, and the time of application and duration of the load. Some techniques of analysis are available for general systems [6], for suddenly loaded structures [13], etc.

4.7 STATIC CRITERION OF STABILITY

This simple criterion uses forces and displacements to assess stability [4]. A static perturbation is introduced in the equilibrium state, and we investigate the static forces generated. If the out-of-balance forces lead the system to the original state, we say that the equilibrium state is stable. On the contrary, if they drive the system to another state, the equilibrium state is unstable.

The static criterion may be stated as follows:

An equilibrium state is stable if, for small static displacements about such state, the static forces tend to restore the system to the original state.

If they tend to drive the system to a different state, the original equilibrium state is said to be unstable. There is also a chance that the resultant forces may be zero. In that case, the original equilibrium state is neutral.

Example 4.9 (Static criterion for a one-degree-of-freedom system). *Let us consider again a one-degree-of-freedom problem, for which static equilibrium takes the form of* (4.41), *i.e.,*

$$[K_0 + K_1(Q^E)]Q^E - \Lambda^E F = 0, \tag{4.45}$$

where again K_0 represents the linear part of stiffness; $K_1(Q^E)$ is the nonlinear part of stiffness and depends on the displacements Q^E; and Λ is the load parameter affecting the loads F of the structure.

A static perturbation q is introduced in the form

$$Q = Q^E + q, \tag{4.46}$$

where Q are now the total displacements. In the new configuration, the system is perhaps no longer in equilibrium, so that the sum of all forces is not zero. This may be found by substitution of the total displacements (4.46), *in the equilibrium condition,* (4.41), *leading to*

$$[K_0 + K_1(Q^E + q)](Q^E + q) - \Lambda^E F \neq 0. \tag{4.47}$$

We expand (4.47) *under the assumption of small static perturbations, so that the nonlinear terms in q can be neglected:*

$$[K_1(q)]q \simeq 0.$$

Because of (4.45), *then there is an out-of-balance force f^* in the form*

$$f^* = -[K_0 + K_1(Q^E)]q. \tag{4.48}$$

To know if the out-of-balance force tends to push the system back to the original state, it may be necessary to calculate the response q^* under f^* using the linear condition

$$K_0\, q^* = f^*. \tag{4.49}$$

If the displacement due to the out-of-balance force q^* is opposite to the perturbation introduced q, then the system is stable. If the direction of q^* is the same as that of q, then the system is at an unstable equilibrium condition. Situations in which $q^* = 0$ represent neutral equilibrium states.

Neutral states are characterized by

$$f^* = 0.$$

Example 4.10 (Linearized one-degree-of-freedom system). *Let us consider again a one-degree-of-freedom system. We want to evaluate stability using the static criterion but starting from a linear equilibrium condition in* (4.41). *Then for a perturbation q we have the force*

$$f^* = -K_0\, q. \tag{4.50}$$

The equilibrium condition for f^* *is*

$$K_0\, q^* = -K_0\, q. \tag{4.51}$$

This is satisfied if

$$q^* = -q. \tag{4.52}$$

The erroneous conclusion from (4.52) *would be that the system is stable. In fact, any state in a system in which the kinematics are simplified to a linear version becomes stable.*

This is an important conclusion and highlights the limitations of a linear analysis: If we do not consider nonlinearity in the original state, then all states seem to be stable. This is in agreement with what was found in the dynamic criterion.

Remarks. We do not employ the static criterion of stability in this book, but it is often employed in the literature. A variant of this approach is used to compute buckling loads under the name of **adjacent equilibrium method**. This, however, is not a stability criteria but a way to compute a critical state.

- A severe limitation of the static criterion of stability is that it cannot be generalized to continuum systems.
- Its generalization to multiple-degrees-of-freedom systems is difficult.
- The use of the static criterion of stability involves the solution of a linear equilibrium problem.
- Kinematic nonlinearity should be included in the analysis; otherwise, all systems appear to be stable.

4.8 FINAL REMARKS

What we have discussed in this chapter covers some of the stability criteria for static systems. But the total potential energy is not the only variational functional employed in solid mechanics. The complementary energy and variational expressions involving two and three fields (displacements, strains, and stresses in a body) have been used for some time [15]. However, the application of such functionals to stability is not easy to find in the literature.

For complementary energy, Koiter [8] has shown that there are several problems, the first one being that there is no universal agreement on the generalization of the complementary energy principle to geometrically nonlinear elastic problems. And, second, for the application of the principle to stability, one must remember that the complementary principle does not involve the condition of a minimum (as the total potential energy does) but of stationary energy. This means that complementary energy cannot be employed directly to assess if a state of equilibrium is stable or unstable. However, Koiter has used complementary energy to identify critical states along a path.

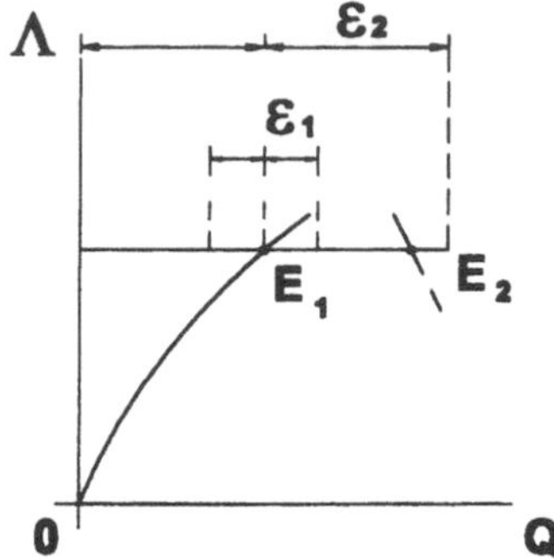

Figure 4.9 Small and large perturbations from an equilibrium state E_1.

Other possibilities of evaluation of stability in systems that are not elastic are based on **thermodynamics** [1, 2, 3] and on the principle of **virtual work**.

Throughout this chapter, the idea of small perturbations has been used, although not in an explicit form. This concept may be better explained with reference to Figure 4.9.

If we investigate the stability of the equilibrium state at E_1 and use small perturbations ϵ_1, what we do is to explore the basin of potential energy that is locally relevant to that state. However, if we allow larger perturbations to occur, such as ϵ_2, then there is a possibility that another state may be reached for the same load level. **Stability in the large** would be concerned with such an approach, typically involving nonlinear dynamic motions. But this is outside the scope of this book, and whenever we employ the words perturbation of an equilibrium state, we actually mean small perturbations, as small as desired.

4.9 PROBLEMS

Review questions. (*a*) Explain in your own words the first stability criterion of Lyapunov. (*b*) What is an autonomous system? (*c*) What is a state of neutral equilibrium? (*d*) Comment on the advantages and disadvantages of each one of the tests based on total potential energy presented in this chapter. (*e*) Distinguish between stability of an equilibrium state and stability of an equilibrium path. (*f*) What is a diagonal form? (*g*) Is it possible to evaluate the stability of a linearized system? (*h*) According to the dynamic criterion, when is a system unstable under static loads? (*i*) Explain the disadvantages of the static criterion of stability.

Problem 4.1 (Two-degrees-of-freedom system). For Example 4.1, evaluate the stability of the unloaded state.

Problem 4.2 (Two-degrees-of-freedom system). Solve Example 4.1 using the energy itself, for the state with $\Lambda = 3,\ Q_1 = Q_2 = 0.022$.

Problem 4.3. Use the determinant of the quadratic form to obtain under what conditions of load and displacements Example 4.1 becomes of critical stability.

Problem 4.4. For the system of Example 4.1 obtain the energy in coordinates so that the quadratic form becomes diagonal.

Problem 4.5. Repeat Problem 4.2 using the diagonal form of Problem 4.4.

Problem 4.6. For an arch structure under central load, with $\theta_0 = 60\circ$, $I/AR^2 = \frac{10}{3} \times 10^{-5}$, the coefficients associated with Q_1 (useful for the fundamental path) are

$$A_1 = -1, \quad A_{11} = 0.810738, \quad A_{111} = -4.29718, \quad A_{1111} = 7.593679.$$

Using a perturbation analysis in terms of Q_1, in which only up to quadratic terms are retained, the solution to the path is of the form

$$\Lambda = \dot{\Lambda} Q_1 + \frac{1}{2} \ddot{\Lambda} Q_1^2,$$

where $\dot{\Lambda} = 0.81073$ and $\ddot{\Lambda} = -4.29716$. The state given by $\Lambda^A = 0.03824$, and $Q_1^A = 0.05526$ satisfies the equation of the path. Use the energy approach to evaluate stability. Use the energy itself.

Problem 4.7 (Circular arch). Consider again Problem 4.6, and repeat the computations for the state given by $\Lambda^B = 0.03824$ and $Q_1^B = 0.322077$. Explain what problems you face in the evaluation of stability and why they arise.

Problem 4.8 (Two-degrees-of-freedom system). Find a diagonal form D_{ij} for the quadratic form of Example 4.2.

Problem 4.9 (Theory). Prove that if one uses the diagonal form, then a state is stable if the stability coefficients are positive.

Problem 4.10 (Circular plate). Consider the circular plate of Problem 2.4 under a load $\Lambda = 0.003$. Is it stable? Repeat the evaluation for $\Lambda = 0.006$.

Problem 4.11 (Plane frame). Investigate the stability of the frame of Problem 2.8 for a state given by $P = \frac{1}{2} EI \left(\frac{\pi}{L}\right)^2$ and $P = \frac{2}{3} EI \left(\frac{\pi}{L}\right)^2$.

Problem 4.12 (Frame). A plane frame under a moment action is shown in Figure 3.8. Evaluate the stability of the equilibrium path.

Problem 4.13 (Frame). Compute the stability determinant for the frame of Figure 3.9.

4.10 BIBLIOGRAPHY

[1] Bazant, Z. P., Stable states and paths of structures with plasticity and damage, *J. Engrg. Mech., ASCE*, 114(12), 2013–2034, 1988.

[2] Bazant, Z. P., Bifurcation and thermodynamic criteria of stable paths of structures exhibiting plasticity and damage propagation, in *Computational Plasticity*, D. Owen, E. Hinton, and E. Oñate, Eds., Pineridge Press, Swansea, UK, 1–26, 1989.

[3] Bazant, Z. P., and Cedolin, L., *Stability of Structures: Elastic, Inelastic, Fracture and Damage Theories*, Oxford University Press, New York, 1991.

[4] Croll, J. G. A., and Walker, A. C., *Elements of Structural Stability*, Macmillan, London, 1972.

[5] Huseyin, K., *Non Linear Theory of Elastic Stability*, Noordhoff, Leyden, 1975.

[6] Huseyin, K., *Vibrations and Stability of Multiple Parameter Systems*, Ivoardhoff, Alphen, 1978.

[7] Koiter, W. T., *On the Stability of Elastic Equilibrium*, Ph.D. thesis, Delft Institute of Technology, Delft, 1945.

[8] Koiter, W. T., Complementary energy, neutral equilibrium and buckling, *Meccanica*, 19, 52–56, 1984.

[9] LaSalle, J., and Lefschetz, S., *Stability by Lyapunov's Direct Method with Applications*, Academic Press, New York, 1961.

[10] Lyapunov, A., *The General Problem of Stability of Motion*, Kharkov, 1892. (There was a French translation in 1907 and an English translation in 1949.)

[11] Prigogine, I., *From Being to Becoming*, W. H. Freeman, San Francisco, 1980.

[12] Reising, R., Liapunov's second method, Chapter 8 in *Stability*, H. H. E. Leipholz, Ed., Waterloo University, Waterloo, 281–320, 1972.

[13] Simitses, G., *Dynamic Stability of Suddenly Loaded Structures*, Springer, New York, 1990.

[14] Thompson, J. M. T., and Hunt, G. W., *A General Theory of Elastic Stability*, Wiley, London, 1973.

[15] Washizu, K., *Variational Methods in Elasticity and Plasticity*, Second edition, Pergamon Press, Oxford, 1975.

CHAPTER

FIVE

CRITICAL STATES

5.1 INTRODUCTION

In the previous chapter it was shown how to evaluate the stability of an equilibrium state, and states were basically classified as stable or unstable. As a special case, a state is at a critical condition in which the quadratic form $V_{ij}x_ix_j$ passes through zero.

Such states are called critical, and important changes in the stability of the structural system occur in them. It will be shown that at the critical state there is an exchange of stability in the sense that the stability changes from stable to unstable. We may even say that the most important states along an equilibrium path are critical for which there is a qualitative change in the nature of the stability of equilibrium. Notice that most practical engineers are not interested in investigating the stability of equilibrium states: They just want to know where the critical states are along a path.

But if these critical states are so important, the first question that we may pose is how we locate them along an equilibrium path. And, furthermore, is it possible to investigate the stability of critical states? Are there many classes of critical states, or just one? Can the energy derivatives help in establishing a criterion to classify critical states?

These and related questions will be addressed in this chapter and will open the door to more elaborate questions to be answered in the following chapters. We start investigating in section 5.2 what conditions are satisfied at a critical equilibrium state and what kinds of mathematical problems arise when critical states are being searched for. The specialized system with linear fundamental path is discussed in section 5.3, and it is shown how the critical state leads to a linearized eigenvalue problem. The specialized system with nonlinear primary path is the subject of section 5.4, leading to a nonlinear eigenvalue problem. Some hints are given as to how to compute the solution of nonlinear eigenvalue problems using iterative techniques. The question of stability of critical states is addressed in section 5.5, to show the computation of such a property.

The remainder of the chapter includes examples of evaluation of critical states. The results are the buckling loads and buckling modes of axially loaded columns (section 5.6), column on an elastic foundation (section 5.7), strut under compression (section 5.8), a two-bar frame (section 5.9), arches under central point load (section 5.10), plates under in-plane loads (section 5.11), rings under radial pressure (section 5.12), and torsional buckling of columns. Some of the examples are classical problems first solved several centuries ago, and we have added some more recent problems of structural components made of composite materials. Section 5.14 discusses the use of finite elements to evaluate critical loads.

Because most of the problems lead to critical states along a linear fundamental path, there are not many plots of equilibrium paths in this chapter. We have plotted results only in selected problems in which a more complex buckling mode should be illustrated or in cases of nonlinear fundamental paths. Of course this is a limited number of examples, and they are included here with the purpose of illustrating the use of the formulation and some basics of the behavior of thin-walled structures. Shell problems will be discussed in Chapter 13.

5.2 DETECTING CRITICAL STATES ALONG AN EQUILIBRIUM PATH

5.2.1 Conditions That Are Satisfied by the Energy at Critical States

In the last chapter it was stated that the quadratic form $V_{ij}^E q_i q_j$ vanishes at a critical state. We denote critical states by an index C, and we keep the index E for noncritical equilibrium states. The vectors q_i indicate directions in the plane of generalized coordinates at a given load level.

At a critical state we assume that the load reaches a critical value, $\Lambda = \Lambda^c$, and that there is only one direction for which the quadratic form vanishes. If x_i is such a direction, it is possible to write

$$V_{ij}^c x_i x_j = 0 \tag{5.1}$$

and

$$V_{ij}^c q_i q_j \geq 0 \tag{5.2}$$

for any vector q_i that is not parallel to x_i.

In the case considered here, in which only one direction x_i exists, the critical state is said to be distinct. If there is more than one direction for which the quadratic form vanishes, the critical state is said to be compound and will be studied in Chapter 14 on mode interaction.

Because the values of the quadratic form are zero in one direction x_i and greater than zero in any other direction $q_i \neq x_i$, it is possible to conclude that x_i minimizes the quadratic form. This leads to the following condition:

$$\left.\frac{\partial(V_{ij}x_ix_j)}{\partial x_i}\right|^c = 2V_{ij}x_j|^c = 0. \tag{5.3}$$

We call x_i the direction of critical stability or the critical mode or eigenmode, and Λ^c the critical load. The term direction of critical stability, or similar ones, are used to denote a specific vector of incremental displacements associated with the vanishing of the quadratic form; this is the *weak* direction out of the critical state.

Since V_{ij} is a function of Λ along the fundamental path, then at the critical state C the above condition is an eigenvalue problem in which Λ^c is the eigenvalue and x_i^c is the eigenvector. We write this in the form

$$\boxed{V_{ij}\left[Q_j, \Lambda\right] x_j|^c = 0 \qquad \Longrightarrow \qquad \text{critical state}} \tag{5.4}$$

This problem has a solution because the determinant of V_{ij} is zero, so that a nontrivial solution of x_i^c must exist for some load level Λ^c.

5.2.2 Critical States in Terms of the Stability Determinant

At the first (lower) critical state, the matrix associated with the quadratic form ceases to be positive definite, and it becomes positive semidefinite

$$V_{ij}^c q_i q_j|^c \geq 0. \tag{5.5}$$

An alternative way to find a critical load is to use the condition that the determinant of the quadratic form is zero, i.e.,

$$\boxed{\det[V_{ij}(\Lambda^c)] = 0 \qquad \Longrightarrow \qquad \text{critical state}} \tag{5.6}$$

Notice, however, that this only provides the critical load Λ^c, not the eigenvector x_j^c. A simple way to find the zeros of the determinant is by evaluation of the determinant along the path, as was done in section 4.5.

5.2.3 Critical States in Terms of the Diagonal Form

Because the quadratic form is diagonal, so that V_{ij} becomes a diagonal matrix D_{ij}, the critical states are reflected by a zero in one of the diagonal elements D_{kk}. The passage through a critical state is identified by the passage through zero of one of the diagonal terms in the energy. This may be written as

$$\boxed{D_{kk} = 0 \qquad \Longrightarrow \qquad \text{critical state}} \tag{5.7}$$

For this reason, the coefficients D_{kk} are called the stability coefficients.

Using the diagonal form is the simplest way to identify a critical state along an equilibrium path. However, if the path is nonlinear the process of diagonalization has to be performed at each load level and this becomes computationally expensive. As mentioned in Chapter 4, the diagonalization of V_{ij} will not be employed in this book. Having said that, we acknowledge that there is a great simplicity in the formulation based on the diagonal form, and this leads to the elegant presentations of [29, 13] and several others.

5.3 THE SPECIALIZED SYSTEM

To illustrate the critical state discussed previously, consider the energy in the specialized system, i.e., a system in which the load potential is linear in Λ

$$V = A_i' \Lambda Q_i + \frac{1}{2!} A_{ij} Q_i Q_j + \frac{1}{3!} A_{ijk} Q_i Q_j Q_k + \frac{1}{4!} A_{ijkl} Q_i Q_j Q_k Q_l, \tag{5.8}$$

where $A_i' = \partial A_i / \partial \Lambda$.

The second derivative is

$$V_{ij} = A_{ij} + A_{ijk} Q_k + \frac{1}{2} A_{ijkl} Q_k Q_l.$$

If the fundamental path is linear, one may write it in the form

$$\boxed{Q_i^F = \Lambda \bar{Q}_i} \tag{5.9}$$

where $\bar{Q}_i$ is the value of the generalized coordinates Q_i^F for a unit load $\Lambda = 1$. For this linear fundamental path we obtain (5.4) in the form

$$\boxed{\left(A_{ij} + \Lambda A_{ijk} \bar{Q}_k + \tfrac{1}{2} \Lambda^2 A_{ijkl} \bar{Q}_k \bar{Q}_l \right) x_j|^c = 0}$$

This is a quadratic eigenvalue problem. Fortunately, if the fundamental path is linear, as assumed in (5.9), then quadratic terms should be neglected for consistency, and we only need to solve the linear eigenvalue problem

$$\boxed{\left(A_{ij} + \Lambda\ A_{ijk} \bar{Q}_k \right) x_j|^c = 0} \tag{5.10}$$

or else, in matrix notation,

$$\{[A_{ij}] + \Lambda\ [A_{ijk} \bar{Q}_k]\} x_j|^c = 0. \tag{5.11}$$

The matrix in the first term $[A_{ij}]$ is called the stiffness matrix of the linear problem, and the second matrix $[A_{ijk} \bar{Q}_k]$ is the load-geometry matrix, also known as the initial stress matrix.

In books on finite elements it is common to denote our matrix $[A_{ij}]$ by K_0; matrix $[A_{ijk}\bar{Q}_k]$ is K_G; and the eigenvector x_j is $\mathbf{x}$, so that the eigenproblem of (5.11) is written as

$$[K_0 + \Lambda\ K_G]\ \mathbf{x} = 0. \tag{5.12}$$

We prefer to employ the index notation as in (5.11) rather than the matrix notation of (5.12), because the former has a direct relation with the derivatives of the energy.

There are many standard computer routines to solve eigenvalue problems of all sizes, and we need not concern ourselves with such aspects of the analysis here. References to effective solution of eigenproblems may be found in [31, 2] and several others.

Example 5.1 (Circular plate: Critical state). *For a circular plate under in-plane loading, the energy coefficients have been derived in Example* 2.4 *of Chapter* 2. *The fundamental path is linear and is given in Example* 2.6.

This is a two-degrees-of-freedom problem, and the critical state satisfies the condition

$$\left[\begin{bmatrix} A_{11} & 0 \\ 0 & A_{22} \end{bmatrix} + \Lambda^c \begin{bmatrix} A_{112}\bar{Q}_2 & A_{112}\bar{Q}_1 \\ A_{112}\bar{Q}_1 & 0 \end{bmatrix}\right] \begin{Bmatrix} x_1^c \\ x_2^c \end{Bmatrix} = \begin{Bmatrix} 0 \\ 0 \end{Bmatrix}.$$

Notice that this is not a diagonal form, because $V_{12} \neq 0$. *Along the fundamental path the displacements are given by*

$$Q_1^F = \Lambda\ \bar{Q}_1 = 0,$$

$$Q_2^F = \Lambda \bar{Q}_2 = -\Lambda \frac{A_2'}{A_{22}} = \Lambda(1 - \nu)\frac{R}{Eh}.$$

Thus the eigenvalue problem reduces to

$$\left[\begin{bmatrix} A_{11} & 0 \\ 0 & A_{22} \end{bmatrix} + \Lambda^c \begin{bmatrix} A_{112}\bar{Q}_2 & 0 \\ 0 & 0 \end{bmatrix}\right] \begin{Bmatrix} x_1^c \\ x_2^c \end{Bmatrix} = \begin{Bmatrix} 0 \\ 0 \end{Bmatrix}.$$

At the critical state the determinant of V_{ij} *is zero, leading to the condition*

$$(A_{11} + \Lambda^c A_{112}\bar{Q}_2)A_{22} = 0.$$

Because $A_{22} \neq 0$, *we must have* $(A_{11} + \Lambda^c A_{112}\bar{Q}_2) = 0$, *from which the critical load is obtained as*

$$\Lambda^c = \frac{A_{11}A_{22}}{A_{112}A_2'} = -D\frac{(1.191 + \nu)}{3.4674}\left(\frac{\pi}{R}\right)^2. \tag{5.13}$$

The displacement along the fundamental path at which a critical state occurs is given by

$$Q_1^c = 0, \qquad Q_2^c = -\frac{A_{11}}{A_{112}} = -\frac{\pi^2}{12.023}\frac{(1.191+\nu)}{1+\nu}\frac{h^2}{R}. \tag{5.14}$$

The direction of critical stability is obtained from the eigenvalue problem, substituting Λ^c and Q_i^c. This leads to

$$\begin{bmatrix} 0 & 0 \\ 0 & A_{22} \end{bmatrix} \begin{Bmatrix} x_1^c \\ x_2^c \end{Bmatrix} = \begin{Bmatrix} 0 \\ 0 \end{Bmatrix}.$$

Of course V_{ij} is singular at C, so that one of the values of x_i^c has to be assumed in order to solve the problem. Say we assume $x_1^c = 1$; then

$$A_{22}x_2^c = 0$$

and from that it follows that

$$x_2^c = 0.$$

Finally, the eigenvector or eigenmode is

$$x_i^c = \begin{Bmatrix} 1 \\ 0 \end{Bmatrix}. \tag{5.15}$$

The results are shown in Figure 5.1. Notice that the eigenmode and the fundamental path are orthogonal and satisfy the orthogonality condition

$$Q_i x_i|^C = \{0, 1\} \begin{Bmatrix} 1 \\ 0 \end{Bmatrix} = 0.$$

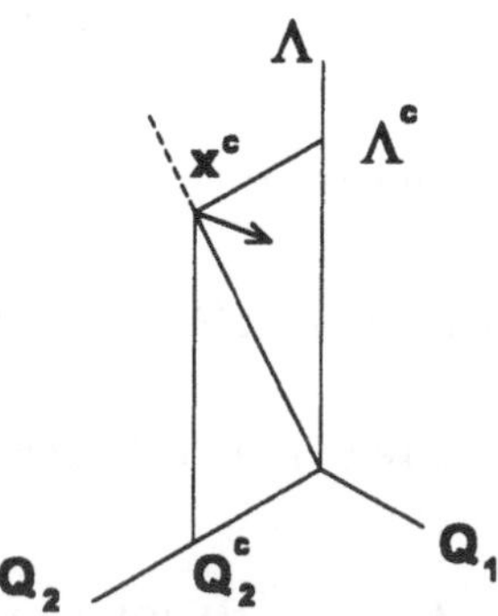

Figure 5.1 Critical state for a circular plate under in-plane compression.

5.4 CRITICAL STATES ALONG NONLINEAR EQUILIBRIUM PATHS

In the previous section the primary path was assumed to be linear, as in many applications that we present in subsequent chapters. However, let us consider here a cubic fundamental path and investigate the consequences of such assumptions. The path is written in terms of some perturbation parameter s in the form

$$Q_i(s) = \dot{Q}_i\, s + \frac{1}{2!}\ddot{Q}_i\, s^2 + \frac{1}{3!}\dddot{Q}_i\, s^3, \tag{5.16}$$

$$\lambda(s) = \dot{\lambda}\, s + \frac{1}{2!}\ddot{\lambda}\, s^2 + \frac{1}{3!}\dddot{\lambda}\, s^3. \tag{5.17}$$

The energy is written in the usual form of a specialized system with quartic terms, (5.8).

The second derivatives V_{ij} now become

$$\begin{aligned} V_{ij}(s) = A_{ij} + s\left(A_{ijk}\dot{Q}_k\right) + \frac{1}{2!}s^2\left(A_{ijk}\ddot{Q}_k + A_{ijkl}\,\dot{Q}_k\dot{Q}_l\right) \\ + \frac{1}{3!}s^3\left(A_{ijk}\,\dddot{Q}_k + 3A_{ijkl}\,\dot{Q}_k\ddot{Q}_l\right) + \frac{1}{4!}s^4\left(3A_{ijkl}\ddot{Q}_k\ddot{Q}_l + 4A_{ijkl}\dot{Q}_k\,\dddot{Q}_l\right) \\ + \frac{1}{5!}s^5\left(10A_{ijkl}\ddot{Q}_k\,\dddot{Q}_l\right) + \frac{1}{6!}s^6\left(10A_{ijkl}\,\dddot{Q}_k\dddot{Q}_l\right). \end{aligned} \tag{5.18}$$

We need not keep such accuracy in $V_{ij}(s)$. Because cubic terms were retained in the original variables we truncate V_{ij} to cubic terms. Thus, the condition of critical state is

$$\begin{aligned} V_{ij}(s)\, x_j|^c = \Big[A_{ij} + s\left(A_{ijk}\dot{Q}_k\right) + \frac{1}{2}s^2\left(A_{ijk}\ddot{Q}_k + A_{ijkl}\,\dot{Q}_k\dot{Q}_l\right) \\ + \frac{1}{3!}s^3\left(A_{ijk}\,\dddot{Q}_k + 3A_{ijkl}\,\dot{Q}_k\ddot{Q}_l\right)\Big]\, x_j|^c = 0. \end{aligned} \tag{5.19}$$

This is a cubic eigenvalue problem, and s^c is the critical load or critical displacement according to the choice of perturbation parameter.

Quadratic and linear approximations can be obtained from (5.19) by elimination of higher-order terms.

5.4.1 Solution of Nonlinear Eigenproblems

The solution of quadratic or cubic eigenproblems is not simple. Some techniques are discussed in the literature, but it is not often that one finds them available in computer packages. A classical text on eigenproblems [31] only mentions one technique: the direct transformation method. Among the few papers on the subject, a combination of sign counting and bisection stages [32] with the Newton–Raphson algorithm have been proposed in [26] to deal with natural frequencies and modes of a linear eigenvalue system and are extended to a transcendental dynamic stiffness matrix in [27]. The flutter boundaries have been calculated by Duncan and Collar in 1932, and updated in [9]; the nonlinear problem that arises in this case is transformed into the search of the zero of a one-dimensional function, and a Newtonian-based solution is used. At least two distinct linear systems of algebraic equations must be solved at each iteration in order to calculate the secant approximation to the one-dimensional function.

But instead of transforming the problem, it is interesting to explore direct possibilities that may be simpler. Iterative techniques were studied in [22] and are summarized in this section. They are attractive because of their simplicity, which leads to an easy implementation in a computer code, but the accuracy and convergence characteristics depend on how the system is linearized at each iteration.

Consider (5.19) in the compact form

$$\left[A_{ij} + s\ B_{ij} + \frac{1}{2!}s^2\ C_{ij} + \frac{1}{3!}s^3\ D_{ij}\right] x_j = 0, \tag{5.20}$$

where the matrices are defined as

$$B_{ij} = A_{ijk}\dot{Q}_k|^c, \tag{5.21}$$

$$C_{ij} = A_{ijk}\ddot{Q}_k + A_{ijkl}\ \dot{Q}_k\dot{Q}_l|^c, \tag{5.22}$$

$$D_{ij} = A_{ijk}\ Q_k + 3A_{ijkl}\ \dot{Q}_k\ddot{Q}_l|^c. \tag{5.23}$$

We begin an iterative procedure by solving the linear part of (5.20), leading to the solution $s = s_0$:

$$\left[A_{ij} + s_0\ B_{ij}\right] x_j = 0. \tag{5.24}$$

At iteration 1, a new matrix is computed

$$E^1_{ij} = B_{ij} + \frac{1}{2}\ s_0\ C_{ij} + \frac{1}{3!}\ s_0^2\ D_{ij} \tag{5.25}$$

and the linear eigenproblem

$$\left[A_{ij} + s_1\ E^1_{ij}\right] x^1_j = 0 \tag{5.26}$$

is solved.

At the nth iteration, we compute

$$E^n_{ij} = B_{ij} + \frac{1}{2}\ s_{n-1}\ C_{ij} + \frac{1}{3!}\ s^2_{n-1}\ D_{ij} \tag{5.27}$$

leading to

$$\left[A_{ij} + s_n\ E_{ij}^n\right] x_j^n = 0. \tag{5.28}$$

Convergence of the iterative procedure is reached when the change in the eigenvalue is smaller than a specified value ϵ

$$\epsilon_s = \left|\frac{s_n - s_{n-1}}{s_n}\right| < \epsilon, \tag{5.29}$$

where ϵ is the accuracy required in the computations.

This is a simple way to solve nonlinear eigensystems, in which the matrix associated with the eigenvalue is updated using information from the previous iteration. Numerical evidence of convergence of this procedure is provided in [22].

5.5 STABILITY OF A CRITICAL STATE

5.5.1 Stability in Terms of the Sign of the Cubic or Quartic Terms

Returning to the discussion on stability of section 4.5 (Chapter 4), an equilibrium state (either critical or regular) is stable if the total potential energy is a minimum in that state. Thus, for a state E to be stable, the following is required:

$$V - V^E = V_i^E q_i + \frac{1}{2!} V_{ij}^E q_i q_j + \frac{1}{3!} V_{ijk}^E q_i q_j q_k + \frac{1}{4!} V_{ijkl}^E q_i q_j q_k q_l. \tag{5.30}$$

Notice that the quadratic form vanishes at a critical state

$$V_i^E = 0,$$

and one has to consider higher-order contributions in the energy expansion. Thus, it is the cubic term $V_{ijk}^c q_i q_j q_k$ that provides the sign of the total change in the energy. We can now say

$$V_{ijk}^c q_i q_j q_k|c \neq 0 \Longrightarrow \text{unstable critical state.} \tag{5.31}$$

On the other hand, if the cubic form itself is zero,

$$V_{ijk}^c q_i q_j q_k|c = 0, \tag{5.32}$$

then it is necessary to investigate the sign of the quartic terms in V to assess the stability of a critical state

$$\text{If} \qquad V_{ijkl} q_i q_j q_k q_l|^c > 0 \Longrightarrow \text{stable critical state,} \tag{5.33}$$

$$\text{If} \qquad V_{ijkl} q_i q_j q_k q_l|^c < 0 \Longrightarrow \text{unstable critical state.} \tag{5.34}$$

The main drawback of the computations in (5.31) to (5.34) is that they have to be done for all possible q_i emerging from the critical state.

Example 5.2 (Circular plate: Stability of the critical state). *Consider the critical load obtained in Example* 5.1 *for a circular plate under in-plane loading.*

It was found that at the critical state the following conditions apply:

$$\Lambda^c = \frac{A_{11}A_{22}}{A_{112}A_2'},$$

$$Q_1^c = 0, \qquad Q_2^c = -\frac{A_{11}}{A_{112}}.$$

Next, we investigate the cubic terms in the V expansion. This has the general expression

$$V_{ijk} = A_{ijk} + A_{ijkl}Q_l.$$

In the present case the cubic terms reduce to

$$V_{111} = A_{111} + A_{1112}Q_2 = 0, \qquad V_{222} = A_{222} + A_{2222}Q_2 = 0,$$

$$V_{122} = A_{122} + A_{1221}Q_1 = 0, \qquad V_{112} = A_{112} + A_{1122}Q_2 = A_{112}.$$

However,

$$\dddot{v} = A_{112}\, x_1^2\, x_2 = 0.$$

It has been easy to show that the cubic terms are zero for this simple problem, but in a more general case, with many degrees of freedom, this may be a more difficult task.

The fourth-order terms are $A_{ijkl}q_iq_jq_kq_l$ and result in

$$A_{1111}(q_1)^4 + 6A_{1122}(q_1)^2(q_2)^2 + A_{2222}(q_2)^4.$$

Since all the A_{ijkl} in this case are positive (see Example 2.6 *in Chapter* 2) *and the increments are affected by even powers, we conclude that*

$$V_{ijkl}q_iq_jq_kq_l > 0$$

for any choice of q_i that is not trivial. This means that the critical state considered is stable.

5.5.2 Parametric Analysis of Directions at the Critical State

The previous example was so simple that there was no need to consider specific directions q_i to evaluate the third- and fourth-order contributions. But in a completely general problem in which the powers of the q_i are not even, it is difficult to evaluate the energy terms in the Taylor expansion of V if one depends on an infinite number of combinations of q_i. Thus, it is desirable to have a more systematic (and simpler) way to evaluate stability of a critical state, without involving the increments q_i.

After reviewing [29], we examine variations of V along all coordinate paths at the critical state considered. A perturbation technique would be most useful in this case, and we write the Q_i as a parametric function of a perturbation parameter denoted by s. It is not necessary to specify s at this stage: It may be an angle, or a scalar which multiplies one of the components, etc. Next, the change in total potential energy $\left(V - V^E\right)$ is expanded in terms of the single parameter s:

$$V - V^c = v(s) = \dot{v}^c s + \frac{1}{2!}\ddot{v}^c s^2 + \frac{1}{3!}\dddot{v}^c s^3 + \frac{1}{4!}\ddddot{v}^c s^4, \tag{5.35}$$

where the ellipses represent derivatives with respect to the perturbation parameter, i.e.,

$$(\dot{v}\,)^c = \left.\frac{dV}{ds}\right|^c, \qquad (\ddot{v})^c = \left.\frac{d^2V}{ds^2}\right|^c,$$

etc. Each one of the required derivatives of v may be evaluated as

$$\dot{v} = V_i \dot{Q}_i, \tag{5.36}$$

$$\ddot{v} = V_i \ddot{Q}_i + V_{ij}\dot{Q}_i\dot{Q}_j, \tag{5.37}$$

$$\dddot{v} = V_i \dddot{Q}_i + V_{ij}\ddot{Q}_i\dot{Q}_j + V_{ijk}\dot{Q}_i\dot{Q}_j\dot{Q}_k + 2V_{ij}\ddot{Q}_i\dot{Q}_j$$

$$= V_i \dddot{Q}_i + 3V_{ij}\dot{Q}_i\ddot{Q}_j + V_{ijk}\dot{Q}_i\dot{Q}_j\dot{Q}_k, \tag{5.38}$$

$$\ddddot{v} = V_i \ddddot{Q}_i + 4V_{ij}\dot{Q}_i \dddot{Q}_j + 3V_{ij}\ddot{Q}_i\ddot{Q}_j + 6V_{ijk}\dot{Q}_i\dot{Q}_j\ddot{Q}_k + V_{ijkl}\dot{Q}_i\dot{Q}_j\dot{Q}_k\dot{Q}_l. \tag{5.39}$$

It is necessary to compute each one of these derivatives at the critical state. Since C is an equilibrium state, it follows that $V_i^c = 0$ and

$$\dot{v}^c = 0. \tag{5.40}$$

The second term reduces to

$$\ddot{v}^c = V_{ij}\dot{Q}_i\dot{Q}_j. \tag{5.41}$$

We have already found this quadratic form, and know that it is positive for all paths chosen (that are stable) except for one: the direction of instability

$$\dot{Q}_i^c = x_i^c. \tag{5.42}$$

Let us investigate stability in the direction of x_i^c, since this will determine stability of the critical state. Thus we consider

$$\ddot{v}^c = 0. \tag{5.43}$$

At this stage, the sign of $V - V^c$ is given by $\dddot{v}^c$. However, a minimum of a function can only be obtained if $v^c > 0$. We write this in the following form: If $v^c \neq 0$ the state C is unstable, so that

$$\dddot{v}^c = V_{ijk} x_i x_j x_k \neq 0 \Rightarrow \text{unstable}. \tag{5.44}$$

On the other hand, if

$$\dddot{v}^c = 0, \tag{5.45}$$

then we have to investigate fourth-order derivatives v^c. We now extend the analysis of [29] and proceed to consider the next derivative:

$$\ddddot{v}^c = 3V_{ij}\ddot{Q}_i\ddot{Q}_j + 6V_{ijk}x_i x_j \ddot{Q}_k + V_{ijkl}x_i x_j x_k x_l. \tag{5.46}$$

There is a problem with the computation of $\ddddot{v}^c$: it depends on $\ddot{Q}_i^c$, and we do not know how to calculate this derivative of displacement. But since a state is stable if $\ddddot{v}^c\,(\ddot{Q}_i) > 0$, then we may consider the value of $\ddot{Q}_i$ for which $\ddddot{v}^c$ attains a minimum. If this minimum is still positive, then the state is stable. Thus, let us minimize $\ddddot{v}^c$ with respect to $\ddot{Q}_j$:

$$\frac{\partial[\ddddot{v}^c\,(\ddot{Q}_j)]}{\partial \ddot{Q}_i} = 0 = V_{ijk}x_j x_k + V_{ij}\ddot{Q}_j|^c$$

from which we get

$$V_{ij}\ddot{Q}_j|^c = -V_{ijk}x_j x_k|^c. \tag{5.47}$$

Here V_{ij}^c is singular and in order to obtain $\ddot{Q}_j^c$ one may have to choose the value of one of its components. This value can be substituted in $\ddddot{v}^c$ and yields

$$\ddddot{v}^c_m = V_{ijkl}x_i x_j x_k x_l + 3V_{ijk}x_i x_j \ddot{Q}_k.$$

Next we define this latter value as a scalar coefficient $\tilde{V}^4$

$$\boxed{\tilde{V}^4 = V_{ijkl}x_i x_j x_k x_l + 3V_{ijk}x_i x_j \ddot{Q}_k} \tag{5.48}$$

The scalar $\tilde{V}^4$ is crucial in most practical cases (symmetric bifurcations, as will be seen in Chapter 7), and the condition of stability of a critical state whenever $v^c = 0$, results in

$$\tilde{V}^4 > 0 \Longrightarrow \text{stable}, \tag{5.49}$$

$$\tilde{V}^4 < 0 \Longrightarrow \text{unstable}. \tag{5.50}$$

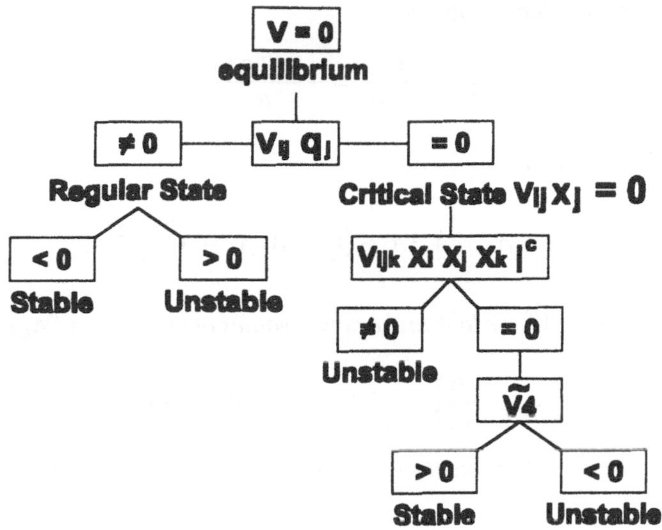

Figure 5.2 Classification of stability of critical states.

A summary of the results of this section is presented in Figure 5.2.

Example 5.3 (Circular plate: Stability and critical directions). *Consider the circular plate under in-plane load at the critical state, following Examples* 5.1 *and* 5.2. *Let us use the above results to investigate stability of the critical state.*

Recall that $x_1^c = 1$, $x_2^c = 0$ *from Example* 5.1. *Then*

$$\dddot{v}^c = V_{ijk} x_i x_j x_k = 0.$$

Notice that using (5.42)

$$\ddot{Q}_1^c = \frac{d(\dot{Q}_1^c)}{ds} = \frac{d(x_1^c)}{ds} = \frac{d(1)}{ds} = 0.$$

Before going into $\tilde{V}^4$ *it is necessary to calculate* $\ddot{Q}_2$ *from*

$$\begin{bmatrix} V_{11}^c & V_{12}^c \\ V_{21}^c & V_{22}^c \end{bmatrix} \begin{Bmatrix} 0 \\ \ddot{Q}_2^c \end{Bmatrix} = \begin{Bmatrix} 0 \\ -V_{211} x_1^2 \end{Bmatrix}$$

or else

$$\begin{bmatrix} 0 & 0 \\ 0 & A_{22} \end{bmatrix} \begin{Bmatrix} 0 \\ \ddot{Q}_2^c \end{Bmatrix} = \begin{Bmatrix} 0 \\ -A_{211} \end{Bmatrix}.$$

From the second line we obtain

$$\ddot{Q}_2^c = -\frac{A_{112}}{A_{22}} = -\frac{0.8668}{R} < 0.$$

Finally, $\tilde{V}^4$ *results in*

$$\tilde{V}^4 = A_{1111} + 3A_{112}\ddot{Q}_2^c = A_{1111} - 3\frac{(A_{112})^2}{A_{22}}.$$

To find the sign of $\tilde{V}^4$ *we look at the actual values of the* A_{ijk} *and* A_{ijkl} *coefficients to find*

$$\tilde{V}^4 = \frac{Eh}{\left(1 - \nu^2\right) R^2}\left[10.55 - 4.50(1 + \nu)\right] > 0.$$

Thus, the circular plate under in-plane compression shows a stable critical point for $\nu < 0.5$.

5.6 THE AXIALLY LOADED COLUMN

This section deals with the straight column with ideal boundary conditions loaded by an axial force through the centroid of the cross section. The example is usually covered in texts on strength of materials and is developed here to show how it is derived from an energy formulation.

5.6.1 Formulation

Total potential energy. Let us first investigate the energy of the column following the notation of Figure 5.3, including membrane as well as bending effects in the energy:

$$V = \frac{1}{2}\int_{x=0}^{L} (N_x \varepsilon_x + M_x \chi_x)\, dx - P\Delta. \tag{5.51}$$

The strain-displacement equations can be assumed including axial deformations and changes in curvature of the middle line

$$\varepsilon_x = \varepsilon'_x + \varepsilon''_x = \frac{du}{dx} + \frac{1}{2}\left(\frac{dw}{dx}\right)^2, \tag{5.52}$$

$$\chi_x = -\frac{d^2w}{dx^2}. \tag{5.53}$$

The end shortening is the displacement at the point of application of the load and may be computed as

$$\Delta = -\int_0^L \frac{du}{dx} dx. \tag{5.54}$$

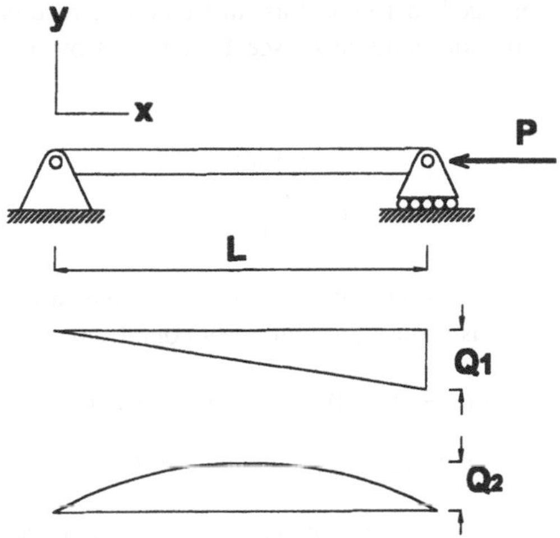

Figure 5.3 Isotropic column under axial load.

The kinematic nonlinearity in this model affects the membrane strains and not the changes in curvature. We shall see that the model of an inextensible column does not have ε_x, so that all nonlinearity is in χ_x.

The constitutive relations for the elastic column made of homogeneous and isotropic material are

$$N_x = AE\varepsilon_x, \qquad M_x = EI\chi_x. \tag{5.55}$$

Thus,

$$V[u, w, P] = \frac{1}{2}EA\int(\varepsilon_x)^2dx + \frac{1}{2}EI\int(\chi_x)^2dx - PL\int\left(\frac{du}{dx}\right)dx.$$

The energy V may also be written in terms of the displacements as

$$V[u, w, \Lambda] = \frac{EA}{2}\int_{x=0}^{L}\left[\left(\frac{du}{dx}\right)^2 + \left(\frac{du}{dx}\right)\left(\frac{dw}{dx}\right)^2 + \frac{1}{4}\left(\frac{dw}{dx}\right)^4\right]dx$$

$$+\frac{EI}{2}\int\left(\frac{d^2w}{dx^2}\right)^2dx - \Lambda\int\left(\frac{du}{dx}\right)dx. \tag{5.56}$$

5.6.2 Simply Supported Column

Next the displacements are approximated by means of trial functions. One possible choice is to employ a polynomial approximation (cubic for w and linear for u), and the reader is encouraged to follow this and compare results. In this section the more classical approximation is adopted (see Figure 5.3) by means of trigonometric functions of w:

$$u = xQ_1,$$
$$w = \left(\sin\frac{n\pi x}{L}\right) LQ_2. \tag{5.57}$$

The displacements are thus written in terms of nondimensional degrees-of-freedom Q_1 and Q_2. Such functions satisfy the boundary conditions:

$$u(x=0)=0, \qquad w(x=0)=0,$$
$$u(x=L)=Q_1, \qquad w(x=L)=0. \tag{5.58}$$

There is no need at this stage to satisfy the natural boundary conditions. Substitution of $u = u(Q_1, x)$ and $w = w(Q_2, x)$ into V leads to

$$V[Q_1, Q_2, \Lambda] = A'_1 \Lambda Q_1$$
$$+\frac{1}{2!}\left(A_{11}Q_1^2 + A_{22}Q_2^2\right) + \frac{1}{3!}(3A_{122})Q_1Q_2^2 + \frac{1}{4!}\left(A_{2222}\right)Q_2^4, \tag{5.59}$$

where

$$\Lambda = \frac{P}{AE}, \qquad A'_1 = EA, \tag{5.60}$$

$$A_{11} = EA, \qquad A_{22} = \frac{1}{2}n^2\pi^2\frac{EI}{L^2}, \qquad A_{122} = \frac{1}{2}n^2\pi^2 AE, \quad A_{2222} = \frac{9}{8}n^4\pi^4 EA.$$

This model of the column has led to a specialized system, in which the load acts on linear terms of displacements. Quartic terms involve only the out-of-plane displacements Q_2.

Fundamental path. The equilibrium conditions are given by

$$V_1 = \Lambda A'_1 + A_{11}Q_1 + \frac{1}{2}A_{122}Q_2^2 = 0,$$
$$V_2 = \left(A_{22} + A_{122}Q_1 + \frac{1}{6}A_{2222}Q_2^2\right)Q_2 = 0,$$

from which we obtain a solution as

$$Q_2^F = 0,$$
$$Q_1^F = -\Lambda\frac{A'_1}{A_{11}} = -\Lambda. \tag{5.61}$$

A compressive load induces a negative axial displacement (shortening) along the fundamental path.

This is a linear path within the approximations of the present model and represents a fundamental path since for $\Lambda = 0$, then $Q_i^F = 0$.

There is another path that satisfies the equilibrium equation for $Q_2 \neq 0$: This new solution is

$$Q_1^S = \frac{3A_{22}A_{122} - \Lambda A_1' A_{2222}}{A_{11}A_{2222} - 3A_{122}A_{122}}, \tag{5.62}$$

$$(Q_2^S)^2 = \frac{6(A_{22} + A_{122}Q_1^S)}{A_{2222}}. \tag{5.63}$$

This is clearly a symmetric path in terms of Q_2^S, but we will only investigate stability along the fundamental path at this stage.

Critical state. To write the eigenvalue problem, the values of second derivatives V_{ij} are required:

$$V_{11} = A_{11}, \qquad V_{22} = A_{22} + A_{221}Q_1 + \frac{1}{2}A_{2222}Q_2^2, \qquad V_{12} = A_{121}Q_2.$$

Along the fundamental path $Q_2^F = 0$, thus

$$V_{11} = A_{11}, \qquad V_{22} = A_{22} + A_{221}Q_1^F, \quad V_{12} = 0$$

is a diagonal form. Clearly, this arises from the simple form of this problem. The eigenproblem results in

$$\left(\begin{bmatrix} A_{11} & 0 \\ 0 & A_{22} \end{bmatrix} - \Lambda \begin{bmatrix} 0 & 0 \\ 0 & A_{221}\bar{Q}_1 \end{bmatrix} \right) \begin{Bmatrix} x_1 \\ x_2 \end{Bmatrix} = \begin{Bmatrix} 0 \\ 0 \end{Bmatrix},$$

where $Q_1^F = \Lambda \, \bar{Q}_1$ has been used. Solution of this problem yields

$$A_{11}(A_{22} - \Lambda \, A_{221}\bar{Q}_1) = 0$$

from which the critical load parameter is

$$\Lambda^c = -\frac{A_{22}}{A_{221}\bar{Q}_1} = \frac{I}{A}\left(\frac{n\pi}{L}\right)^2 \tag{5.64}$$

and, in the more classical form, the critical load becomes

$$\boxed{P^{cr} = EI\left(\tfrac{n\pi}{L}\right)^2} \tag{5.65}$$

The lowest critical load occurs for $n = 1$, and we say that this is the buckling load of the column. The displacements along the fundamental path for this load level are given by

$$Q_i^c = -\frac{n^2}{L^2}\frac{I}{A}\begin{Bmatrix} 1 \\ 0 \end{Bmatrix}. \tag{5.66}$$

The eigenvector associated with Λ^c is obtained from

$$\begin{bmatrix} A_{11} & 0 \\ 0 & 0 \end{bmatrix}\begin{Bmatrix} x_1 \\ x_2 \end{Bmatrix} = \begin{Bmatrix} 0 \\ 0 \end{Bmatrix}.$$

Setting $x_2 = 1$, we obtain $x_1 = 0$ and

$$x = \begin{Bmatrix} 0 \\ 1 \end{Bmatrix}. \tag{5.67}$$

Notice that the choice $x_1 = 1$ would lead to the condition $A_{11} \times 1 = 0$, which cannot be satisfied for $A_{11} \neq 0$.

Stability of the critical state. To investigate the stability of the critical state itself, we compute the third-order stability coefficient

$$\dddot{v} = 0. \tag{5.68}$$

Because $\dddot{v}$ is zero, we must investigate the fourth-order coefficient of stability. For the evaluation of $\tilde{V}_4$ we need the vector $\ddot{Q}_j$, which is computed from

$$V_{ij}^c \ddot{Q}_j = -V_{ijk}^c x_j x_k.$$

The solution yields

$$\ddot{Q}_j = \begin{Bmatrix} -A_{122}/A_{11} \\ 0 \end{Bmatrix}, \tag{5.69}$$

and the stability coefficient becomes

$$\tilde{V}_4^c = \left(V_{2222} x_2^2 + 3V_{221}\ddot{Q}_1\right) x_2^2 = \frac{3}{8}(\pi n)^4 EA > 0. \tag{5.70}$$

The stability of the critical state is given by the sign of $\tilde{V}_4^c$, and this is clearly positive from (5.70). Thus, the critical state of the column is stable.

The present study of the extensible column (including axial deformations) yields a buckling load identical to Euler's load, obtained for the inextensional strut. This latter case is discussed below in section 5.8. According to the eigenvector obtained in the analysis, the direction of instability involves out-of-plane displacements but not axial displacements.

5.6.3 Influence of Boundary Conditions

The buckling load of the column depends on the boundary conditions of the structure. The case studied in section 6.2 is simply supported at both ends. For other boundary conditions, the following critical loads apply:

Clamped-clamped

$$P^{cr} = 4EI\left(\frac{\pi}{L}\right)^2. \tag{5.71}$$

Clamped simply supported

$$P^{cr} = 2.04EI\left(\frac{\pi}{L}\right)^2. \tag{5.72}$$

Clamped-free

$$P^{cr} = \frac{1}{4}EI\left(\frac{\pi}{L}\right)^2. \tag{5.73}$$

Because of the similarity between the form of the above equations and the critical load of the simply supported case, it is common to say that the clamped-clamped case is equivalent to the simply supported case provided an effective length $L' = 0.7L$ is taken into account. For the clamped simply supported case, $L' = 0.5L$, while for the clamped-free case, $L' = 2L$.

5.7 COLUMN ON AN ELASTIC FOUNDATION

The same column of section 5.6 is considered here again, but with the new feature of an elastic foundation distributed along the column (Figure 5.4), i.e., a Wrinkler foundation. This problem was investigated by Timoshenko, and is reported, for example, in [28].

We include a new term due to the effect of the foundation. If K is the stiffness of the foundation, then the energy becomes

$$V[u, w, P] = \frac{1}{2}\int\left(EA(\varepsilon_x)^2 + EI(\chi_x)^2 + K\,w^2 - P\frac{du}{dx}\right)dx. \tag{5.74}$$

The same interpolation function employed for the column (5.57) is applied to the discretization of the column on an elastic foundation.

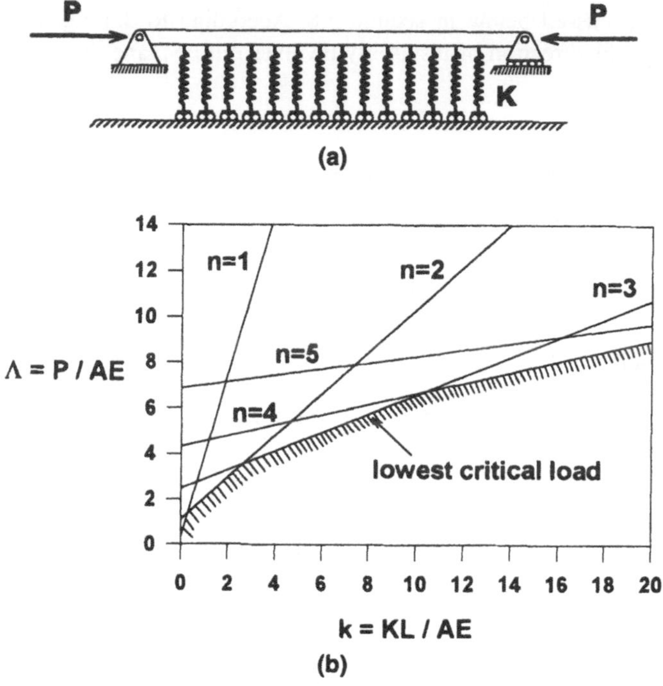

Figure 5.4 (a) Column on an elastic foundation; (b) critical loads for the column on an elastic foundation.

The only coefficient of the energy affected by the elastic foundation is A_{22}

$$A_{22} = \frac{1}{2}\left(n^4\pi^4\frac{EI}{L^2} + KL^3\right). \tag{5.75}$$

Critical state. The fundamental path is the same as in the column studied in the previous section. The critical state leads to

$$\Lambda^c = -\frac{A_{22}}{A_{221}\bar{Q}_1^F} = \frac{1}{AE}\left(n^2\pi^2\frac{EI}{L^2} + \frac{KL^3}{n^2\pi^2}\right). \tag{5.76}$$

The first term is the critical load of a column, and the solution is modified by the second term due to the foundation. It is not possible to say that the lower values of critical loads occur for $n = 1$, and computations or sensitivity analysis have to be carried out. Typical results are presented in Figure 5.4.b, in which the critical load is plotted versus the ratio of stiffness between the foundation and the column. For example, the results of Fiure. 5.4.b show that at $k \leq 0.4$, buckling occurs for $n = 1$. For $0.4 \leq k \leq 3$, the lowest buckling mode is $n = 2$. An increment in the stiffness of the foundation produces an increase in the number of waves in the buckling mode (or a decrease in the wavelength of the buckling mode).

The eigenvector remains the same as in the column (section 5.6) and is given by

$$x = \begin{Bmatrix} 0 \\ 1 \end{Bmatrix}. \tag{5.77}$$

It turns out that sensitivity of the critical load with respect to the stiffness of the foundation decreases as the lowest buckling mode number increases. This is an interesting introductory example that shows how a critical state changes with a design parameter. We shall investigate this topic in more detail in Chapter 7.

5.8 THE AXIALLY LOADED STRUT

This is perhaps the best known example of buckling in structures. This kind of analysis is first attributed to Euler, although the problem that attracted Euler's attention was different.

The columns in section 5.6 and 5.7 were assumed to deform with axial strains. Let us consider the influence of neglecting axial strains in the total potential energy, i.e., an inextensional column, or strut:

$$\varepsilon_x = 0, \tag{5.78}$$

$$V = \frac{1}{2}\int_{x=0}^{L} M_x \chi_x \, dx - P\Delta. \tag{5.79}$$

Our source of nonlinearity in section 6.2 was the axial strain, but such term is not present now. This means that it is necessary to include higher-order terms in the definition of the curvature χ_x, and the refinement takes the form

$$\chi_x = \frac{d^2w}{dx^2}\left[1 - \left(\frac{dw}{dx}\right)^2\right]^{-1/2}. \tag{5.80}$$

Thus, by refining our definition of the curvature we recover nonlinear terms in the kinematics.

The displacement Δ is given by

$$\Delta = L - \int_{x=0}^{L}\left[1 - \left(\frac{dw}{dx}\right)^2\right]^{1/2} dx.$$

Expanding Δ we obtain

$$\Delta = \int_{x=0}^{L}\left[\frac{1}{2}\left(\frac{dw}{dx}\right)^2 + \frac{1}{8}\left(\frac{dw}{dx}\right)^4 + \cdots\right] dx. \tag{5.81}$$

Substitution of (5.81) and (5.80) into V and expanding the strain energy leads to

$$V[w,P] = \frac{1}{2}EI\int_{x=0}^{L}\left[\left(\frac{d^2w}{dx^2}\right)^2 + \left(\frac{dw}{dx}\right)^2\left(\frac{d^2w}{dx^2}\right)^2 + \cdots\right]dx$$

$$+P\int_{x=0}^{L}\left[\frac{1}{2}\left(\frac{dw}{dx}\right)^2 + \frac{1}{8}\left(\frac{dw}{dx}\right)^4 + \cdots\right]dx. \tag{5.82}$$

Notice that the axial displacement u is not present in the energy V of the strut. Next possible displacement fields are assumed in the form

$$w(x) = Q_n \sin\left(\frac{n\pi x}{L}\right), \tag{5.83}$$

where $n = 1, 2, 3, \ldots.$

We substitute (5.83) into (5.80) and compute the energy coefficients as

$$A_{11} = \frac{L}{2}\left(\frac{n\pi}{L}\right)^2\left[EI\left(\frac{n\pi}{L}\right)^2 - P\right],$$

$$A_{1111} = -\frac{3L}{8}\left(\frac{\pi}{L}\right)^4\left[2EI\left(\frac{\pi}{L}\right)^2 - 3P\right]. \tag{5.84}$$

In an inextensible column there are quadratic and quartic terms associated with the load parameter.

The fundamental path is trivial, with no deformations. The critical state is given by

$$V_{11} = A_{11} = 0$$

leading to

$$\boxed{P^c = EI\left(\frac{n\pi}{L}\right)^2} \tag{5.85}$$

The lowest value of the critical load occurs for $n = 1$. Notice that this is the same critical load that we obtained for the column with axial strains. However, differences between the two cases arise in the postcritical path.

5.9 A TWO-BAR FRAME

The next problem is a frame formed with two bars, denoted by bar a and bar b in Figure 5.5.a. The load acts in the x direction and is negative with respect to the assumed positive direction x. Each member has the same length L, area A, and modulus $E\,I$.

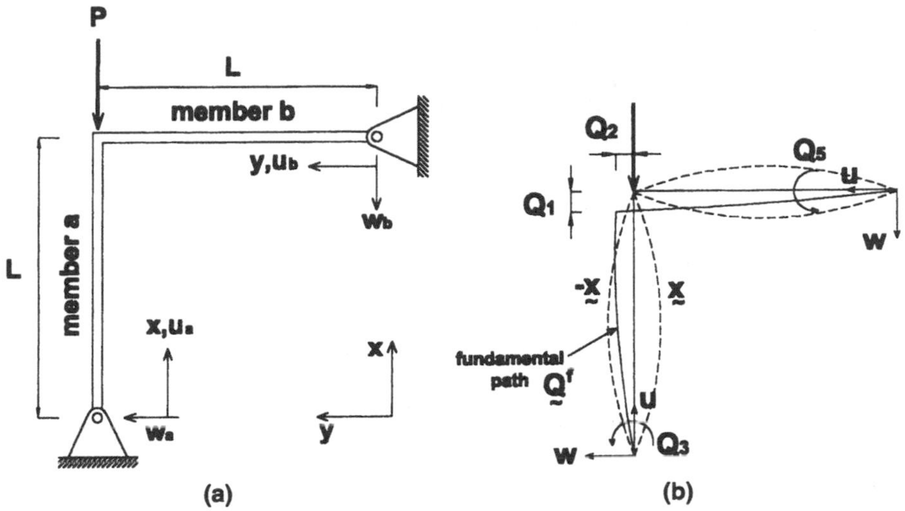

Figure 5.5 (a) Two-bar frame; (b) fundamental path and critical state.

At the common joint of the bars they can rotate an angle θ, but the angle between the bars continues to be $\pi/2$.

Interest in this model comes from the experimental work done by Roorda [24] to illustrate asymmetric bifurcation behavior, in his search to show that the types of instability predicted by Koiter could in fact be observed in practice. And he was successful!

Solutions to this problem of considering inextensible bars were derived by Koiter [18] in analytical form and in [29, 7] using finite elements. An analytical solution including axial deformations of the members was discussed in [3]. We employ a finite element analysis in this section, in which axial deformation of the members is included (see Appendix B). In Chapter 7 we study the same problem of neglecting the membrane energy of the frame, and we shall see that the deformation of the members has no consequences on the critical load, but they are very important for the postbuckling behavior of the frame.

Total potential energy. The energy V has contributions from each one of the bars, with membrane strains and curvatures, so that it may be written as

$$V = \frac{1}{2}\int_{x=0}^{L} (\varepsilon_a N_a + \chi_a M_a)dx$$

$$+\frac{1}{2}\int_{y=0}^{L} (\varepsilon_b N_b + \chi_b M_b)dy - P\Delta. \tag{5.86}$$

The following kinematics apply:

$$\varepsilon_a = \frac{du_a}{dx} + \frac{1}{2}\left(\frac{dw_a}{dx}\right)^2, \qquad \chi_a = \frac{d^2 w_a}{dx^2},$$

$$\varepsilon_b = \frac{du_b}{dy} + \frac{1}{2}\left(\frac{dw_b}{dy}\right)^2, \qquad \chi_b = \frac{d^2 w_b}{dy^2}. \tag{5.87}$$

The constitutive relations are

$$N_a = EA\varepsilon_a, \qquad M_a = EI\chi_a,$$

$$N_b = EA\varepsilon_b, \qquad M_b = EI\chi_b. \tag{5.88}$$

A key point in the model is the satisfaction of the boundary conditions

$$u_a(x=0) = w_a(x=0) = 0, \qquad at x = 0,$$

$$u_b(y=0) = w_b(y=0) = 0, \qquad at y = 0,$$

while compatibility should be satisfied at the joint between the two bars

$$u_a(x=L) = -w_b(y=L),$$

$$u_b(y=L) = w_a(x=L),$$

$$\frac{dw_a}{dx}(x=L) = \frac{dw_b}{dy}(y=L).$$

A discrete model of the frame. As indicated in Appendix B, linear polynomials for the in-plane displacements and cubic interpolation for out-of-plane displacements can be employed for the discretization of frames. Discretization of the structure considering only two elements (one in each member) requires 12 degrees of freedom. However, there are 7 boundary or compatibility conditions, and this leaves us with 5 degrees of freedom. This is a rather crude discretization of the stability problem, but it is shown in detail because the equations become cumbersome as the number of degrees of freedom of the frame increases.

The interpolation functions are written in a more compact form with the aid of nondimensional coordinates

$$\xi = \frac{x}{L}, \qquad \eta = \frac{y}{L}. \tag{5.89}$$

The displacements in the vertical bar a are written in terms of nondimensional generalized coordinates Q_j,

$$u_a = \xi Q_1,$$
$$w_a = Q_2(3\xi^2 - 2\xi^3) + LQ_3(\xi - 2\xi^2 + \xi^3) + LQ_4(\xi^3 - \xi^2).$$

In the horizontal element b, the displacements are assumed in the form

$$u_b = \eta Q_2,$$
$$w_b = -Q_1\left(3\eta^2 - 2\eta^3\right) + LQ_5\left(\eta - 2\eta^2 + \eta^3\right) + LQ_4\left(\eta^3 - \eta^2\right). \quad (5.90)$$

At the joint, we have

$$\Delta = -Q_1. \quad (5.91)$$

The above functions satisfy all the boundary conditions of the problem.

The energy of the discrete system is written next in terms of the generalized coordinates, and the linear energy coefficient is

$$A'_1 = -AE.$$

The quadratic coefficients are

$$A_{11} = A_{22} = \frac{AE}{L} + 12\frac{EI}{L^3}, \qquad A_{33} = A_{55} = 4\frac{EI}{L}, \qquad A_{44} = 8\frac{EI}{L},$$
$$A_{14} = A_{15} = 6\frac{EI}{L^2}, \qquad A_{23} = A_{24} = -6\frac{EI}{L^2}, \qquad A_{34} = A_{45} = 2\frac{EI}{L}. \quad (5.92)$$

The cubic coefficients are

$$A_{112} = A_{122} = \frac{6}{5}\frac{EA}{L^2}, \qquad A_{123} = -\frac{1}{10}\frac{EA}{L} = -A_{125},$$
$$A_{133} = A_{144} = A_{244} = A_{255} = \frac{2AE}{15}, \qquad A_{134} = A_{245} = -\frac{1}{30}\frac{EA}{L}. \quad (5.93)$$

The quartic coefficients are

$$A_{1111} = A_{2222} = \frac{216}{35}\frac{AE}{L^3}, \qquad A_{3333} = A_{5555} = \frac{6}{35}AEL, \qquad A_{4444} = \frac{12}{35}AEL,$$
$$A_{1144} = A_{2244} = A_{2233} = A_{1155} = \frac{9}{35}AE, \qquad A_{3344} = A_{4455} = \frac{AEL}{70},$$
$$A_{1114} = A_{1115} = \frac{27}{35}\frac{AE}{L^2}, \qquad A_{4441} = A_{5551} = -\frac{3}{140}AE,$$
$$A_{4555} = A_{5444} = A_{4443} = A_{3334} = -\frac{3}{140}AEL, \qquad A_{2223} = A_{2224} = -\frac{18}{35}\frac{AE}{L^2},$$
$$A_{1445} = A_{1455} = \frac{3}{140}AE, \qquad A_{2333} = A_{2444} = \frac{3}{140}AE, \quad (5.94)$$

where

$$\Lambda = \frac{P}{AE}. \tag{5.95}$$

Fundamental path. Equilibrium of the above system for a unit load $\Lambda = 1$ takes the form

$$\begin{bmatrix} A_{11} & 0 & 0 & A_{14} & A_{15} \\ & A_{22} & A_{23} & A_{24} & 0 \\ & & A_{33} & A_{34} & 0 \\ & & & A_{44} & A_{45} \\ & & & & A_{55} \end{bmatrix} \begin{Bmatrix} Q_1 \\ Q_2 \\ Q_3 \\ Q_4 \\ Q_5 \end{Bmatrix} = \Lambda \begin{Bmatrix} AE \\ 0 \\ 0 \\ 0 \\ 0 \end{Bmatrix}. \tag{5.96}$$

This leads to the following solution:

$$Q_1^F = -\frac{1}{2}\left(\frac{2AL^2+3I}{AL^2+3I}\right), \qquad Q_2^F = \frac{3}{2}\left(\frac{I}{AL^2+3I}\right),$$

$$Q_3^F = -\frac{1}{4L}\left(\frac{AL^2-I}{AL^2+3I}\right), \qquad Q_4^F = \frac{1}{2L}, \qquad Q_5^F = \frac{1}{4L}\left(\frac{5AE+6I}{AE+3I}\right). \tag{5.97}$$

This fundamental path involves bending of the frame, as can be seen from the presence of the bending stiffness coefficient EI and the rotations Q_3, Q_4, and Q_5.

An approximate membrane solution for the fundamental path, useful for the W formulation in Chapter 13, is in this case

$$Q_1^F = -1,$$

$$Q_j^F = 0 \qquad \text{for} \qquad j = 2, 3, 4, 5. \tag{5.98}$$

This solution may be found, for example, in [5, p. 244].

Example 5.4. *Compute the critical state of the frame using the following data (taken from* [24]*):* $A = 1.6 \times 25.4mm^2$, $E = 200GPa$, $I = 8.466mm^4$, $L = 580.03mm$. *The shape of the fundamental path is shown in Figure* 5.5.b *and involves end shortening as well as lateral displacement of the joint.*

To compute the critical state we need to evaluate the term $\Lambda A_{ijk} Q_k^F$*.The determinant of* V_{ij} *is a quintic function of the load parameter* Λ *if bending terms are included in the fundamental path. This reduces to a cubic polynomial for membrane fundamental path. The lowest root of the polynomial is*

$$\Lambda^c = \frac{P^c}{AE} = 8.932 \times 10^{-6}$$

from which the critical load is $P^c = 72.62N$ (16.32 *lb*).

The eigenvector (Figure 5.5) has only rotations and takes the form

$$x_j = \left\{ \begin{array}{c} 0 \\ 0 \\ 1 \\ -0.5826 \\ 0.3459 \end{array} \right\}.$$

Notice that other eigenvalues can be computed from the analysis, and they are associated with other mode shapes. The other eigenvalues are detected at loads $P = 395.47$, $P = 928.9$, $P = 356,008.9$, *and* $P = 356,016.6$.

Because of the interpolation functions employed, this solution is an approximation, and to improve it with the same functions one should have a discretization of the frame using more elements. A better approximation can be obtained if each bar is divided in two elements. The number of degrees of freedom increases from 12 *to* 24, *and the number of unknowns increases from* 5 *to* 11. *This can be handled with a small computer program, and the critical load results in*

$$P^c = 82.1N.$$

The solution obtained by Roorda [24] *is* $P^c = 80.5N$, *and the difference is approximately* 2%.

5.10 CIRCULAR ARCH UNDER CENTRAL POINT LOAD

The circular arch considered in Chapter 2 is shown schematically in Figure 5.6.a; this is an elastic arch with pinned supports and a load applied at the center. The geometry

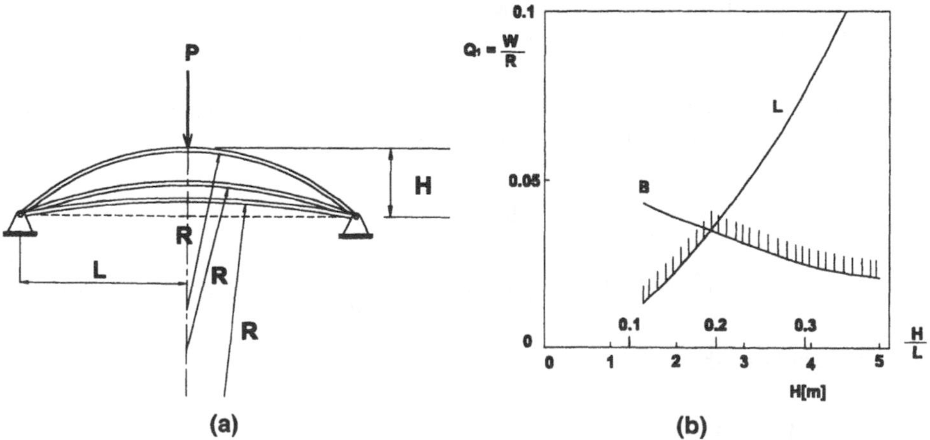

Figure 5.6 (a) Geometry of a circular arch; (b) influence of the height of the arch on the critical state.

of this symmetric arch is written with the help of an angular coordinate θ (with values in the range $-\theta_0 \leq \theta \leq \theta_0$) and the radius R.

The displacement field is represented in terms of two degrees of freedom in the form

$$w(\theta) = R\left[Q_1 \cos(c\theta) + Q_2 \sin(2c\theta)\right], \tag{5.99}$$

where

$$c = \frac{\pi}{2\theta_0}.$$

The energy and equilibrium equations for this arch have been obtained in Chapter 2 in terms of two generalized coordinates. The nonzero coefficients of the energy for this arch are

$$A'_1 = -1, \qquad A_{11} = \frac{8}{\pi^2} + \frac{I}{AR^2}\,c^4, \qquad A_{22} = 16\,\frac{I}{AR^2}\,c^4,$$

$$A_{111} = -\frac{6}{\pi}c^2, \qquad A_{122} = -\frac{8}{\pi}c^2,$$

$$A_{1111} = \frac{3}{2}c^4, \qquad A_{2222} = 24c^4, \qquad A_{1122} = 2c^4, \tag{5.100}$$

where the normalizing coefficients employed are

$$\Lambda = \frac{P}{EA\theta_0}. \tag{5.101}$$

The fundamental path can be considered with different levels of approximation, and they have importance on the critical loads that are computed from them. A detailed analysis of the nonlinear fundamental path of the arch has been carried out in Chapters 2 and 3, and such results are employed in the following.

Cubic fundamental path. If cubic terms are retained in the fundamental path (and this is the exact solution), then the path satisfies the relations

$$Q_2^F = 0,$$

$$\Lambda^F = A_{11}Q_1^F + \frac{1}{2}A_{111}(Q_1^F)^2 + \frac{1}{6}A_{1111}(Q_1^F)^3. \tag{5.102}$$

The matrix of the quadratic form, V_{ij}, becomes

$$V_{ij} = \begin{bmatrix} A_{11} + A_{111}Q_1^F + \frac{1}{2}A_{1111}(Q_1^F)^2 & 0 \\ 0 & A_{22} + A_{122}Q_1^F + \frac{1}{3}A_{1122}(Q_1^F)^2 \end{bmatrix}. \tag{5.103}$$

This is a case in which V_{ij} is diagonal, not only at the level of critical loads but also all along the path. The condition of critical state

$$V_{ij}x_j|^C = 0$$

is a nonlinear eigenvalue problem.

At the critical state, the determinant of V_{ij} vanishes, i.e.,

$$V_{11}V_{22}\,|^c = 0.$$

There are two possibilities: either $V_{11}^c = 0$ or $V_{22}^c = 0$. Let us start exploring the first case, i.e.,

$$V_{11}^c = 0 = A_{11} + A_{111}Q_1 + \frac{1}{2}A_{1111}Q_1^2\Big|^c \tag{5.104}$$

from which we obtain two roots:

$$Q_1^L = \frac{-A_{111} \pm \sqrt{(A_{111})^2 - 2A_{11}A_{1111}}}{A_{1111}}, \tag{5.105}$$

$$Q_2^L = 0.$$

These roots represent the critical displacements for $V_{11}^c = 0$, and the associated critical load is given by (5.102).

Equation (5.105) can also be written in terms of the parameters of the arch as

$$Q_1^L = \frac{4}{\pi c^2}\left(1 \pm \sqrt{\frac{1}{3} - \frac{\pi^2}{12}\frac{I}{AR^2}\,c^4}\right). \tag{5.106}$$

Let us now explore the critical conditions that arise if $V_{22}^c = 0$ is satisfied:

$$V_{22}^c = 0 = A_{22} + A_{122}Q_1^F + \frac{1}{3}A_{1122}(Q_1^F)^2|^c. \tag{5.107}$$

This leads to two roots:

$$Q_1^B = \frac{-A_{122} \pm \sqrt{(A_{122})^2 - \frac{4}{3}A_{22}A_{1122}}}{\frac{2}{3}\ A_{1122}}, \tag{5.108}$$

$$Q_2^B = 0,$$

and the associated critical load is given by (5.102). Equation (5.108) can also be written in terms of the parameters of the arch as

$$Q_1^B = \frac{4}{\pi c^2}\left(1 \pm \sqrt{1 - \pi^2 \frac{I}{AR^2}\, c^4}\right). \tag{5.109}$$

In summary, we have found four critical states along the cubic fundamental path of the arch, with the critical displacements given by (5.105) and (5.108), and the corresponding critical load given by (5.102).

The critical load results in

$$\Lambda^c = \left[A_{11} + \frac{1}{2}A_{111}(Q_1^c) + \frac{1}{6}A_{1111}(Q_1^c)^2\right] Q_1^c.$$

To compute the eigenvectors we go back to the condition of critical state $V_{ij}x_j|^c = 0$. For the two critical states that satisfy $V_{11}^c = 0$ we now have

$$\begin{bmatrix} 0 & 0 \\ 0 & V_{22} \end{bmatrix} \begin{Bmatrix} x_1 \\ x_2 \end{Bmatrix} = \begin{Bmatrix} 0 \\ 0 \end{Bmatrix}. \tag{5.110}$$

If we normalize the eigenvector using $x_1^c = 1$, then $V_{22}x_2|^c = 0$; therefore, $x_2^c = 0$ and

$$x_i^L = \begin{Bmatrix} 1 \\ 0 \end{Bmatrix}. \tag{5.111}$$

Next we consider the case $V_{22}^c = 0$, leading to

$$\begin{bmatrix} V_{11} & 0 \\ 0 & 0 \end{bmatrix} \begin{Bmatrix} x_1 \\ x_2 \end{Bmatrix} = \begin{Bmatrix} 0 \\ 0 \end{Bmatrix} \tag{5.112}$$

from which

$$x_i^B = \begin{Bmatrix} 0 \\ 1 \end{Bmatrix}. \tag{5.113}$$

Engineers are mainly interested in the lowest critical load of a system. In the present case the lowest critical displacements will be associated with the (−) sign in (5.106) and (5.109). Which one will occur first? It may be shown that

$$\text{If} \quad \frac{I}{AR^2}\left(\frac{\pi}{2\theta_0}\right)^4 > 0.074 \quad \Rightarrow \quad Q_1^B > Q_1^L, \tag{5.114}$$

while if

$$\text{If} \quad \frac{I}{AR^2}\left(\frac{\pi}{2\theta_0}\right)^4 < 0.074 \quad \Rightarrow \quad Q_1^B < Q_1^L. \tag{5.115}$$

From the geometric parameters in (5.115), it follows that for a given cross section of the arch, shallow arches with small θ_0 find Q_1^L before Q_1^B. With large values of θ_0 (deep arches), the reverse occurs and the lowest critical state is defined by Q_1^B. We shall see in the following chapters that the critical points associated with Q_1^L are limit points, while those associated with Q_1^B are bifurcation points.

Linear fundamental path. In the previous analysis cubic terms were retained along the fundamental path, leading to four critical states. Let us consider in this section a simplification of the above analysis by retaining only linear terms in (5.102). This means that the fundamental path is represented by

$$Q_2^F = 0, \qquad \Lambda^F = A_{11}\, Q_1^F . \tag{5.116}$$

The matrix V_{ij} now becomes

$$V_{ij} = \begin{bmatrix} A_{11} + A_{111} Q_1^F & 0 \\ 0 & A_{22} + A_{122} Q_1^F \end{bmatrix}.$$

We may rewrite the second equation of (5.116) in the form

$$Q_1^F = \frac{1}{A_{11}} \Lambda^F,$$

so that V_{ij} is an explicit function of Λ:

$$V_{ij}(\Lambda) = \left(\begin{bmatrix} A_{11} & 0 \\ 0 & A_{22} \end{bmatrix} + \Lambda \begin{bmatrix} A_{111}/A_{11} & 0 \\ 0 & A_{22}/A_{11} \end{bmatrix} \right). \tag{5.117}$$

The condition $V_{ij}(\Lambda) x_j|^c = 0$ is now a linear eigenvalue problem in Λ.
For $V_{11}^c = 0$ we get

$$\Lambda^L = -\frac{(A_{11})^2}{A_{111}}, \qquad Q_1^L = -\frac{A_{11}}{A_{111}}, \tag{5.118}$$

$$x_i^L = \{1, 0\} .$$

For $V_{22}^c = 0$ the following solution is obtained:

$$\Lambda^B = -\frac{A_{22} A_{11}}{A_{122}}, \qquad Q_1^B = -\frac{A_{22}}{A_{122}}, \tag{5.119}$$

and

$$x_i^B = \{0, 1\}.$$

In both cases, the critical states Q_2^L and Q_2^B computed for the cubic fundamental path are not present in the linearized path. The solutions of critical states along a linear path have now reduced to two.

The solution presented in this section can be acceptable in deep arches for which bifurcation buckling occurs first. The path in the region of Q_1^B in this case is approximately linear, and the results may be a good estimate with respect to computing critical states from a cubic fundamental path.

Example 5.5. *To illustrate the buckling behavior of a circular arch structure, we consider in this section the following data:* $A = 7.74 \times 10^{-3} m^2$, $I = 2.49 \times 10^{-4} m^4$, $E = 200 GPa$, *a fixed span of* $2L = 26m$ *and a radius* $R = 19.4m$, *leading to a central angle* $2\theta_0 = 1.46 rad$. *The height of the arch is* $H = 5m$. *Using the above dimensions, we find*

$$\frac{I}{AR^2}\left(\frac{\pi}{2\theta_0}\right)^4 = 0.0179.$$

This is a value below 0.07, *so that we would expect to find bifurcation occurring before a limit point.*

The results show that there is bifurcation at a critical displacement $Q_1^B = 0.025$, *while a limit point occurs for* $Q_1^L = 0.12$, *which is a much higher value.*

To investigate the influence of the elevation of the arch on the stability problem, parametric studies were carried out for $0.1 \leq H/L \leq 0.8$, *and we obtained the diagram of Figure* 5.6.b. *For small values of* H, *that is, for shallow arches, the unstable behavior is a limit point. For* $H \approx 2.5m$ $(H/L = 0.19)$ *the first bifurcation and the limit point are coincident, and we should be careful because the eigenvalues are no longer distinct. This may lead to mode coupling. Finally, for deep arches, in which* $H/L \geq 0.2$, *the behavior is dominated by bifurcation.*

We have identified two classes of critical states in the problem of a circular arch under a point load. In one of them the eigenvector has the same direction as the primary path, while in the other case the eigenvector is orthogonal to the primary path. For the discretization adopted, the primary path results in the form of a function between the load Λ^F and the Q_1^F component, with $Q_2^F = 0$.

Which one of the critical loads occurs first depends on the geometry of the arch. In shallow arches, the first eigenvector has the direction of the primary path, while in deep arches the first eigenvector is orthogonal to the primary path. Linearization of the fundamental path may have small consequences on the first critical load, if it occurs at a load level for which the path is moderately nonlinear. The limit point, on the other hand, occurs at a highly nonlinear part in the path, and the error in the identification by means of a linear analysis is very large.

An interesting nonclassical application of an arch structure, with buckling studies as design considerations, is reported in [15].

5.11 RECTANGULAR PLATE UNDER IN-PLANE LOADS

The buckling of laminated composite plates was reviewed by Leissa in [19] including 350 references. Such a review includes orthotropic plates, anisotropic plates, unsymmetrical laminates, and additional effects such as holes, nonlinear constitutive relations, local effects, influence of shear on buckling, postbuckling, and imperfection sensitivity. Critical loads of isotropic rectangular plates for various boundary and loading conditions may be found in [5, 4] and several others. The buckling of composite rectangular plates is studied, for example, in [30, 16, 1].

In this section we present one example to illustrate orthotropic rectangular plates under in-plane loadings. Boundary conditions are assumed to be simply supported to be able to carry out a simple analysis.

5.11.1 Orthotropic Plate under Biaxial Loading

The kinematic equations of the rectangular plate in Cartesian coordinates are

$$\varepsilon_{ij} = \frac{1}{2}\left(\frac{\partial u_i}{\partial x_j} + \frac{\partial u_j}{\partial x_i}\right) + \frac{1}{2}\frac{\partial u_3}{\partial x_i}\frac{\partial u_3}{\partial x_j}, \tag{5.120}$$

$$\chi_{ij} = -\frac{\partial^2 u_3}{\partial x_i \partial x_j}. \tag{5.121}$$

The plate considered is made of a composite material, which is a specially orthotropic laminate, with the constitutive relations taken from [12]:

$$N_{11} = K_{11}\varepsilon_{11} + K_{12}\varepsilon_{22}, \qquad N_{22} = K_{12}\varepsilon_{11} + K_{22}\varepsilon_{22},$$
$$N_{12} = K_{66}(2\varepsilon_{12}). \tag{5.122}$$

Here K_{66} is the constitutive coefficient associated with the engineering shear strain $2\varepsilon_{12}$. The constitutive relations for bending are

$$M_{11} = D_{11}\chi_{11} + D_{12}\chi_{22}, \qquad M_{22} = D_{12}\chi_{11} + D_{22}\chi_{22},$$
$$M_{12} = (2D_{66})\chi_{12}. \tag{5.123}$$

The total potential energy of the plate may be written as

$$V = \frac{1}{2}\int_{x1=0}^{a}\int_{x2=0}^{b}(N_{ij}\varepsilon_{ij} + M_{ij}\chi_{ij})dx_1dx_2 + \Lambda\oint f_iu_ids, \tag{5.124}$$

where s is the boundary coordinate, and it can be either x_1 or x_2; and a, b are the dimensions of the plate. It is assumed that all the loads are increased by the same load parameter Λ.

Raleigh–Ritz solution. Next we assume a solution to the displacement field in the form

$$u_1(x_1, x_2) = Q_1 \left(\frac{2x_1}{a} - 1 \right) + Q_3 (\sin \mu x_1 \cos \eta x_2),$$

$$u_2(x_1, x_2) = Q_2 \left(\frac{2x_2}{b} - 1 \right) + Q_4 (\cos \mu x_1 \sin \eta x_2),$$

$$u_3(x_1, x_2) = Q_5 \left(\sin \mu x_1 \sin \eta x_2 \right), \tag{5.125}$$

where $\mu \equiv m\pi/a$, $\eta = n\pi/b$, and $m, n = 1, 3, 5, 7$. The degrees of freedom Q_1 and Q_2 are associated with the end shortening on the two coordinate directions, and Q_3, Q_4, and Q_5 represent the displacement field after buckling.

Substitute (5.125) into (5.124), and after integration one obtains the quadratic energy coefficients

$$A_1' = 2bf_1, \qquad A_2' = 2af_2,$$

$$A_{11} = 4\frac{b}{a}K_{11}, \qquad A_{22} = 4\frac{a}{b}K_{22},$$

$$A_{33} = \frac{ab}{4}\left(\mu^2 K_{11} + \eta^2 K_{66}\right), \qquad A_{44} = \frac{ab}{4}\left(\eta^2 K_{22} + \mu^2 K_{66}\right),$$

$$A_{55} = \frac{ab}{4}\left[\mu^4 D_{11} + 2\mu^2\eta^2 (D_{12} + 2D_{66}) + \eta^4 D_{22}\right],$$

$$A_{12} = 4K_{12}, \qquad A_{34} = \frac{ab}{4}(K_{66} + K_{12})\,\eta\mu.$$

The cubic energy coefficients result in

$$A_{155} = \frac{b}{2}\left(\mu^2 K_{11} + \eta^2 K_{12}\right), \qquad A_{255} = \frac{a}{2}\left(\mu^2 K_{12} + \eta^2 K_{22}\right),$$

$$A_{355} = 0, \qquad A_{455} = 0.$$

The higher-order coefficients will not be required in this section, but they are computed because they will be necessary in the postbuckling analysis:

$$A_{5555} = 3\frac{ab}{64}\left[9\mu^4 K_{11} + 2\mu^2\eta^2 (K_{12} + 2K_{66}) + 9\eta^4 K_{22}\right]. \tag{5.126}$$

The total potential energy now results in the usual form

$$V[Q_i, \Lambda] = A_i \Lambda Q_i + \frac{1}{2!}A_{ij} Q_i Q_j + \frac{1}{3!}A_{ijk} Q_i Q_j Q_k + \cdots \tag{5.127}$$

with $i, j, k, l = 1, 2, 3, 4, 5$ in the present case.

Primary path. Since all loads are increased by the same parameter Λ, then there is a fixed relation between f_1 and f_2, i.e.,

$$f_2 = \gamma\ f_1, \tag{5.128}$$

where γ indicates the relation between loads in the x and in the y coordinate directions.

From the condition of equilibrium $V_i = 0$, one obtains the following linear fundamental path:

$$Q_1^F = -\frac{1}{2} a f_1 \frac{K_{22} - \gamma K_{12}}{K_{11}K_{22} - K_{12}^2},$$

$$Q_2^F = -\frac{1}{2} b f_1 \frac{\gamma K_{11} - K_{12}}{K_{11}K_{22} - K_{12}^2},$$

$$Q_3^F = Q_4^F = Q_5^F = 0. \tag{5.129}$$

Critical state. The condition of critical state, $V_{ij}\ x_j\ |^c = 0$, results in the form

$$\begin{bmatrix} A_{11} & A_{12} & 0 & 0 & 0 \\ & A_{22} & 0 & 0 & 0 \\ & & A_{33} & A_{34} & 0 \\ & & & A_{44} & 0 \\ & & & & A_{55} + A_{551} Q_1^F + A_{552} Q_2^F \end{bmatrix} \begin{Bmatrix} x_1 \\ x_2 \\ x_3 \\ x_4 \\ x_5 \end{Bmatrix} = \begin{Bmatrix} 0 \\ 0 \\ 0 \\ 0 \\ 0 \end{Bmatrix}. \tag{5.130}$$

The eigenvalue Λ^c is given by the last row

$$\Lambda^c f_1 = -\frac{A_{55}}{A_{551} Q_1^F + A_{552} Q_2^F} \tag{5.131}$$

or else

$$\Lambda^c f_1 = \frac{\mu^4 D_{11} + 2\mu^2\eta^2 (D_{12} + 2D_{66}) + \eta^4 D_{22}}{\gamma\eta^2 + \mu^2}. \tag{5.132}$$

Equation (5.132) represents a set of values of critical loads as a function of the specific mode considered (μ, η) and the relation γ between load components.

The eigenvector can be computed from (5.130). Although V_{ij} has dimensions 5×5, the first two rows and columns are uncoupled from the rest. We choose the value of one of the components of the eigenvector to solve the problem, say, $x_1 = 1$. This leads to

$$A_{33}x_3 + A_{34}x_4 = 0,$$

$$A_{34}x_3 + A_{44}x_4 = 0,$$

from which the solution is

$$x_j = \begin{Bmatrix} 0 \\ 0 \\ 0 \\ 0 \\ 1 \end{Bmatrix}. \tag{5.133}$$

In the present example, the eigenvector has a component only in the out-of-plane direction.

Example 5.6. *Let us compute the critical state for a rectangular plate with $b = 250mm$, and $a = 2b$. The coefficients of the constitutive equations are assumed to be $D_{11} = 0.856 GPa.mm^3$, $D_{12} = 0.824 GPa.mm^3$, $D_{22} = 2.88 GPa.mm^3$, and $D_{66} = 0.972 GPa.mm^3$.*

The critical loads for different values of m, and for $n = 1$, have been plotted in Figure 5.7.a. The figure also shows the dependence of Λ^c on the relation between the load components, as reflected by the coefficient γ.

For this particular plate geometry and material properties, the lowest critical load does not occur for $m = 1$ but for a higher value of m.

It is also clear that the critical load decreases with increasing values of γ. For $\gamma = 0$ (uniaxial load), the lowest critical load is $f_1\Lambda^c = 1.46 \times 10^{-3} GPa.mm$, while for $\gamma = 1.0$, we get $f_1\Lambda^c = 0.73 \times 10^{-3} GPa.mm$.

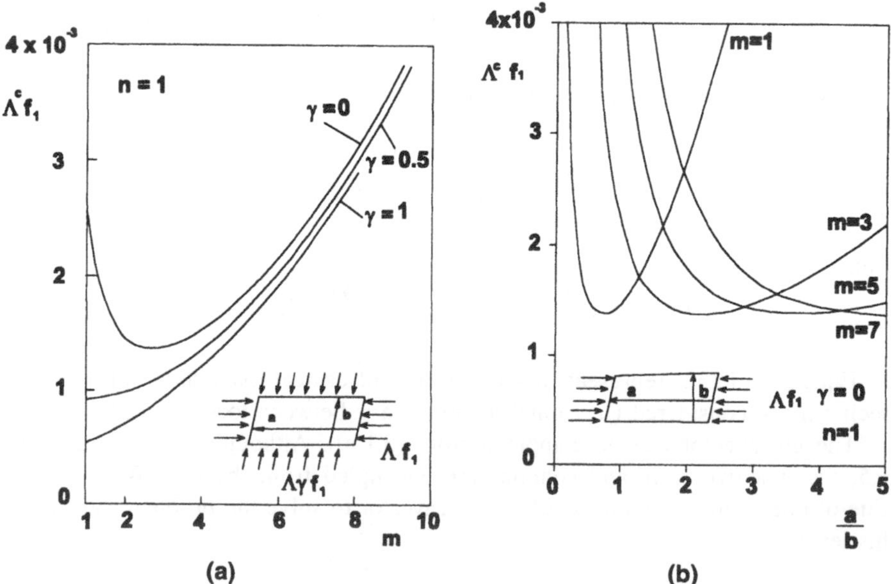

Figure 5.7 Composite plate. (a) Critical loads for different mode shapes with $n = 1$; (b) critical loads for different plates.

5.11.2 Orthotropic Plate under Uniaxial Loading

For the special case in which $\gamma = 0$, there is only one load component f_1 acting on two edges of the plate. The critical load becomes a function of η and μ:

$$\Lambda^c = \frac{1}{f_1}\left[\mu^2 D_{11} + 2\eta^2 (D_{12} + 2D_{66}) + \frac{n^4}{\mu^2} D_{22}\right]. \tag{5.134}$$

In this particular case it is possible to find the minimum value of Λ^c in terms of η and μ.

The minimum of Λ^c in terms of η occurs for $n = 1$, in which case $\eta_{min} = \pi/b$. The minimum in terms of μ is not evident, and we find it from the condition

$$\frac{\partial \Lambda^c}{\partial \mu} = 0 = 2\mu D_{11} - 2\left(\frac{\pi}{b}\right)^4 \frac{1}{\mu^3} D_{22}$$

from which

$$\mu^2_{min} = \left(\frac{\pi}{b}\right)^2 \sqrt{\frac{D_{22}}{D_{11}}}.$$

Substitution into Λ^c leads to

$$f_1 \Lambda^c_{min} = 2\left(\frac{\pi}{b}\right) 2\left[\sqrt{D_{11} D_{22}} + (D_{12} + 2D_{66})\right]. \tag{5.135}$$

Example 5.7. *A plot of Λ^c versus the aspect ratio of the plate, a/b, is shown in Figure* 5.7.b *for different values of m and for $n = 1$. It is interesting to notice that the minimum of each curve is the same value,* $(\Lambda^c f_1)$ min $= 1.37 \times 10^{-3}$, *because we have fixed b and increased a.*

5.11.3 Isotropic Plate

In a single layer isotropic plate, the constitutive coefficients reduce to

$$A_{12} = \nu K, \qquad A_{22} = K, \qquad A_{16} = A_{26} = 0, \qquad A_{66} = \frac{1-\nu}{2} K,$$

$$D_{12} = \nu D, \qquad D_{22} = D, \qquad D_{16} = D_{26} = 0,$$

$$D_{66} = \frac{1-\nu}{4} D, \qquad D_{12} + 4D_{66} = D, \tag{5.136}$$

where

$$D = \frac{Et^3}{12\left(1 - \nu^2\right)}, \qquad K = \frac{Et}{1 - \nu^2}.$$

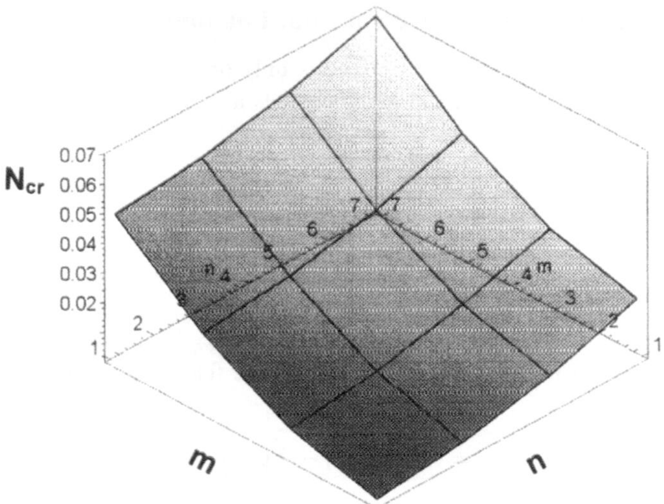

Figure 5.8 Aluminum plate, critical loads for different modes.

The critical load for biaxial loading results in

$$f_1 \Lambda^c = D \frac{\left(\mu^2 + \eta^2\right)^2}{\left(\mu^2 + \gamma \eta^2\right)}. \tag{5.137}$$

If the load has the same value along all edges of the plate, then $\gamma = 1$ and

$$f_1 \Lambda^c = D\left(\mu^2 + \eta^2\right). \tag{5.138}$$

Finally, for uniaxial loading, $\gamma = 0$ and

$$f_1 \Lambda^c_{\min} = 4\left(\frac{\pi}{b}\right)^2 D_{11}.$$

Example 5.8. *Let us consider an aluminum plate with* $E = 70,000\ N/mm^2$ *and* $\nu = 0.3$. *The geometry of the plate is defined by* $a = 250mm$, $b = 400mm$, *and thickness* $h = 0.1mm$. *For a value of* $\gamma = 1$ *(uniform compression on all boundaries), the critical load for different values of m and n is shown in Figure 5.8. The lowest load occurs for* $m = n = 1$ *and is given by* $\Lambda^c f_1 = 1.407\ N/mm^2$.

5.12 CIRCULAR RING

Discrete energy of a ring. The problem of a circular ring under a uniform radial load is interesting from the point of view of its buckling behavior and also because of the large number of applications of rings as part of structures. The case studied is shown in Figure 5.9, where the load always acts in the radial direction. Other load cases are considered by several authors, including follower pressures, which act normally on the deformed surface of the ring.

The total potential energy was already calculated in Problem 2.1 and can be written as

$$V = \frac{1}{2}\int_0^{2\pi} (N_\theta \varepsilon_\theta + M_\theta \chi_\theta)\, R d\theta - \int_0^{2\pi} pw R d\theta. \tag{5.139}$$

We employ the kinematic relations

$$\varepsilon_\theta = \varepsilon'_\theta + \varepsilon''_\theta = \left(\frac{du}{Rd\theta} + \frac{w}{R}\right) + \frac{1}{2}\left(\Gamma\frac{\bar{u}}{R} - \frac{dw}{Rd\theta}\right)^2,$$

$$\chi_\theta = \frac{1}{R}\left(\Gamma\frac{du}{Rd\theta} - \frac{d^2w}{Rd\theta^2}\right). \tag{5.140}$$

The value of Γ in (5.140) is 0 or 1 depending on the ring theory employed. In Donnell's theory, we take $\Gamma = 0$, while in the improved theory, $\Gamma = 1$ is assumed.

The elastic constitutive laws are as in (5.55). Substitution into (5.139) leads to the energy $V = V(\bar{u}, \bar{w}, \Lambda)$.

The displacements are next assumed in the convenient form

$$u(\theta) = RQ_3 \sin(n\theta),$$

$$w(\theta) = R\left[Q_1 + Q_2\cos(n\theta)\right] \tag{5.141}$$

for $n = 2, 3, 4, 5, \ldots$. If the energy is divided by AE, the coefficients become as follows:

$$\Lambda = \frac{R}{AE}p, \qquad A_1 = -2\Lambda, \qquad A_{11} = 2, \qquad A_{22} = 1 + n^4\frac{I}{AR^2},$$

$$A_{23} = n\left(1 + n^2\Gamma\frac{I}{AR^2}\right), \qquad A_{33} = n^2\left(1 + \Gamma^2\frac{I}{AR^2}\right), \tag{5.142}$$

$$A_{122} = n^2, \qquad A_{123} = \Gamma n, \qquad A_{133} = \Gamma^2, \qquad A_{2222} = \frac{9}{4}n^4,$$

$$A_{2223} = \frac{9}{4}\Gamma n^3, \qquad A_{2233} = \frac{9}{4}\Gamma^2 n^2, \qquad A_{2333} = \frac{9}{4}\Gamma^3 n, \qquad A_{3333} = \frac{9}{4}\Gamma^4,$$

Notice that there are only 7 coefficients in Donnell's approximation ($\Gamma = 0$), while in the improved theory the number of terms is 13. The properties of the cross section only enter up to quadratic terms, while cubic and quartic terms are only a function of the number n of waves in the buckling mode.

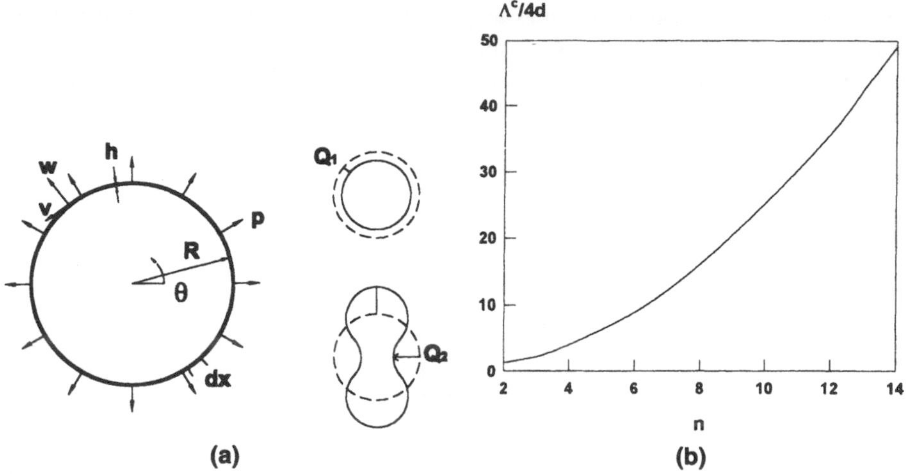

Figure 5.9 (a) Circular ring under radial load; (b) buckling mode for a circular ring under radial load.

Critical state. The fundamental path is shown to be

$$Q_1^F = \Lambda, \qquad Q_2^F = 0, \qquad Q_3^F = 0. \tag{5.143}$$

The Hessian of the quadratic form is given in this case by

$$V_{ij} = \begin{bmatrix} 2 & 0 & 0 \\ 0 & 1 + n^2(n^2 \frac{I}{AR^2} + \Lambda) & n(1 + \Gamma n^2 \frac{I}{AR^2}) + \Gamma n \Lambda \\ 0 & n(1 + \Gamma n^2 \frac{I}{AR^2}) + \Gamma n \Lambda & n^2(1 + \Gamma^2 \frac{I}{AR^2}) + \Gamma^2 \Lambda \end{bmatrix}. \tag{5.144}$$

The critical load results in

$$\Lambda^c = -\left(\frac{n}{R}\right)^2 \frac{I}{A}. \tag{5.145}$$

The lowest value of critical load occurs for $n = 2$,

$$\Lambda^c_{\min} = -4 \frac{I}{AR^2},$$

and the associated mode shape is illustrated in Figure 5.9.b. It is interesting to notice that the critical load in this case is independent of the value of Γ, so that both theories of the ring yield identical values of critical load.

The displacements at the critical load are given by

$$Q_1^c = -n^2 \frac{I}{AR^2}, \qquad Q_2^c = 0, \qquad Q_3^c = 0. \tag{5.146}$$

The eigenmode can be computed as

$$x_1^c = 0, \qquad x_2^c = 1, \qquad x_3^c = -\frac{1}{n}. \tag{5.147}$$

Finally, the stability coefficient evaluated at the critical state results in

$$\tilde{V}_4^c = \frac{3}{4}\left(n - \frac{\Gamma}{n}\right)^4 > 0. \tag{5.148}$$

The value of $\tilde{V}_4^c$ depends on the theory employed (i.e., on the value of Γ). But in both theories the ring has a stable critical point because of the fourth power in (5.148). Because the stability coefficient is positive, the critical state of a circular ring under a radial load is stable.

Results are plotted in Figure 5.9.b for a ring with $I/AR^2 = \frac{1}{12} \times 10^{-4}$.

5.13 TORSIONAL BUCKLING OF COMPRESSED ELEMENTS

Thin-walled columns may have a form of buckling in which the cross section remains almost undeformed but with a rotation in a torsional mode. Such deflection is shown in Figure 5.10 for a thin-walled open section with arbitrary shape. The centroid of the section is denoted by C, and the shear center is S. If the section is loaded at the shear center, then there is bending without twisting of the column. Details on the behavior of such cross sections can be found in books on mechanics of solids.

The cross section has a translation u along axis x; a displacement v in the direction of axis y; and a rotation β, as shown in Figure 5.10.

The energy due to bending in the x direction is

$$V = \frac{1}{2}\int_0^L EI_y \left(\frac{d^2u}{dz^2}\right)^2 dz.$$

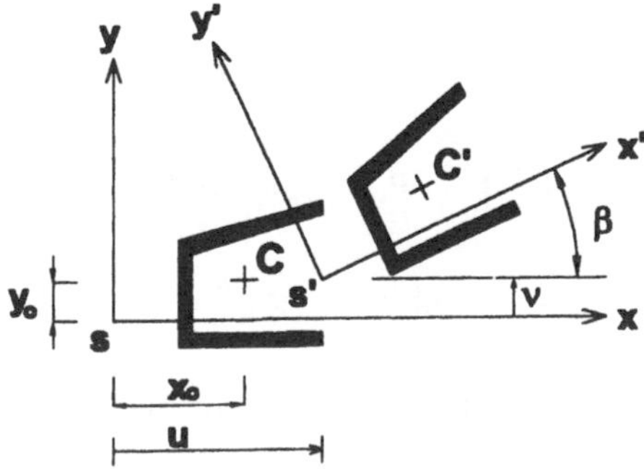

Figure 5.10 Torsional buckling of thin-walled elements under compression.

Bending in the y direction leads to

$$V = \frac{1}{2}\int_0^L EI_x \left(\frac{d^2v}{dz^2}\right)^2 dz.$$

St. Venant torsion introduces the energy term

$$V = \frac{1}{2}\int_0^L GJ \left(\frac{d\beta}{dz}\right)^2 dz.$$

Finally, the energy produced by longitudinal stresses due to warping torsion is given by

$$V = \frac{1}{2}\int_0^L E\Gamma \left(\frac{d^2\beta}{dz^2}\right)^2 dz,$$

where Γ is the warping constant of the cross section. The column was assumed to be inextensional in the axial direction, so that for the calculation of the end shortening necessary for the load potential, the equations of the strut should be used.

Next, assume boundary conditions at $z = 0$ and $z = L$:

$$u = v = \beta = 0,$$

$$\frac{d^2u}{dz^2} = \frac{d^2v}{dz^2} = \frac{d^2\beta}{dz^2} = 0.$$

We neglect deflections in the fundamental path and consider that the buckling mode of this problem may be written as

$$u = Q_1 \sin\left(\frac{\pi z}{L}\right), \qquad v = Q_2 \sin\left(\frac{\pi z}{L}\right), \qquad \beta = Q_3 \sin\left(\frac{\pi z}{L}\right).$$

Including kinematics up to quadratic terms in the energy leads to

$$V = \frac{1}{L}\left(\frac{\pi}{2}\right)^2 \left(A_{11}Q_1^2 + A_{22}Q_2^2 + A_{33}Q_3^2 + 2A_{13}Q_1Q_3 + 2A_{23}Q_2Q_3\right),$$

where

$$A_{11} = EI_y\left(\frac{\pi}{L}\right)^2 - P, \qquad A_{22} = EI_x\left(\frac{\pi}{L}\right)^2 - P,$$

$$A_{33} = GJ + E\Gamma\left(\frac{\pi}{L}\right)^2 - r_0^2 P, \qquad A_{13} = y_0 P, \qquad A_{23} = -x_0 P.$$

The radius of gyration is r_0, and (x_0 and y_0) are the coordinates of the distance between S and C.

Torsional buckling of the member can now be studied, and this is left to the reader as a problem at the end of this chapter.

5.14 FINITE ELEMENTS FOR BUCKLING LOADS

The examples presented in this chapter are representative of what can be solved using simplified models and just a few degrees of freedom. Of course, more complex problems are faced by the engineer because of the geometry, the boundary conditions, and the loads that lead to complex nonlinear structural systems. Solution of such systems is only possible by means of numerical methods, and the finite element method is at present the most common choice in structural engineering. A convenient feature of the finite element method in stability problems is that both can be based on the total potential energy. Thus, the equations derived in this chapter to identify critical states can be approximated at a finite element level and then assembled to have the energy of a structural system.

Let us consider the energy of one element in a finite element discretization

$$V = \frac{1}{2}\int \sigma^T \varepsilon d\mathcal{V} - q^T \int \phi^T b d\mathcal{V} - q^T \int \phi^T p d\mathcal{S},$$

where the stress vector is σ, the strain vector is ε, q are the degrees of freedom of the element, b are the boundary forces, and p are the surface forces. When the structure has large strains, σ is the second Piola–Kirchhoff stress tensor, and ε is the Green–Lagrange strain tensor.

The displacements are interpolated by means of shape functions ϕ and modal unknowns, i.e.,

$$u = \phi q.$$

The strains are assumed to be the linear term in q plus a quadratic term, represented in the form

$$\varepsilon = B_0 q + B_1(q)q.$$

B_0 and B_1 are differential operators acting on the interpolation functions ϕ.

For linear elastic material, the constitutive equations can be written as

$$\sigma = C\varepsilon.$$

With the above information, the energy can be constructed and assembled. Use of the theory of elastic stability can then proceed in the normal form with the energy thus derived.

5.15 CONCLUDING REMARKS

In the classical approach of the theory of elastic stability, in the Timoshenko tradition, the computation of critical loads is the end of the subject, so that critical displacements and directions are not considered. A limited number of examples were presented in this chapter, mainly for the purpose of clarification of the ideas as they were presented.

Critical equilibrium states were shown in this chapter to be either stable or unstable, and the stability is computed by investigation of the higher-order derivatives of the total potential energy with respect to the generalized coordinates of the system. Only the energy criterion has been employed to identify a critical state, but there are other ways to do that. For example, the dynamic approach is developed in [14]; the static approach in [28, 5]; the complementary energy principle in [21, 17]; the principle of Castigliano in [11, 10]; and the principle of virtual work in [4].

5.16 PROBLEMS

Review questions. (*a*) Explain how a critical state can be detected by means of an eigenvalue problem. (*b*) What is the load-geometry matrix? (*c*) Under what conditions is the eigenproblem nonlinear? (*d*) Why is it difficult to identify the stability of a critical state from the sign of the cubic or quartic coefficients?

Problem 5.1. For a two-degrees-of-freedom problem, the stiffness matrix has been computed as

$$A_{ij} = \begin{bmatrix} 90.16 & 71.92 \\ 71.92 & 90.16 \end{bmatrix}.$$

The load-geometry matrix is in this case

$$A_{ijk}\bar{Q}_k = \begin{bmatrix} -14.83 & 0.005 \\ 0.005 & -14.83 \end{bmatrix}.$$

Calculate the critical load, the displacements at the critical state, and the eigenvectors. Perform the computations for all eigenvalues that you find along the linear fundamental path, not just for the first critical load. Draw the eigenvectors in the $Q_1 - Q_2$ space and show that they are orthogonal.

Problem 5.2. Evaluate the stability of the first critical state, using the coefficient $\tilde{V}^4$. Use the same data as in Problem 5.1 and consider the following energy coefficients:

$$A_{111} = -2402.4, \qquad A_{112} = 0.4 = A_{122},$$

$$A_{1111} = A_{2222} = 24040, \qquad A_{1112} = A_{1222} = -12026, \qquad A_{1122} = 12024.$$

Problem 5.3. At a given load level Λ, the total potential energy of a system takes the form

$$V[Q_1, Q_2] = \alpha\, Q_1^4 + Q_1^2 Q_2 + Q_2^2.$$

Consider α as a parameter associated with the properties of the structure. Evaluate the stability of that state.

Problem 5.4. For the case studied in Example 5.1 show that it was not possible to assume $x_2^c = 1$.

Problem 5.5 (Two-bar truss) [23]. A simple two-bar truss is shown in Figure 5.14. It may be shown that if the displacement components are approximated by

$$w = Q_1 \sin\left(\frac{\pi x}{L_1}\right), \qquad v = Q_2 \sin\left(\frac{\pi y}{L_2}\right),$$

then the total potential energy of the discrete system is as follows:

$$\begin{aligned} V = {} & E\frac{\pi^4}{4}\left(\frac{I_1}{L_1^3}Q_1^2 + \frac{I_2}{L_2^3}Q_2^2\right) - P\frac{\pi^2}{4S}\left(\frac{S2}{L_1}Q_1^2 + \frac{S1}{L_2}Q_2^2\right) \\ & + \frac{3\pi^8}{8L_1^7}Q_1^4\left[\frac{1}{2}EI_1 - \frac{1}{8S}P\,(S1C + C1S)\right] \\ & + \frac{3\pi^4}{64L_1^4}Q_1^4\left[\frac{2CS1}{S} + \frac{S1C^3}{S^3} + \frac{L_1}{L_2}\frac{S1}{S^3} + \frac{C1C^2}{S^2} - \frac{2S1C}{S}\right] \\ & + \frac{3\pi^8}{8L_2^7}Q_2^4\left[\frac{1}{2}EI_2 - \frac{1}{8S}P\frac{S1}{S}\right] \\ & + \frac{3\pi^4}{64L_2^3L_1S^2}Q_2^4\left[\frac{S1C}{S}\left(1 + C\frac{L_1}{L_2}\right) + C1\right] \\ & + \frac{\pi^4}{16L_1^2L_2S}Q_1^2Q_2^2\left[S1\left(1 + \frac{C^2}{S^2} + \frac{L_1}{L_2}\frac{C}{S^2}\right) - \frac{C1C}{S} + S1\right], \end{aligned}$$

where

$$S = \sin(\theta_1 + \theta_2), \qquad S1 = \sin(\theta_1), \qquad S2 = \sin(\theta_2),$$

$$C = \cos(\theta_1 + \theta_2), \qquad C1 = \cos(\theta_1), \qquad C2 = \cos(\theta_2).$$

Find the critical states along the trivial fundamental path. Comment on the results that you obtain. Is the critical state stable or unstable?

Problem 5.6 (Theory). Show that (5.114) and (5.115) are correct.

Problem 5.7 (Column). The Euler buckling of a column was studied in section 5.6 using trigonometric functions to approximate the displacement field. However, other functions can also be employed to obtain a discretization of the structure. In this problem you should use a cubic polynomial interpolation for the out-of-plane displacement w and a linear interpolation for the axial displacement component u, as in Appendix B. Compute the critical load and the eigenvector. Compare this solution with the formulation of section 5.6. Explain the differences.

Problem 5.8 (Column). One can think of the analysis of bars using cubic polynomials as a finite element discretization of the column. Solve again Problem 5.7 using a discretization with two elements. Compare with the results of Problem 5.7.

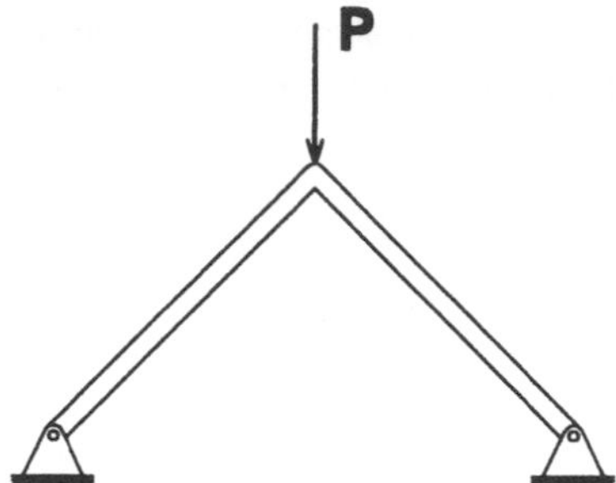

Figure 5.11 Problem 5.9.

Problem 5.9 (Plane frame). The frame of Figure 5.11 is formed by two members at an angle $\alpha = \pi/2$. Use the formulation of Appendix B to evaluate the critical state of the frame. Compute the results using the same data as in section 5.9.

Problem 5.10 (Plane frame). Repeat Problem 5.9 using two elements per member and compare the critical results with those of Problem 5.9. Comment on the convergence of the solution.

Problem 5.11 (Plane frame). Repeat Problem 5.9 considering $\alpha = 115°$, and compare the solution with the results of Problem 5.9.

Problem 5.12 (Plane frame). The steel structure shown in Figure 5.12 is a frame made with solid circular bars. The diameter of the cross section is $40mm$, and $E = 200GN/m^2$. Neglect axial strains in the analysis, and use polynomial functions to approximate the displacements. Compute the load and displacements at the critical state.

Problem 5.13 (Plane frame). A steel frame, shown in Figure 5.13, is made with four tubular members. Each vertical member is a hollowed tube with $75mm$ diameter

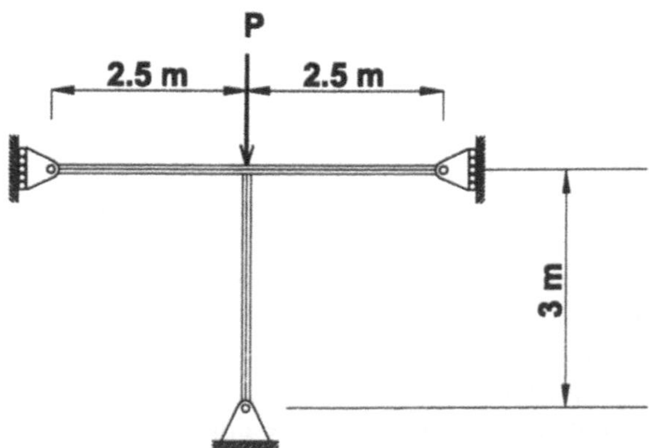

Figure 5.12 Problem 5.12.

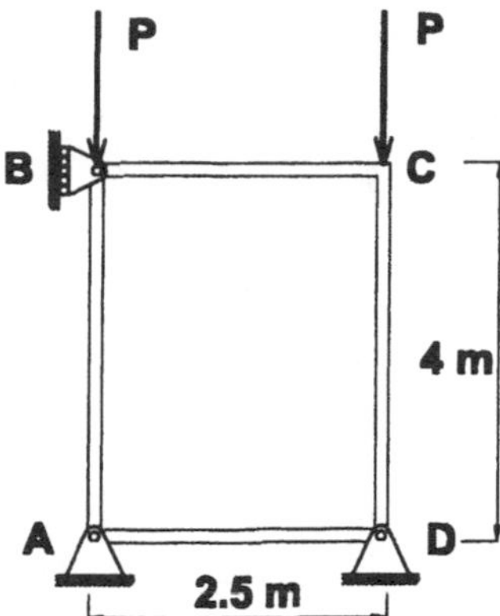

Figure 5.13 Problem 5.13.

and 10mm thickness, while horizontal members have the same diameter but 7mm thickness. Buckling is assumed to occur in the plane of the drawing. Use polynomial functions to approximate the buckling modes.

Problem 5.14 (Plane truss). The truss shown in Figure 5.14 and Problem 5.5 is loaded at the joint between the two members. The dimensions are $L_1 = 271mm$, $L_2 = 194mm$, $I_1 = I_2 = 1.165mm^4$. The bars are inextensional, and trigonometric functions can be used to model the buckled shape of the structure. Find the critical state.

Problem 5.15 (Plane truss). Figure 5.15 shows a plane truss. The fundamental path is trivial because the load P is taken by member BC to the support C. Find the critical state.

Problem 5.16 (Plane frame). Find the critical load of the frame illustrated in Figure 2.13. Assume only vertical loads P_1.

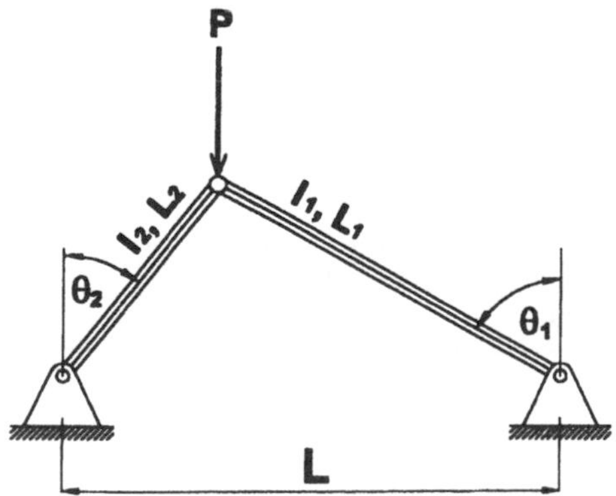

Figure 5.14 Problem 5.14.

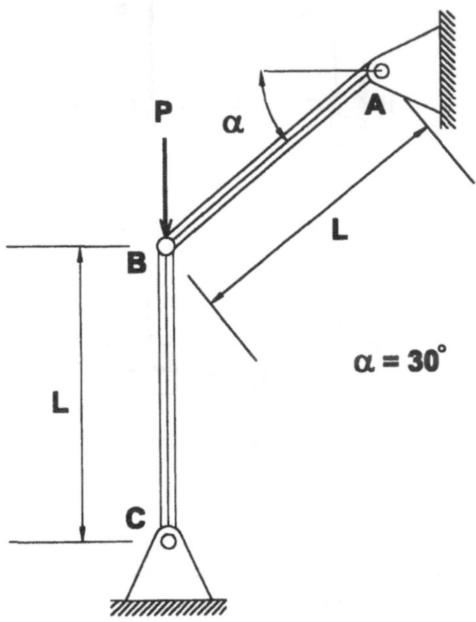

Figure 5.15 Problem 5.15.

Problem 5.17 (Column with bracing). The column shown in Figure 5.16 has two elements that provide partial bracing. Find the critical state.

Problem 5.18 (Plate). A square plate is loaded by pure shear N_{12} at the boundaries (Figure 5.17). Compression in this case arises in planes at 45° with respect to the coordinate directions x_1 and x_2. If the origin of the coordinates is at one of the corners, then the out-of-plane displacement can be approximated by

$$w = Q_1 \sin\left(\frac{\pi x_1}{a}\right) \sin\left(\frac{\pi x_2}{a}\right) + Q_2 \sin\left(\frac{2\pi x_1}{a}\right) \sin\left(\frac{2\pi x_2}{a}\right),$$

where a is the length of the side of the plate. Evaluate the critical state.

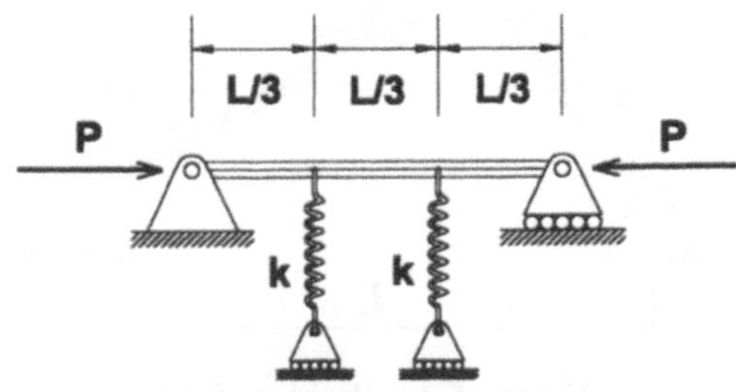

Figure 5.16 Problem 5.17.

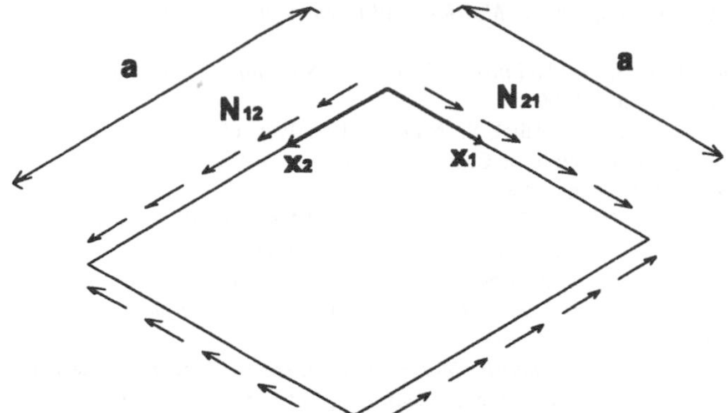

Figure 5.17 Problem 5.18.

Problem 5.19 (Thin-walled column) [8]. Evaluate the critical state for the thin-walled element under compression, assuming a torsional mode. Use results from section 5.13.

Problem 5.20 (I-section column). A thin-walled composite column (made with fiber-reinforced plastic) with $L = 3.5m$ has an I-shaped cross section with flange, web, and thickness dimensions $6 \times 6 \times 1/4in$ (or $152.4 \times 152.4 \times 6.35mm$). Consider $E = 17,245N/mm^2$, $G = 6,624N/mm^2$, and find the critical state in the torsional mode under axial compression. Use the results of Problem 5.19. Remember that this section has two axes of symmetry.

Problem 5.21 (Angle section column). A column has an angle-shaped cross section and is loaded under axial compression. The flanges have equal angles $127mm$ width and $9.53mm$ thickness, with length $L = 3m$. Find the critical torsional mode of buckling using the results of Problem 5.19. Consider $E = 17,245N/mm^2$, $G = 6,624N/mm^2$.

5.17 BIBLIOGRAPHY

[1] Barbero, E., *Introduction to Composite Materials Design*, Taylor & Francis, Philadelphia, PA, 1998.

[2] Bathe, K. J., *Finite Element Procedures*, Prentice-Hall, Englewood Cliffs, NJ, 1996.

[3] Batista, R. C., *Estabilidade Elastica de Sistemas Mecanicos Estruturais*, LNCC/CNPq, Rio de Janeiro, 1982.

[4] Bazant, A., and Cedolin, L., *Stability of Structures: Elastic, Inelastic, Fracture, and Damage Theories*, Oxford University Press, New York, Oxford, 1991.

[5] Brush, D. O., and Almroth, B. O., *Buckling of Bars, Plates and Shells*, McGraw-Hill-Kogakusha, Tokyo, 1975.

[6] Casciaro, R., Lanzo, A. D., and Salerno, G., *A New Algorithm for Nonlinear Symmetric Eigenvalue Problems*, Report 122, Department of Structures, University of Calabria, Italy, 1990.

[7] Casciaro, R., Di Carlo, A., and Pignataro, M., A finite element technique for bifurcation analysis, 14th *IUTAM Congress*, Delft, 1976.

[8] Chajes, A., *Principles of Structural Stability Theory*, Prentice-Hall, Englewood Cliffs, NJ, 1974, Chapter 5.

[9] Collar, A. R., and Simpson, A., *Matrices and Engineering Dynamics*, Ellis Horwood, Chichester, 1987.

[10] El Naschie, M. S., *Stress, Stability and Chaos in Structural Engineering: An Energy Approach*, McGraw-Hill, London, 1990.

[11] El Naschie, M. S., and Al Athel, S., Numerical analysis of eigenvalue and initial post-buckling problems using the principle of Castigliano, in *Proc. Int. Congress on Numerical Methods for Engineering*, Dunod, Paris, 1980.

[12] Gibson, R. F., *Principles of Composite Materials*, McGraw-Hill, New York, 1994.

[13] Huseyin, K., *Nonlinear Theory of Elastic Stability*, Noordhoff, Leyden, 1975.

[14] Huseyin, K., *Vibrations and Stability of Multiple Parameter Systems*, Ivoardhoff, Alphen, 1978.

[15] Iyengar, H., Baker, W., and Sinn, R., The Broadgate Exchange House, London, UK, *Structural Engineering International*, 4, 1993, 214–216.

[16] Jones, R., *Mechanics of Composite Materials*, Hemisphere Publishing, New York, 1975.

[17] Koiter, W. T., Complementary energy, neutral equilibrium and buckling, *Meccanica*, 19, 52–56, 1984.

[18] Koiter, W. T., Post-buckling of a simple two-bar frame, in *Progress in Applied Mechanics*, The Folke Odqvist Volume, Almqvist & Wiksell, Stockholm, 1967.

[19] Leissa, A. W., A review of laminated composite plate buckling, *Appl. Mech. Rev.*, 40(5), 1987, 575–591.

[20] Numaier, A., Residual inverse iteration for the nonlinear eigenvalue problem, *SIAM J. Numer. Anal.*, 22(5), 914–923, 1985.

[21] Masur, E. F., and Popelar, C. H., On the use of the complementary energy in the solution of buckling problems, *Internat. J. Solids Structures*, 12, 203–216, 1976.

[22] Mirasso, A. E., and Godoy, L. A., Iterative techniques for non-linear eigenvalue buckling problems, *Comm. Appl. Numer. Methods*, 8, 311–317, 1992.

[23] Reis, A. J., and Roorda, J., Post-buckling behavior under mode interaction, *J. Engrg. Mech., ASCE*, 105(4), 609–621, 1979.

[24] Roorda, J., *Instability of Imperfect Elastic Structures*, Ph.D. thesis, University of London, London, 1965.

[25] Ruhe, A., Algorithms for the nonlinear eigenvalue problem, *SIAM J. Numer. Anal.*, 10(4), 674–689, 1973.

[26] Simpson, A., A Newtonian procedure for the solution of $Ex = \lambda Ax$, *J. Sound Vibration*, 82, 161–170, 1982.

[27] Simpson, A., On the solution of $S(\omega)x = 0$ by a Newtonian procedure, *J. Sound Vibration*, 97, 153–164, 1984.

[28] Timoshenko, S., and Gere, J. M., *Theory of Elastic Stability*, Second edition, McGraw-Hill, New York, 1965.

[29] Thompson, J. M. T., and Hunt, G. W., *A General Theory of Elastic Stability*, Wiley, London, 1973, pp. 93–94.

[30] Vinson, J. R., and Sierrakowski, R. L., *The Behavior of Structures Composed of Composite Materials*, Kluwer, Dordrecht, 1987.

[31] Wilkinson, J. H., *The Algebraic Eigenvalue Problem*, Clarendon Press, Oxford, 1965.

[32] Wittrick, W. H., and Williams, F. W., A general algorithm for computing natural frequencies of elastic structures, *Quart. J. Mech. Appl. Math.*, 34, 263–284, 1971.

CHAPTER

SIX

THE LIMIT POINT

6.1 INTRODUCTION

Let us consider an equilibrium path in which, by increasing the load parameter Λ, a critical state has been reached. This was expressed in Chapter 5 by means of the condition

$$V_{ij}\left[Q_j, \Lambda\right] x_j \mid^c = 0 \tag{6.1}$$

with the eigenvector normalized in the form

$$x_1^c = 1.$$

The (6.1) allows evaluation of the critical load Λ^c, the displacement field along the fundamental path for that load level Q_i^c, and the eigenvector x_i^c.

Until a few decades ago, the critical load was the only information available to the designer regarding buckling of a structure. However, there is also a need to know many other related questions:

- Is there only one or several different types of critical states?
- If there were several types, would they be different in nature? Perhaps it is possible that some critical states are stable and others are unstable.
- Is it possible to classify critical states?
- Are there equilibrium paths that a structure can follow beyond a critical state?
- Can a structure take loads higher than the first critical load?
- What is the relation between a critical state and the collapse of a structure?

These are profound and important questions, with practical consequences. Behind them, there is also a common behavior in humans: If we know of a situation that can be categorized as "critical," we also want to know what are the possible alternatives that we may have to face whenever we encounter such a situation. We would like to classify these alternatives and understand which one is the worst. Knowledge may give us some feeling of control of a situation, and this is particularly true if we are facing a critical situation. And, finally, we want to know what is beyond the critical situation.

This chapter continues in section 6.2 with the perturbation equations that arise from the equilibrium condition, but evaluating them at the critical state. These equations are supplemented with the contracted equations. In section 6.3 we obtain the conditions that are satisfied at a limit point, leading to an algorithm to carry out the computations. The formulation is also simplified to the case of linear load terms, in which explicit solutions are easy to obtain (section 6.4). An example is given to illustrate the use of the algorithm. Finally, the shallow arch is investigated in section 6.5 using two levels of approximation for the fundamental equilibrium path.

6.2 PERTURBATION EQUATIONS

Once a critical state has been detected along an equilibrium path, a technique of nonlinear analysis may be employed to follow any paths emerging from the critical point. Several techniques could be used here, but we prefer to employ perturbations, because they require and use valuable information at the critical state itself: the derivatives of the energy and the state variables at the critical state. In doing so, we are able to produce a classification of critical states based on the signs and values of such derivatives. The first general presentation of this topic known to the author is that of [3], but we have followed [2] here.

6.2.1 Expansion of Displacement and Load Variables

Anticipating the existence of postcritical states, we expand the displacement and load parameters as a function of a perturbation parameter s, in the form

$$Q_i(s) = Q_i^c + q_i(s), \tag{6.2}$$

$$\Lambda(s) = \Lambda^c + \lambda(s), \tag{6.3}$$

where q_i and λ are incremental values in the generalized coordinates and the load parameter measured with respect to the critical state. They are given as

$$s = q_1,$$

$$q_i(s) = \frac{dq_i}{ds}\, s + \frac{1}{2}\frac{d^2q_i}{ds^2}\, s^2 + \cdots, \qquad i \neq 1, \tag{6.4}$$

$$\lambda(s) = \frac{d\lambda}{ds}\, s + \frac{1}{2}\frac{d^2\lambda}{ds^2}\, s^2 + \cdots. \tag{6.5}$$

Here s is an adequate perturbation parameter. By "adequate," we mean a parameter that has a significant value (nonzero or nonnegligible) along the path being investigated to serve as a one-dimensional coordinate to describe the path. The load itself is not an adequate parameter for limit point behavior, for reasons that will be clear later in this chapter.

An arbitrary component of the generalized coordinates Q_k (where k identifies only one of the coordinates) can be used as a perturbation parameter, provided that $x_k^c \neq 0$. For convenience of notation, one may choose the increment in Q_1 as the perturbation parameter, associated with the condition $x_1^c = 1$.

The increments in Q_i and Λ measured from the critical state were termed q_i and λ and now become

$$q_1 = s,$$

$$q_i(q_1) = \left.\frac{dq_i}{dq_1}\right|^c q_1 + \frac{1}{2}\left.\frac{d^2q_i}{dq_1^2}\right|^c q_1^2 + \cdots, \tag{6.6}$$

$$\lambda(q_1) = \left.\frac{d\lambda}{dq_1}\right|^c q_1 + \frac{1}{2}\left.\frac{d^2\lambda}{dq_1^2}\right|^c q_1^2 + \cdots. \tag{6.7}$$

A more compact notation of the above can be done as follows:

$$q_i(q_1) = q_i^{(1)} q_1 + \frac{1}{2} q_i^{(2)} q_1 + \cdots, \tag{6.8}$$

$$\lambda(q_1) = \lambda^{(1)} q_1 + \frac{1}{2}\lambda^{(2)} q_1^2 + \cdots, \tag{6.9}$$

where the first derivative with respect to q_1 is denoted by $(\#)^{(1)}$, the second derivative by $(\#)^{(2)}$, etc., all evaluated at the critical state C.

6.2.2 Perturbation of Equilibrium Equations

Equilibrium may now be written in the form

$$V_i = V_i[Q_j(q_1), \Lambda(q_1)] = 0. \tag{6.10}$$

The derivative of (6.10) with respect to the perturbation parameter q_1 leads to the first-order perturbation equations

$$\boxed{V_{ij} q_j^{(1)} + V_i' \lambda^{(1)} = 0} \tag{6.11}$$

where $(\#)'$ denotes derivatives with respect to λ, so that $V_i' = \partial^2 V / \partial Q_i \partial \Lambda$.

Differentiating a second time, one gets

$$\boxed{V_{ijk}q_j^{(1)}q_k^{(1)} + 2V'_{ij}q_j^{(1)}\lambda^{(1)} + V_{ij}q_j^{(2)} + V''_i\lambda^{(1)2} + V'_i\lambda^{(2)} = 0} \tag{6.12}$$

The third derivative yields

$$3[V'_{ijk}q_j^{(1)}q_k^{(1)}\lambda^{(1)} + V_{ijk}q_j^{(1)}q_k^{(2)} + V''_{ij}q_j^{(1)}\lambda^{(1)2} + V'_{ij}(q_j^{(2)}\lambda^{(1)} + q_j^{(1)}\lambda^{(2)})$$
$$+V''_i\lambda^{(1)}\lambda^{(2)}] + V_{ijkl}q_j^{(1)}q_k^{(1)}q_l^{(1)} + V'''_i\lambda^{(1)3} + V_{ij}q_j^{(3)} + V'_i\lambda^{(3)} = 0. \tag{6.13}$$

Differentiating (6.13), one obtains the fourth derivative

$$V_{ijklm}q_j^{(1)}q_k^{(1)}q_l^{(1)}q_m^{(1)} + 4V'_{ijkl}q_j^{(1)}q_k^{(1)}q_l^{(1)}\lambda^{(1)} + 6V_{ijkl}q_j^{(1)}q_k^{(1)}q_l^{(2)}$$
$$+3[2V''_{ijk}q_j^{(1)}q_k^{(1)}\lambda^{(1)2} + 4V'_{ijk}q_j^{(1)}q_k^{(2)}\lambda^{(1)} + 2V'_{ijk}q_j^{(1)}q_k^{(1)}\lambda^{(2)} + V_{ijk}q_j^{(2)}q_k^{(2)}$$
$$+2V''_{ij}q_j^{(2)}\lambda^{(1)2} + 4V''_{ij}q_j^{(1)}\lambda^{(1)}\lambda^{(2)} + 2V'_{ij}q_j^{(2)}\lambda^{(2)} + 2V''_i\lambda^{(1)2}\lambda^{(2)} + V''_i\lambda^{(2)2}]$$
$$+4V''_i\lambda^{(1)}\lambda^{(3)} + 4V''_{ij}q_j^{(1)}\lambda^{(1)3} + V''''_i\lambda^{(1)4} + 4V_{ijk}q_j^{(1)}q_k^{(3)} + 4V'_{ij}q_j^{(3)}\lambda^{(1)}$$
$$+V_{ij}q_j^{(4)} + 4V'_{ij}q_j^{(1)}\lambda^{(3)} + V'_i\lambda^{(4)} = 0. \tag{6.14}$$

These perturbation equations are next evaluated at the critical state. This means that the coefficients of (6.8) and (6.9) are the same as those present in the evaluated perturbation equations (6.11)–(6.14). We investigate the solution of the evaluated perturbation equations in this and the next chapter, and in doing so, it is possible to establish a classification of behavior depending on the values of the derivatives of V, from which limit and bifurcation points are identified.

6.2.3 Contracted Perturbation Equations

In the solution of the perturbation systems it will be clear that another scalar equation is necessary for each perturbation system. An additional equation can be provided by means of the **contraction mechanism** (see, for example, [4]). This operation was described in Chapter 3 and plays a central role in the analysis. A contracted perturbation equation is obtained by premultiplication of each set of perturbation equation by the eigenvector and addition of the results. This provides a new scalar condition for each order of perturbation equation.

The contracted first-order equation is

$$\boxed{\underbrace{V_{ij}x_i}_{0}\, q_j^{(1)} + V'_i x_i \lambda^{(1)}|^c = 0} \tag{6.15}$$

Notice that the first term is zero because it is the condition of critical state.

The contracted second-order perturbation equation is

$$V_i' x_i \lambda^{(2)} + V_{ijk} x_i q_j^{(1)} q_k^{(1)} + 2V_{ij}' x_i q_j^{(1)} \lambda^{(1)} + V_i'' x_i (\lambda^{(1)})^2 |^c = 0. \tag{6.16}$$

The contracted third-order perturbation equation becomes

$$V_{ijkl} x_i q_j^{(1)} q_k^{(1)} q_l^{(1)} + 3[V_{ijk}' q_j^{(1)} q_k^{(1)} \lambda^{(1)} + V_{ijk} q_j^{(1)} q_k^{(2)} + V_{ij}'' q_j^{(1)} \lambda^{(1)2}$$

$$+ V_{ij}' (q_j^{(2)} \lambda^{(1)} + q_j^{(1)} \lambda^{(2)}) + V_i'' \lambda^{(1)} \lambda^{(2)}] x_i + V_i''' x_i \lambda^{(1)3} + V_i' x_i \lambda^{(3)} |^c = 0, \tag{6.17}$$

and higher-order contracted equations are obtained in a similar way. Notice that the product $x_i V_{ij} q_j^{(n)} |^c = (V_{ij} x_j) q_i^{(n)} |^c$, which is present in all contracted equations, is zero. This is because of the symmetry $V_{ij} = V_{ji}$, and because the contracted equations are also evaluated at the critical state in which $(V_{ij} x_j)^c = 0$.

6.3 THE CONDITIONS AT A LIMIT POINT

6.3.1 Solution of Perturbation Equations

The first-order perturbation equations (6.11) are N conditions (where N is the number of degrees of freedom of the problem) and contain $(N + 1)$ unknowns (the values of $\lambda^{(1)}$ and $q_j^{(1)}$). It is then necessary to supplement (6.11) with a scalar equation, i.e., the first-order contracted (6.15)

$$\lambda^{(1)} (V_i' x_i) |^c = 0. \tag{6.18}$$

There are two possibilities to satisfy 6.18: either $x_i V_i' \neq 0$ and $\lambda^{(1)} = 0$, or else $x_i V_i' = 0$. The first possibility is called limit point and is studied in this chapter. The second possibility (bifurcation points) is studied in the next chapter.

Let us consider here the case in which the system has load components in the direction of the critical mode x_i; that is, the load vector and the eigenmode are not orthogonal. In this case,

$$V_i' x_i |^c \neq 0. \tag{6.19}$$

Under this condition, (6.18) admits a unique solution in the form

$$\boxed{\lambda^{(1)} |^c = 0} \tag{6.20}$$

and the critical equilibrium state is called a limit point. The above condition may also be written as

$$\left. \frac{d\lambda}{dq_1} \right|^c = 0, \tag{6.21}$$

so that the limit point represents a maximum (or a minimum) in the $\Lambda - Q_i$ space.

We now consider the first-order perturbation equation. This reduces to

$$V_{ij}q_j^{(1)}|^c = 0.$$

But this is the same condition employed to obtain the critical state, which is (6.1). Thus,

$$\boxed{q_j^{(1)} = x_j^c} \tag{6.22}$$

which means that the first-order component of displacements in the postbuckling path is the direction of the eigenvector.

The second-order contracted equations of (6.16) now become

$$V_{ijk}x_ix_jx_k + V_i'x_i\lambda^{(2)}|^c = 0. \tag{6.23}$$

Let us define a new coefficient C in the form

$$\boxed{C = V_{ijk}x_ix_jx_k|^c} \tag{6.24}$$

In the limit point, the following condition is satisfied:

$$\boxed{C \neq 0} \tag{6.25}$$

Because of (6.25), it is possible to obtain $\lambda^{(2)}$ from (6.23) in the form

$$\boxed{\lambda^{(2)c} = -\frac{C}{x_iV_i'}\,|^c} \tag{6.26}$$

The second-order components of the displacement vector in the postbuckling path are obtained from the second-order perturbation equation, that is,

$$V_{ij}q_j^{(2)}\,|^c = -V_{ijk}x_jx_k - \lambda^{(2)}V_i'\,|^c \tag{6.27}$$

or

$$\boxed{V_{ij}q_j^{(2)}\,|^c = -V_{ijk}x_jx_k + \left(\frac{C}{x_iV_i'}\right)V_i'\,|^c} \tag{6.28}$$

One of the components of vector $q_j^{(2)}$ is known, i.e., for $j = 1$,

$$q_1^{(2)c} = \frac{d}{dq_1}(q_1^{(1)c}) = \frac{d}{dq_1}(x_1^c) = \frac{d}{dq_1}(1) = 0.$$

The above condition allows solution of (6.27) and (6.25), which is associated with a singular matrix V_{ij}. Notice that because of (6.24), the limit point is always unstable.

The analysis can be extended to include higher-order terms using a similar procedure to the one presented in this section.

The postbuckling path for the limit point results in

$$Q_j(s) = Q_j^c + x_j\, s + \frac{1}{2} q_j^{(2)}\, s^2 + \cdots, \qquad j \neq 1, \tag{6.29}$$

$$\Lambda(s) = \Lambda^c + 0\, s + \frac{1}{2}\lambda^{(2)}\, s^2 + \cdots, \tag{6.30}$$

where s, the perturbation parameter, is the generalized coordinate used to normalize the eigenvector x, i.e., $s = Q_1$, $x_1 = 1$, $q_1^{(2)} = 0$.

6.3.2 An Algorithm of Solution for the Limit Point

Let us summarize the conditions required to have a limit point

$$x_i V_i' |^c \neq 0$$

and

$$C = V_{ijk} x_i x_j x_k |^c \neq 0.$$

An algorithm for the solution of the limit point should proceed as follows:

1. Define
$$\lambda^{(1)} |^c = 0.$$
2. Define
$$q_j^{(1)c} = x_j^c.$$
3. Compute
$$C = V_{ijk} x_i x_j x_k\ |^c.$$
4. Compute
$$x_i V_i'.$$
5. Calculate
$$\lambda^{(2)c} = -\frac{C}{x_i V_i'}\bigg|^c.$$
6. Solve $q_j^{(2)c}$ using
$$V_{ij} q_j^{(2)} |^c = -V_{ijk} x_j x_k + \left(\frac{C}{x_i V_i'}\right) V_i' |^c$$
with $q_1^{(2)c} = 0$.

6.4 THE SPECIALIZED SYSTEM

Let us consider the energy expansion for the specialized system, defined in Chapter 2, in which the load parameter Λ is linear. This was extensively developed in previous chapters, and we have employed the explicit form

$$V[Q_i, \Lambda] = A'_i Q_i \Lambda + \frac{1}{2!} A_{ij} Q_i Q_j + \frac{1}{3!} A_{ijk} Q_i Q_j Q_k + \frac{1}{4!} A_{ijkl} Q_i Q_j Q_k Q_l. \quad (6.31)$$

Because of (6.31), the coefficient C is now

$$C = (A_{ijk} + A_{ijkl} Q_l) x_i x_j x_k |^c. \quad (6.32)$$

The term $x_i V'_i$, which is useful in the identification of a limit or bifurcation state, is given by

$$x_i V'_i |^c = A'_i x_i |^c, \quad (6.33)$$

and the curvature of the postcritical path results in

$$\lambda^{(2)c} = - \left. \frac{(A_{ijk} + A_{ijkl} Q_l) x_i x_j x_k}{A'_i x_i} \right|^c. \quad (6.34)$$

Equations (6.31)–(6.34) are similar to those obtained for the general system, but now the derivatives have been carried out and are given in explicit form.

Example 6.1 (Two-degrees-of-freedom system). *Let us consider the energy V of a given two-degrees-of-freedom system in the form of* (6.31). *The coefficients of V are those employed in Example* 4.1:

$$A'_1 = A'_2 = -1, \qquad A_{11} = A_{22} = 90.16, \qquad A_{12} = 71.92,$$

$$A_{111} = A_{222} = -2402.4, \qquad A_{112} = A_{122} = 0.4,$$

$$A_{1111} = A_{2222} = 24040, \qquad A_{1112} = A_{2221} = -12026, \qquad A_{1122} = 12024.$$

A critical state was found to be given by $\Lambda^c = 6.332$, $Q^c = \{0.08599 \quad 0.08599\}$, and the eigenvector by $x_i^c = \{1 \quad 1\}$. Let us examine the nature of this state.

The value of C results in

$$C = -2737.33.$$

The computed value of the projection of the load vector on the eigenvector becomes

$$V'_i x_i |^c = -2,$$

which means that the problem is a limit point.

The curvature of the postbuckling path results in

$$\lambda^{(2)c} = -1368.67$$

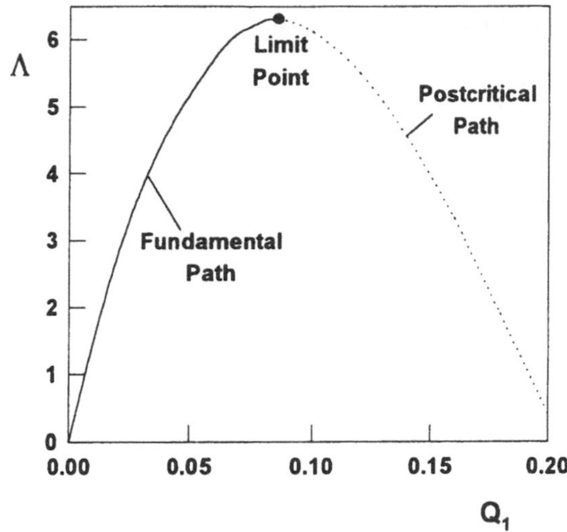

Figure 6.1 Path emerging from a limit point.

and

$$q_i^{(2)c} = 0.$$

The load may be written in the approximate form

$$\Lambda(q_1) = 6.3322 - 1368.67q_1^2.$$

Figure 6.1 *shows the descending path emerging from the limit point considered.*

6.5 THE SHALLOW ARCH

This is the most commonly used example of limit point behavior in structures [5]. Let us investigate two levels of approximation in the fundamental path, leading to linear and cubic equilibrium paths.

Example 6.2 (Shallow arch with cubic fundamental path). *The energy and equilibrium conditions of the shallow arch were already obtained in Chapter 2. We concentrate here on the nonlinear fundamental path and investigate a critical state along this path.*

The equations governing the fundamental path of the arch take the form

$$Q_i^F = s,$$

$$Q_2^F = 0,$$

so that there are only displacements Q_1 before buckling. The load-displacement relation is given by

$$\Lambda^F = \dot{\Lambda} Q_1^F + \frac{1}{2}\ddot{\Lambda}(Q_1^F)^2 + \frac{1}{3!}\dddot{\Lambda}(Q_1^F)^3,$$

where the derivatives of the load with respect to the perturbation parameter are

$$\dot{\Lambda} = A_{11}, \qquad \ddot{\Lambda} = A_{111}, \qquad \dddot{\Lambda} = A_{1111}.$$

It was found in Chapter 5 that one of the critical states is given by

$$Q_1^c = -\left\{A_{111} + \frac{[(A_{111})^2 - 2A_{11}A_{1111}]^{1/2}}{A_{1111}}\right\},$$

$$Q_2^c = 0,$$

$$\Lambda^c = A_{11}Q_1^c + \frac{1}{2}A_{111}(Q_1^c)^2 + \frac{1}{3!}A_{1111}(Q_1^c)^3.$$

The eigenvector for this state was found to be

$$x_i = \begin{Bmatrix} 1 \\ 0 \end{Bmatrix}.$$

We want to obtain a classification of this state and find the postbuckling response.

Let us start by looking at the relation between the load vector and the eigenvector. The load vector in this problem is

$$V_i' = \begin{Bmatrix} -1 \\ 0 \end{Bmatrix},$$

so that the product

$$V_i' x_i|^c = A_i' x_i|^c = -1.$$

According to our criterion of classification, we say that this is a limit point. In fact, the load vector and the eigenvector have the same direction.

Next, we apply the results already obtained in section 6.3 for the secondary path of a limit point, i.e.,

$$\lambda^{(1)c} = 0,$$

$$q_i^{(1)c} = x_i = \begin{Bmatrix} 1 \\ 0 \end{Bmatrix}.$$

The important coefficient C needs the computation of the following evaluated derivatives:

$$V_{111} = A_{111} + A_{1111} Q_1, \qquad V_{111}^c = A_{111} + A_{1111} Q_1^c,$$

$$V_{222} = A_{222} Q_2, \qquad V_{222}^c = 0,$$

$$V_{112} = \frac{2}{3} A_{1122} Q_2, \qquad V_{112}^c = 0,$$

$$V_{122} = A_{122} + \frac{2}{3} A_{1122} Q_1, \qquad V_{122}^c = A_{122} + \frac{2}{3} A_{1122} Q_1^c.$$

Substitution in C, because $x_2 = 0$, leads to the following value:

$$C = V_{111} x_1 x_1 x_1 = A_{111} + A_{1111} Q_1^c \neq 0.$$

This is clearly nonzero, as expected for a limit point.

The curvature of the postbuckling path at the limit state, according to section 6.3, is

$$\lambda^{(2)c} = -\frac{C}{V_i' x_i} = C = A_{111} + A_{1111} Q_1^c,$$

$$\lambda^{(2)c} = -\sqrt{A_{111}^2 - 2A_{11} A_{1111}}. \tag{6.35}$$

The above expression can also be written in terms of the coefficients of the arch

$$\lambda^{(2)c} = C = -\frac{3\pi}{2\theta_0^2} \sqrt{1 - \frac{I}{A}\left(\frac{\pi}{R}\right)^2 \left(\frac{\pi}{2\theta_0}\right)^4}.$$

The next term in the perturbation expansion of the incremental displacement results in

$$V_{ij} q_j^{(2)}|^c = -V_{ijk} x_j x_k - \lambda^{(2)} V_i'$$

with $q_1^{(2)} = 0$. Computation of the right-hand side leads to

$$\begin{bmatrix} 0 & 0 \\ 0 & V_{22} \end{bmatrix} \begin{Bmatrix} 0 \\ q_2^{(2)} \end{Bmatrix} = -\begin{Bmatrix} V_{111} x_1 x_1 \\ 0 \end{Bmatrix} - \begin{Bmatrix} V_1' \lambda^{(2)c} \\ 0 \end{Bmatrix},$$

where

$$V_{22}^c = A_{22} + A_{122} Q_1^c + \frac{1}{3} A_{1122} (Q_1^c)^2 \neq 0.$$

Solution of this system is only possible if

$$q_2^{(2)c} = 0,$$

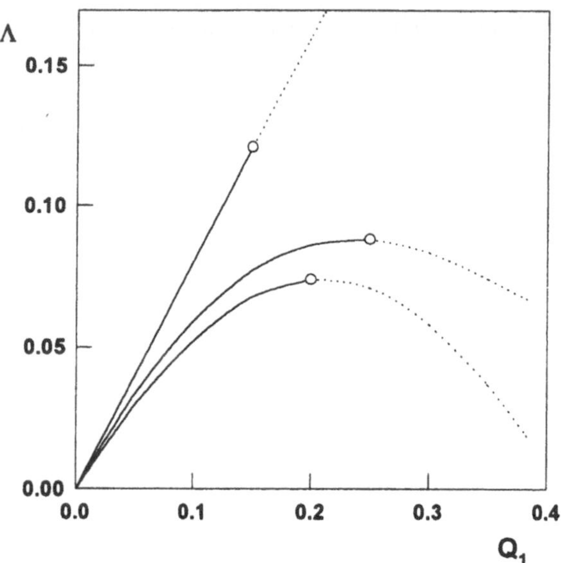

Figure 6.2 Postbuckling behavior of a shallow arch.

so that

$$q_i^{(2)c} = \begin{Bmatrix} 0 \\ 0 \end{Bmatrix}. \tag{6.36}$$

An improved calculation of the postbuckling displacements shows a zero second-order contribution: This means that the path emerging from this limit point follows the direction of the eigenvector.

The results can be summarized as follows:

$$q_j = x_j q_1,$$

$$\lambda(q_1) = (A_{111} + A_{1111} Q_1^c)(q_1)^2 + \cdots.$$

For example, for $\theta_0 = 60°$, $A = 1$, $I = 0.0833$, $R = 50$, we get

$$\lambda^{(2)c} = -2.48.$$

The critical load is positive, while the curvature of the postbuckling path is negative; this shows that the path, plotted in Figure 6.2, is descendent.

Example 6.3 (Shallow arch with linearized fundamental path). *The fundamental path is next approximated by a linear model, i.e.,*

$$\Lambda^F = \dot{\Lambda} Q_1^F = A_{11} Q_1^F,$$

$$Q_2^F = 0.$$

Consider the critical state found in section 5.10 *as*

$$\Lambda^c = -\frac{(A_{11})^2}{A_{111}},$$

$$x_i = \begin{Bmatrix} 1 \\ 0 \end{Bmatrix}.$$

We want to classify this critical state and compute the postcritical path. Similar to what was found in the previous section, the state is characterized by

$$V_i' x_i = A_1' x_1 = -1,$$

so that it represents a limit point. Notice that this is found even in the presence of a linear fundamental path, in which there is no possibility of computing a curved path with zero slope. Here, the linear eigenvalue problem "approximates" the situation by detecting an eigenvector contained in the same plane as the fundamental path. The situation is shown in Figure 6.2.

6.6 FINAL REMARKS

A typical form of the first limit point along an equilibrium path starting at the origin in the $\Lambda - q_i$ space is illustrated in Figure 6.3. This is an unstable critical state, followed by a descendent postcritical path. Thus, there are no stable states in the neighborhood of the limit point, and the system **jumps** to a far-away stable state. This jump is known as **snap through**. It may happen that there is no stable state to which the system can jump following a limit point, and a catastrophic failure occurs.

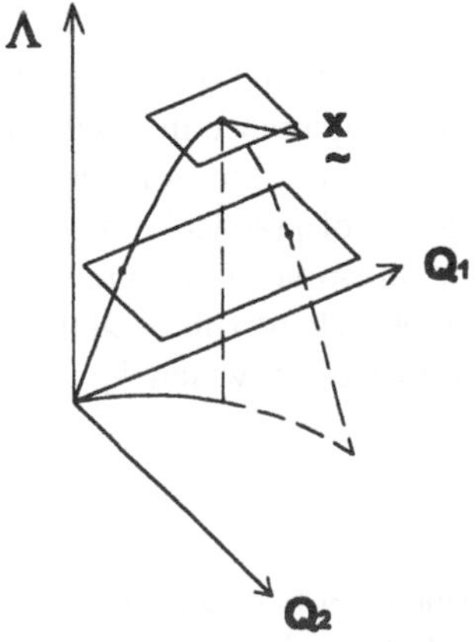

Figure 6.3 A limit point.

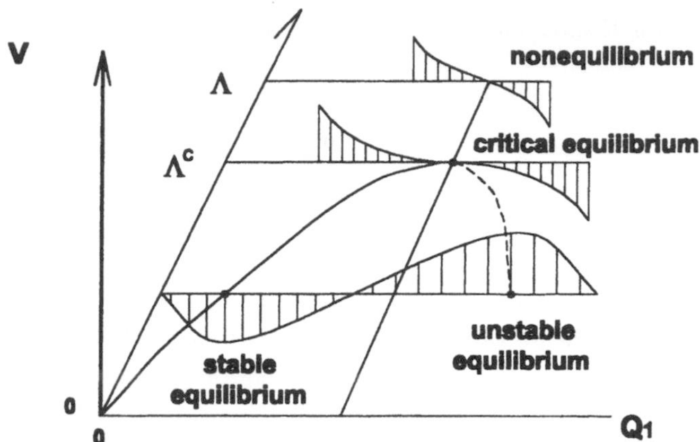

Figure 6.4 Energy levels as a path approaches a limit point.

Notice that there are at least two states of equilibrium for a load level such as Λ^E. Thus, the load is not an adequate choice as a perturbation parameter because it does not define states along the path in a unique way.

A plot of the energy for a constant load level is shown in Figure 6.4. For a load level $\Lambda = \Lambda_1$, the energy has a minimum on the rising (stable) path, and a maximum on the descending (unstable) path. At the limit point the two stationary points are coincident and form an inflection point in the energy surface. The limit point is neither a maximum nor a minimum. Finally, if the energy is computed at a load level $\Lambda = \Lambda_2 > \Lambda_c$, then a stationary point is not found, meaning that equilibrium is not satisfied for that load level.

The limit point has been investigated for a general system and also for a specialized system in which some linear assumptions are made about the load terms. The chapter was concerned with the identification of limit points and with developing an algorithm of computations to obtain the postcritical path in terms of perturbation expansion.

In real structures, it is not often found that limit point behavior is reached. "The snapping situation is not of great importance in practical structures, because the snapping condition is reached only after very considerable deformation; such large deformation would not be characteristic of practical structures in their working loads. The more interesting and important situation from a practical point of view is the state of bifurcation" [1, p. 46]. In the next chapter the solution of the perturbation equations is reconsidered, and bifurcation cases, in which $V_i' x_i|^c = 0$, are studied.

6.7 PROBLEMS

Review questions. (*a*) Describe what is meant by a limit point. (*b*) What perturbation parameter can be chosen to detect a limit point? (*c*) Why do we need contracted perturbation equations in the analysis of limit points? (*d*) Explain the meaning of

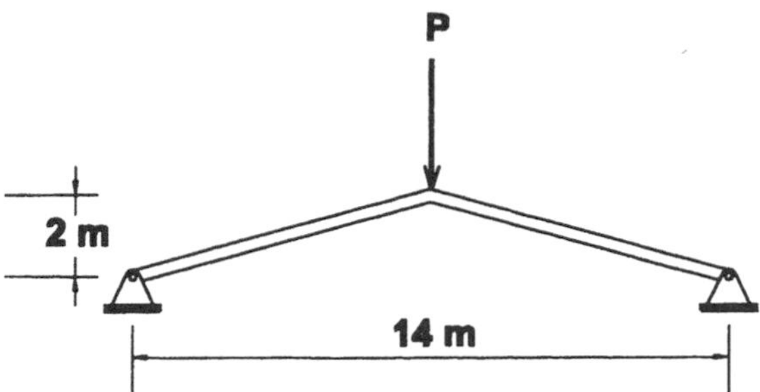

Figure 6.5 Frame of Problem 6.2.

the condition $V_i' x_i \neq 0$. (*e*) Explain the consequences of $q_j^{(1)c} = x_j^c$. (*f*) Why is it necessary to have $C \neq 0$ in a limit point? (*g*) Discuss why jumps occur in limit points.

Problem 6.1 (Circular arch). Consider the arch with cubic fundamental path.
(a) Plot the curvature of the postbuckling path as a function of the angle θ_0.
(a) What is the range of validity of the results?

Problem 6.2 (Plane frame). A two-bar frame is shown in Figure 6.5 for which the properties are $E = 200kN/mm^2$, $\nu = 0.3$, $\rho = 7.7 \times 10^{-8} kN/mm^3$, $L = 7.28m$, $A = 6451.6mm^2$, $I = 286.62 \times 10^8 mm^4$, and the height of the frame is $2m$. Find the limit point load of the frame and the initial postcritical path.

Problem 6.3 (Theory). Obtain third-order terms in the perturbation expansion of a limit point.

Problem 6.4. Find the critical states of an arch under a uniform distributed load acting perpendicular to the midline of the arch. Assume simply supported boundary conditions.

6.8 BIBLIOGRAPHY

[1] Chilver, A. H., The elastic stability of structures, Chapter 3 in *Stability*, H. H. E. Leipholz, Ed., University of Waterloo, Waterloo, 1972.

[2] Flores, F. G., and Godoy, L. A., Elastic post buckling analysis via finite element and perturbation techniques: Part 1, Formulation, *Internat. J. Numer. Methods Engrg.*, 33, 1775–1794, 1992.

[3] Thompson, J. M. T., Basic principles in the general theory of elastic stability, *J. Mech. Phys. Solids*, 11, 1963, 13.

[4] Thompson, J. M. T., and Hunt, G. W., *A General Theory of Elastic Stability*, Wiley, London, 1973.

[5] Walker, A. C., A nonlinear finite element analysis of shallow circular arches, *Internat. J. Solids Structures*, 5, 97, 1969.

CHAPTER

SEVEN

BIFURCATION STATES[27]

7.1 INTRODUCTION

In the last chapter, a class of behavior was identified in which a fundamental path has significant nonlinearity. This nonlinearity induces a decrease in the stiffness of the structure along the path, until the tangent to the path in the load-displacements space becomes horizontal. Such a critical state was identified as a limit point and was associated with a vanishing derivative of the load with respect to the displacements.

There is also a possibility of reaching a critical state along a path without large displacements. This occurs because an initial path crosses a new one, called a secondary path, and this situation is illustrated in Figure 7.1. Even a linear fundamental path can have critical states of this kind. This peculiar behavior is hidden in the initial path, in the sense that the response prior to the appearance of the critical state is quite under control and with small displacements associated with increments in the load parameter.

Because there is now more than one equilibrium path, we have to distinguish between them. The path starting from the unloaded situation is called the **primary** path, sometimes also called the fundamental, initial, or precritical path. The new path is the secondary path, also known as postcritical or postbuckling path. There may be other paths crossing the primary path, and they are all known as **secondary** paths. New paths that cross the secondary paths are known as **tertiary** paths. The state at which the primary path crosses the secondary path is called bifurcation state or bifurcation point, and it will be shown in this chapter that it is a critical state. A state

[27]This chapter was written with Dr. Fernando G. Flores, National University of Córdoba, Argentina.

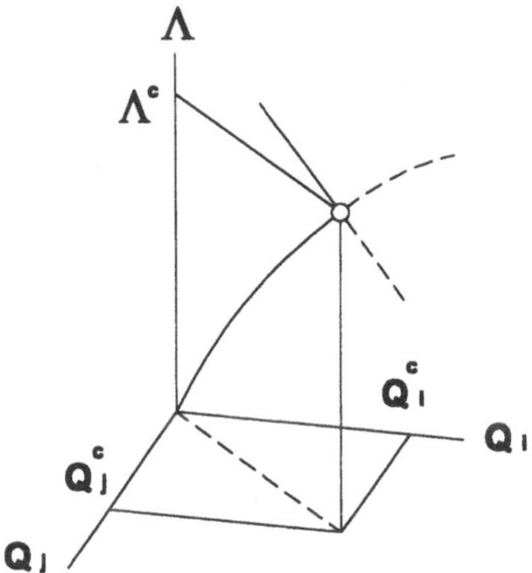

Figure 7.1 Example of bifurcation behavior.

at which a secondary path crosses a tertiary path is called secondary bifurcation point. Higher-order bifurcations are not studied in this book, but they may be of interest in the study of chaotic motions (see, for example, [6]). Another term used in this context is branching, meaning that a secondary path may be seen as a branch of the primary path. Figure 7.2 illustrates branching exploited for aesthetic reasons.

The study of critical points using bifurcation theory in the theory of elastic stability started with the work of Koiter [13] for continuous systems and has been followed by Budianski and Hutchinson (see, for example, [3]). For systems specified by a finite number of generalized coordinates, the work of Thompson [20] was one of the earliest to provide general information for branching behavior. Sewell (see, for example, [19] and his earlier work cited there) and a number of other researchers were also pioneers in this field.

There are several ways to present a bifurcation analysis, depending on the generalized coordinates employed to describe the postcritical behavior. The first possibility is to use the same generalized coordinates that were employed in the primary path. This does not require any transformation in the energy, and as such it can be called the V-**formulation**. In the context of nonlinear analysis, this is a Lagrangian formulation.

The first work known to the author in which the V-approach was followed is due to Sewell, which reads:

> This accounts for the presence of some extra terms here as compared with Thompson's formulae [the W-approach], but the penalty of carrying them ... is light and felt to be outweighed by the advantages. [19]

A second choice would be to rewrite the energy so that the generalized coordinates are measured with respect to the primary path. This involves a $V \to W$ transformation, called the W-**formulation**, and is presented in Chapter 13. This transformation

Figure 7.2 From an exhibition by Frans Krajcberg on trees and fire at the Museum of Modern Art in Rio de Janeiro, 1992. (Photograph by the author.)

represents a Lagrangian formulation using the primary path as an intermediate configuration.

A third possibility is to transform the energy V or W into a new functional D, involving a $V \rightarrow D$ transformation. We may call this the D-**formulation**; however, it will not be discussed in this book. We mentioned in Chapter 4 that this transformation requires the computation of the full set of eigenvalues and eigenvectors of the problem, and is not a practical technique for general multiple degree-of-freedom systems.

This chapter is an extended version of the analysis of perfect systems using the V-formulation, presented in [8]. To avoid duplication of efforts, we will make frequent use of the perturbation equations obtained in the last chapter on limit point analysis. The starting point of the analysis of bifurcation is the same as that of limit points, and they differ in the values of certain parameters. Thus, the same general theory is used to study both limit and bifurcation points. In section 7.2 the basic conditions that distinguish bifurcation from limit points are identified, and it is shown that when these conditions are met there is always more than one path. Symmetric bifurcations are studied in section 7.3, and an algorithm of analysis is presented to carry out the computations. A similar analysis is developed in section 7.4 for the case of asymmetric bifurcation. The simplifications associated with a specialized system are described in section 7.5, leading to explicit equations in terms of the energy coefficients. Two theorems are presented in section 7.6: they relate stability of the path with critical states. The rest of the chapter includes examples of bifurcation behavior.

The column problem in section 7.7 is an example of symmetric bifurcation with stable postbuckling response. This is a classical problem, and is perhaps the first buckling problems studied in the history of mechanics. The two-bar frame was studied in Chapter 5 and is reviewed in section 7.8 to investigate postcritical behavior. It is found that the problem leads to an asymmetric bifurcation. Section 7.9 deals with a column on an elastic foundation, already considered in Chapter 5, for which unstable bifurcation is found to occur. The circular ring under radial pressure is another example of stable symmetric bifurcation, and the solution is discussed in section 7.10. The equations of a plate made of a composite material, under in-plane biaxial compression, is studied in section 7.11 and represents a problem with two independent load parameters. The use of finite elements to evaluate postbuckling equilibrium paths using perturbations is discussed in section 7.12.

7.2 CONDITIONS FOR BIFURCATION POINTS

In our previous chapters it was shown that a critical state satisfies the following conditions:

$$V_i = 0 \quad \Longrightarrow \quad \text{equilibrium state,} \tag{7.1}$$

$$V_{ij} x_j |^c = 0 \quad \Longrightarrow \quad \text{critical state.} \tag{7.2}$$

The postcritical path was written in Chapter 6 taking q_1 as a perturbation parameter (i.e., $s = q_1$) and expanding the variables that define the path in the form

$$q_i(q_1) = q_i^{(1)} q_1 + \frac{1}{2} q_i^{(2)} q_1^2 + \cdots, \tag{7.3}$$

$$\lambda(q_1) = \lambda^{(1)} q_1 + \frac{1}{2} \lambda^{(2)} q_1^2 + \cdots. \tag{7.4}$$

To solve the evaluated derivatives in these expansions, $q_i^{(n)}$, $\lambda^{(n)}$, we employ a perturbation analysis of the equilibrium equations. It was shown in Chapter 6 that these conditions are not enough to solve the problem, and one scalar equation should be added to each perturbation set. This scalar equation is a contracted equation obtained by premultiplication of the perturbation equations by the eigenvector.

Let us again consider the contracted form of the first-order perturbation equation, that is,

$$\underbrace{V_{ij}x_i}_{0} q_j^{(1)} + V_i' x_i \lambda^{(1)}|^c = 0, \tag{7.5}$$

which reduces to

$$V_i' x_i \lambda^{(1)}|^c = 0.$$

Unlike our previous analysis of the limit point, we shall now consider the case in which

$$\boxed{V_i' x_i|^c = 0} \tag{7.6}$$

This condition implies that the load does not produce work when it acts on a displacement field that has the direction of the eigenvector. Incorporation of (7.6) into (7.5) satisfies the contracted perturbation equation for any value of $\lambda^{(1)}$, and thus it provides no new information.

The first-order perturbation equation is in this case

$$V_{ij}q_j|^c = -V_i' \lambda^{(1)}|^c. \tag{7.7}$$

Notice that there are fewer independent equations than unknowns because V_{ij} is singular at the critical state.

It is clear from the first-order perturbation equation (7.7) that the displacements $q_i^{(1)}$ should be a function of the load parameter $\lambda^{(1)c}$. We construct the complete solution of $q_i^{(1)}$ as the sum of a solution of the homogeneous equation associated with (7.7) plus a particular solution of the complete (7.7). The homogeneous equation associated with (7.7) is $V_{ij}q_j|^c = 0$, and the solution is the eigenvector x_j (since the homogeneous equation **is** the eigenproblem). Let us call y_j the particular solution, obtained for $\lambda^{(1)} = 1$. Then one can write

$$\boxed{q_j^{(1)c} = x_j + \lambda^{(1)}\, y_j|^c} \tag{7.8}$$

Substitution of (7.8) into (7.7) leads to

$$\boxed{V_{ij}\, y_j|^c = -V_i'|^c} \tag{7.9}$$

This new vector $y_j|^c$ evaluated at the critical state is the slope of the fundamental path at the critical state (the derivatives of the displacements with respect to the load parameter). The proof of this is left to the reader as Problem 7.1.

To solve (7.9), which is a system with singularity of order one, we need to know the value of one of the components of y_j . From (7.8) we know that $x_1 = 1$; and if $q_i^{(1)}$ is normalized in the same way, it is possible to write

$$q_1^{(1)} = x_1 + \lambda^{(1)}\, y_1 = 1 + \lambda^{(1)}\, y_1.$$

But since $q_1^{(1)} = 1$, it follows that

$$y_1 = 0. \tag{7.10}$$

Thus, (7.9) contains $(N-1)$ unknowns and has a unique solution.

The solution of $\lambda^{(1)}$ is achieved by contraction of the evaluated second-order perturbation equation, which yields

$$V_{ijk}x_i q_j^{(1)} q_k^{(1)} + 2V'_{ij}x_i q_j^{(1)}\lambda^{(1)} + V''_i x_i \lambda^{(1)2}|^c = 0. \tag{7.11}$$

Substitution of (7.8) into (7.11) leads to

$$(V_{ijk}x_i y_j y_k + 2V'_{ij}x_i y_j + V''_i x_i)\lambda^{(1)2} + 2(V_{ijk}y_k + V'_{ij})x_i x_j \lambda^{(1)}$$
$$+(V_{ijk}x_i x_j x_k)|^c = 0. \tag{7.12}$$

This is a quadratic equation in $\lambda^{(1)c}$, and the solution has two roots of $\lambda^{(1)}$. Thus, there are two tangents; in other words, there are two paths emerging from the critical state. Thus we may say that the bifurcation point is a good name to identify this critical state.

Equation (7.12) can be written as

$$\boxed{A\,\lambda^{(1)2} + B\,\lambda^{(1)} + C|^c = 0} \tag{7.13}$$

in which

$$A = V_{ijk}x_i y_j y_k + 2V'_{ij}x_i y_j + V''_i x_i|^c,$$
$$B = (V_{ijk}y_k + V'_{ij})x_i x_j|^c, \tag{7.14}$$
$$C = (V_{ijk}x_i x_j x_k)|^c.$$

Equation (7.8) may be employed for both primary and secondary paths. However, if the generalized coordinate chosen as perturbation parameter does not have a component in the primary path ($A = 0$), then $\lambda^{(1)} = \infty$, and (7.8) is no longer useful to define this path. Notice that this would not be a severe limitation, since at this stage we are only interested in the secondary path.

Two cases can be identified, depending on the solution of (7.13): symmetric and asymmetric bifurcations.

7.3 SYMMETRIC BIFURCATION

The first case to be considered is when the coefficient C vanishes at the critical state. Mathematically this case is not very general in the sense that it requires a further condition ($C = 0$). However, it is the most common case found in simple structures.

7.3.1 First-Order Perturbation Coefficients

The nature of the bifurcation depends on the value of the scalar $C = V_{ijk}x_i x_j x_k|^c$ at the critical point in (7.13). If

$$\boxed{C = 0} \tag{7.15}$$

then (7.13) becomes

$$(A\,\lambda^{(1)} + 2\,B)\,\lambda^{(1)}|^c = 0. \tag{7.16}$$

One of the solutions of (7.16) is the tangent to the fundamental path

$$A\,\lambda^{(1)} + 2\,B|^c = 0$$

leading to

$$\lambda^{(1)} = -\frac{2B}{A}. \tag{7.17}$$

The other solution of (7.16) is the tangent to the secondary path; i.e.,

$$\boxed{\lambda^{(1)c} = 0} \tag{7.18}$$

The solution of (7.8) reduces to

$$\boxed{q_j^{(1)c} = x_j^c} \tag{7.19}$$

This means that the tangent to the secondary or postcritical path has the same direction as the eigenvector.

Up to this point, we have found the first-order coefficients in the expressions of the postcritical path. However, one of the derivatives is zero, according to (7.18), and it is necessary to continue our analysis to find the first nonzero derivative in order to define the path.

7.3.2 Second-Order Perturbation Coefficients

The procedure for second-order derivatives is similar to what we have already followed and should lead to values of $\lambda^{(2)c}$ and $q_j^{(2)c}$. Let us start from the second-order perturbation equation

$$V_{ij}q_j^{(2)}|^c = -V_{ijk}x_j x_k - V_i'\lambda^{(2)}|^c. \tag{7.20}$$

The values of $q_j^{(2)}$ may be written as before in the form

$$q_j^{(2)c} = z_j + \lambda^{(2)}\,y_j|^c. \tag{7.21}$$

Notice that the condition $V_{ij}q_j^{(2)}|^c = -V_i'\lambda^{(2)}|^c$ is similar to (7.9), so we have again made use of the solution y_j. Vector z_j is the solution of the homogeneous equation (with $\lambda^{(2)} = 0$), and y_j is a particular solution to the complete equation. To show the conditions that each vector should satisfy, let us substitute (7.21) into (7.20) to obtain

$$V_{ij}(z_j + \lambda^{(2)}\, y_j)|^c = -V_{ijk}\, x_j x_k - V_i'\, \lambda^{(2)}|^c$$

or else

$$V_{ij}\, z_j + V_{ijk}\, x_j x_k|^c = -(V_i' + V_{ij}\, y_j)\, \lambda^{(2)}|^c.$$

It is then necessary to solve the equation

$$\boxed{V_{ij}\, z_j|^c = -V_{ijk}\, x_j x_k|^c} \tag{7.22}$$

Next we notice that

$$q_1^{(2)c} = z_1 + y_1 \lambda^{(2)}|^c$$

with $q_1^{(2)c} = 0$ and $y_1 = 0$; this means that

$$z_1 = 0, \tag{7.23}$$

and (7.22) has a unique solution.

The contraction mechanism is next applied to the third-order perturbation equation, leading to

$$V_{ijkl} x_i x_j x_k x_l + 3 V_{ijk} x_i x_j z_k + 3(V_{ijk} y_k + V_{ij}') x_i x_j \lambda^{(2)}|^c = 0. \tag{7.24}$$

This is a linear equation in $\lambda^{(2)}$ and may be written as

$$3B\, \lambda^{(2)} + \tilde{V}^4|^c = 0, \tag{7.25}$$

where

$$\tilde{V}^4 = V_{ijkl} x_i x_j x_k x_l + 3 V_{ijk} x_i x_j z_k|^c. \tag{7.26}$$

The scalar $\tilde{V}^4$ is the stability coefficient, which was already defined in Chapter 5. It is interesting to notice that since $\tilde{V}^4$ is involved in the computations, we can now distinguish between stable symmetric bifurcations and unstable symmetric bifurcations:

$$\tilde{V}^4 > 0 \Longrightarrow \text{stable symmetric bifurcation,} \tag{7.27}$$

$$\tilde{V}^4 < 0 \Longrightarrow \text{unstable symmetric bifurcation.} \tag{7.28}$$

The solution of (7.25) is

$$\boxed{\lambda^{(2)c} = -\frac{\tilde{V}^4}{3B}} \tag{7.29}$$

Notice that the solution (7.29) requires that

$$B \neq 0. \tag{7.30}$$

Substitution of (7.29) into (7.21) yields

$$\boxed{q_j^{(2)} = z_j - \frac{\tilde{V}_4}{3B} \quad y_j} \tag{7.31}$$

The perturbation analysis could stop at this point, since we have a nonzero value for all derivatives involved.

7.3.3 Third-Order Perturbation Coefficients

Let us continue with the new set of coefficients. The third-order perturbation equations are

$$V_{ij}q_j^{(3)} + V_i'\lambda^{(3)}|^c = -3[V_{ijk}'q_j^{(1)}q_k^{(1)}\lambda^{(1)} + V_{ijk}q_j^{(1)}q_k^{(2)} + V_{ij}''q_j^{(1)}\lambda^{(1)2}$$
$$+V_{ij}'(q_j^{(2)}\lambda^{(1)} + q_j^{(1)}\lambda^{(2)}) + V_i''\lambda^{(1)}\lambda^{(2)}] - V_{ijkl}q_j^{(1)}q_k^{(1)}q_l^{(1)} - V_i''\lambda^{(1)3}. \tag{7.32}$$

The solution $q_j^{(3)}$ may be written in terms of $\lambda^{(3)}$ as

$$q_j^{(3)c} = w_j + y_j\,\lambda^{(3)}|^c. \tag{7.33}$$

The new vector w_j results from

$$V_{ij}w_j|^c = -3[V_{ijk}'q_j^{(1)}q_k^{(1)}\lambda^{(1)} + V_{ijk}q_j^{(1)}q_k^{(2)} + V_{ij}''q_j^{(1)}\lambda^{(1)2}$$
$$+V_{ij}'(q_j^{(2)}\lambda^{(1)} + q_j^{(1)}\lambda^{(2)}) + V_i''\lambda^{(1)}\lambda^{(2)}] - V_{ijkl}q_j^{(1)}q_k^{(1)}q_l^{(1)} - V_i''\lambda^{(1)3}|^c \tag{7.34}$$

with $w_1 = 0$.

The contracted fourth-order perturbation equation is

$$[4(V_{ijk}q_k^{(3)} + V_{ij}'\lambda^{(3)}) + V_{ijklm}x_kx_lx_m + 6V_{ijkl}x_kq_l^{(2)} + 6V_{ijk}'x_k\lambda^{(2)}]x_ix_j$$
$$+(3V_{ijk}q_j^{(2)}q_k^{(2)} + 6V_{ij}'\,q_j^{(2)}\lambda^{(2)} + 3V_i''\lambda^{(2)2})x_i|^c = 0. \tag{7.35}$$

Replacement of (7.33) into (7.35) leads to

$$\lambda^{(3)c} = -\frac{[(V_{ijklm}x_kx_lx_m + 6V_{ijkl}x_kq_l^{(2)} + 6V_{ijk}'x_k\lambda^{(2)})x_ix_j}{}$$
$$\left.\frac{+(3V_{ijk}q_j^{(2)}q_k^{(2)} + 6V_{ij}'\,q_j^{(2)}\lambda^{(2)} + 3V_i''\lambda^{(2)2})x_i]}{4(V_{ijk}y_k + V_{ij}')x_ix_j}\right|^c \tag{7.36}$$

in which all the variables are now known.

Higher-order derivatives may be obtained in a similar way, but it must be noticed that fifth-order derivatives of V are required in (7.36), and they will automatically be zero if only fourth-order terms have been retained in the original energy.

7.3.4 An Algorithm for Symmetric Bifurcation

It was shown in the last section that the conditions to have a symmetric bifurcation are

$$V_i' x_i = 0, \qquad C = 0, \qquad B \neq 0.$$

An algorithm for the solution of symmetric bifurcation should proceed as follows:

1. Define
$$\lambda^{(1)c} = 0.$$
2. Define
$$q_j^{(1)c} = x_j^c.$$
3. Solve $y_j{}^c$ using
$$V_{ij}\, y_j|^c = -V_i'|^c$$
with $y_1 = 0$.
4. Compute
$$B = (V_{ijk} y_k + V_{ij}') x_i x_j|^c.$$
5. Solve $z_j{}^c$ using
$$V_{ij} z_j|^c = -V_{ijk} x_j x_k|^c$$
with
$$z_1 = 0.$$
6. Compute
$$\tilde{V}^4 = V_{ijkl} x_i x_j x_k x_l + 3 V_{ijk} x_i x_j z_k.$$
7. Characterize the bifurcation
$$\tilde{V}^4 > 0 \quad \Longrightarrow \quad \text{stable symmetric bifurcation,}$$
$$\tilde{V}^4 < 0 \quad \Longrightarrow \quad \text{unstable symmetric bifurcation.}$$
8. Compute
$$\lambda^{(2)c} = -\frac{\tilde{V}^4}{3B}.$$
9. Compute
$$q_j^{(2)} = z_j - \frac{\tilde{V}_4}{3B}\, y_j.$$

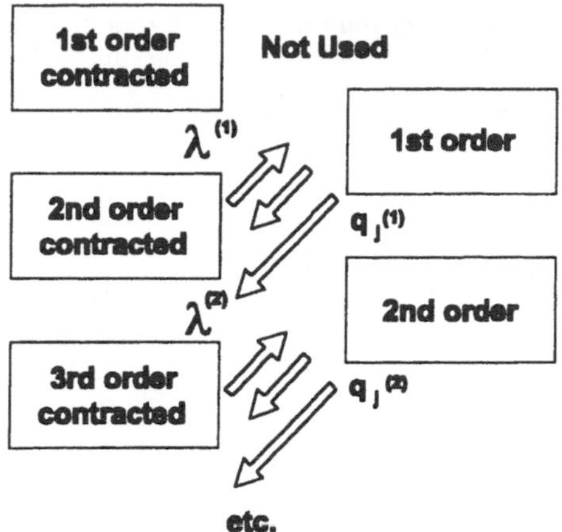

Figure 7.3 Sequence of perturbation analysis in bifurcation states.

10. The postcritical path is

$$\Lambda(q_1) = \lambda^c + \frac{1}{2}\lambda^{(2)c}\, q_1^2 + \cdots,$$

$$Q_i(q_1) = Q_i^c + x_i q_1 + \frac{1}{2} q_i^{(2)} q_1^2 + \cdots.$$

The scheme of perturbation analysis employed in this case is shown in Figure 7.3.

Example 7.1. *The circular plate is a simple example of symmetric bifurcation. Let us consider the discretization of the plate using two degrees of freedom, as was done in Chapters* 2 *and* 5. *Here we follow the results obtained in Examples* 5.1, 5.2, *and* 5.3. *The load vector and the eigenvector are*

$$V_i'|^c = \begin{Bmatrix} 0 \\ A_2' \end{Bmatrix}, \qquad x_i = \begin{Bmatrix} 1 \\ 0 \end{Bmatrix},$$

so that

$$x_i V_i'|^c = 0 \qquad \Rightarrow \qquad \textit{bifurcation.}$$

Next, consider the value of the third derivatives, $C = V_{ijk} x_i x_j x_k$. *From Example* 5.2 *we know that the only nonzero term is* V_{112}. *However,*

$$C = V_{112} x_1 x_1 \underbrace{x_2}_{0} = 0 \qquad \Rightarrow \qquad \textit{symmetric bifurcation.}$$

Finally, the value of $\tilde{V}_4$ *computed in Example* 5.3 *is positive*

$$\tilde{V}_4 > 0 \qquad \Rightarrow \qquad \textit{stable symmetric bifurcation.}$$

Let us compute the secondary path using perturbation analysis and

$$s = Q_1.$$

The vector y_j *is*

$$\begin{bmatrix} 0 & 0 \\ 0 & V_{22} \end{bmatrix} \begin{Bmatrix} 0 \\ y_2 \end{Bmatrix} = - \begin{Bmatrix} 0 \\ A_2' \end{Bmatrix} \quad \Rightarrow \quad y_j = -\frac{A_2'}{A_{22}} \begin{Bmatrix} 0 \\ 1 \end{Bmatrix}.$$

The vector z_j *is*

$$\begin{bmatrix} 0 & 0 \\ 0 & V_{22} \end{bmatrix} \begin{Bmatrix} 0 \\ z_2 \end{Bmatrix} = - \begin{Bmatrix} 0 \\ V_{211} x_1 x_1 \end{Bmatrix} \quad \Rightarrow \quad z_j = -\frac{A_{211}}{A_{22}} \begin{Bmatrix} 0 \\ 1 \end{Bmatrix}.$$

Next, the scalar B results in

$$B = \left(V_{112} y_2 + V_{11}'\right) x_1 x_1 = -A_{112} \frac{A_2'}{A_{22}}.$$

The stability coefficient is

$$\tilde{V}_4 = (A_{1111} + 3A_{112} z_2)\, x_1^4 = A_{1111} - 3\frac{(A_{112})^2}{A_{22}}.$$

Finally, the curvature of the postbuckling path becomes

$$\lambda^{(2)c} = -\frac{\tilde{V}_4}{3B} = \frac{A_{1111} A_{22} - 3\,(A_{112})^2}{3A_2' A_{112}}.$$

The displacements results in

$$q_j^{(2)} = \begin{Bmatrix} 0 \\ q_2^{(2)} \end{Bmatrix} \quad \text{with} \quad q_2^{(2)} = z_2 + \lambda^{(2)c}\, y_2.$$

The example illustrates that the computations required to obtain the postcritical path using the present perturbation approach are simple and can be done with little effort.

7.4 ASYMMETRIC BIFURCATION

The results of the previous section were obtained under the assumption that the coefficient C vanished at the critical state. However, this is not the only possibility, and we shall investigate in this section cases in which $C \neq 0$. The analysis starts with the solution of first-order perturbation equations.

7.4.1 First-Order Perturbation Coefficients

We return to (7.13), under the condition

$$C = (V_{ijk}x_i x_j x_k)|^c \neq 0. \tag{7.37}$$

It is assumed that $B < 0$ and $A \neq 0$; then the explicit form of the roots is

$$\boxed{\lambda^{(1)c} = \frac{-B \pm \sqrt{B^2 - AC}}{A}} \tag{7.38}$$

One of these conditions is the tangent to the fundamental path, and the other is tangent to the secondary path. Another possibility of identification of the roots [17], employed when the analysis is carried out by means of a step-by-step algorithm, is to compare both solutions of (7.8) with the last value of the tangent to the primary path (obtained, for example, using a continuation method).

The displacement derivatives are written in the form

$$q_j^{(1)c} = x_j + y_j \lambda^{(1)}|^c, \tag{7.39}$$

where y_j is solved as in (7.9). Notice that now

$$q_j^{(1)c} \neq x_j^c.$$

This is an important difference between symmetric and asymmetric bifurcations.

7.4.2 Second-Order Perturbation Coefficients

The second-order equations developed in the last chapter provide us with $(N - 1)$ independent relations

$$V_{ij}q_j^{(2)} + V_i' \lambda^{(2)}|^c = -(V_{ijk}q_j^{(1)}q_k^{(1)} + 2V_{ij}' q_j^{(1)}\lambda^{(1)} + V_i'' \lambda^{(1)2})|^c. \tag{7.40}$$

The other equation needed is the contracted third-order perturbation equation, i.e.,

$$3\left(V_{ijk}x_i q_j^{(1)} + V_{ij}' x_i \lambda^{(1)}\right) q_j^{(2)} + 3\left(V_{ij}' x_i q_j^{(1)} + V_i'' x_i \lambda^{(1)}\right) \lambda^{(2)}|^c$$

$$= -\left[V_{ijkl}q_j^{(1)}q_k^{(1)}q_l^{(1)} + V_i''' \lambda^{(1)3} + 3V_{ijk}' q_j^{(1)}q_k^{(1)}\lambda^{(1)} + 3V_{ij}'' q_j^{(1)}\lambda^{(1)2}\right] x_i|^c. \tag{7.41}$$

To obtain an explicit form for $\lambda^{(2)c}$ it is necessary to write the values of $q_j^{(2)c}$ in terms of $\lambda^{(2)c}$ in the second-order perturbation equations. This is left as a problem at the end of this chapter.

7.4.3 An Algorithm for Asymmetric Bifurcation

The conditions required to have asymmetric bifurcation are

$$V_i' x_i = 0, \qquad C \neq 0, \qquad B < 0, \qquad A \neq 0.$$

An algorithm for the solution of asymmetric bifurcation should proceed as follows:

1. Solve y_j^c using

$$V_{ij}\, y_j|^c = -V_i'|^c$$

with

$$y_1 = 0.$$

2. Compute

$$A = V_{ijk} x_i y_j y_k + 2V_{ij}' x_i y_j + V_i'' x_i|^c,$$

$$B = (V_{ijk} y_k + V_{ij}') x_i x_j|^c,$$

$$C = (V_{ijk} x_i x_j x_k)|^c.$$

3. Compute

$$\lambda^{(1)c} = -\frac{C}{2B}.$$

4. Compute

$$q_j^{(1)} = x_j + y_j \lambda^{(1)}|^c.$$

5. Postcritical path

$$\Lambda(q_1) = \lambda^c + \lambda^{(1)c} q_1 + \cdots,$$

$$Q_i(q_1) = Q_i^c + q_i^{(1)} q_1^1 + \cdots.$$

The asymmetric bifurcation is a more general case than the symmetric bifurcation. In fact, symmetric bifurcations may be seen as a special case of asymmetric bifurcations, and this is shown in a section below.

7.4.4 The Tangent to the Asymmetric Secondary Path

Let us identify the tangents to the paths emerging from an asymmetric bifurcation. We proceed to perform a geometric interpretation of the solutions: The two tangents associated with (7.38) may be written as

$$\lambda^{(1)c} = \beta \pm \alpha. \tag{7.42}$$

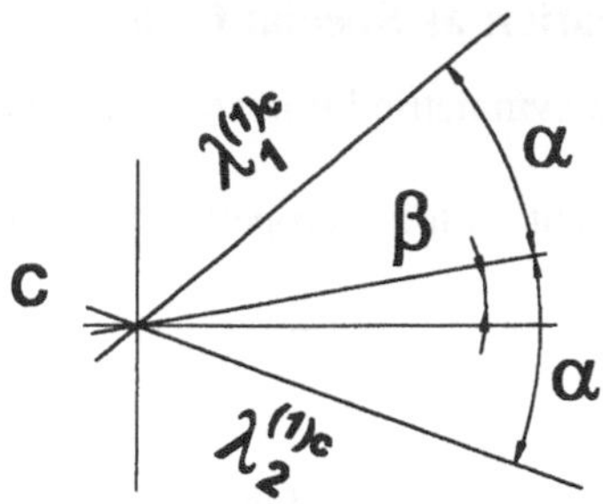

Figure 7.4 Tangent to the secondary path.

In Figure 7.4 we show two paths emerging from C, an asymmetric point of bifurcation. One of the paths is descendent, and its tangent is identified as $\lambda_2^{(1)c}$. Notice that

$$2\alpha \neq \frac{\pi}{2}.$$

A comparison between (7.38) and (7.42) shows that

$$\beta = -\frac{B}{A}, \tag{7.43}$$

$$\alpha = \frac{\sqrt{B^2 - AC}}{A} = \frac{B}{A}\sqrt{1 - \frac{AC}{B^2}}. \tag{7.44}$$

We may expand the root of α in the form

$$\alpha = \frac{B}{A}\left(1 - \frac{1}{2}\frac{AC}{B^2} - \cdots\right) = \frac{B}{A} - \frac{1}{2}\frac{C}{B} - \cdots, \tag{7.45}$$

The tangent $\lambda_1^{(1)c}$ results in

$$\lambda_2^{(1)c} = \beta - \alpha \approx -2\frac{B}{A}. \tag{7.46}$$

This is the same as (7.17), and represents the primary path.
The other tangent, $\lambda_2^{(1)c}$ must be associated with the secondary path and is

$$\lambda_1^{(1)c} = \beta + \alpha = -\frac{1}{2}\frac{C}{B}, \tag{7.47}$$

showing that it requires values of the coefficients B and C.

7.4.5 Symmetric Bifurcation as Special Case of Asymmetric Bifurcation

In this section we show that symmetric bifurcations can be obtained as a special case of asymmetric bifurcations.

Let us consider (7.38), but with the property of symmetric bifurcations, (7.15):

$$C = 0.$$

Thus,

$$\lambda^{(1)c} = \frac{-B \pm \sqrt{B^2}}{A}.$$

One of the roots is the tangent to the fundamental path.

$$\lambda_1^{(1)c} = -\frac{2B}{A}.$$

The other root yields

$$\lambda_1^{(1)c} = 0$$

as it should be in symmetric bifurcations.

It is left to the reader to prove that the other coefficients of the perturbation equations also result from specialization of the asymmetric formulation.

7.5 THE SPECIALIZED SYSTEM

We have seen that an important case in structures occurs when the energy takes the form

$$V = A_i Q_i + \frac{1}{2!} A_{ij} Q_i Q_j + \frac{1}{3!} A_{ijk} Q_i Q_j Q_k + \frac{1}{4!} A_{ijkl} Q_i Q_j Q_k Q_l. \tag{7.48}$$

In this section we derive the bifurcation equations for the energy of (7.48) from the general analysis presented in sections 7.3 and 7.4.

A bifurcation state is identified by (7.6), now in the form

$$\left(A_i + A_{ij} Q_j\right) x_i |^c = 0. \tag{7.49}$$

The nature of bifurcation is given by the coefficient C, which can be computed from (7.14) and (7.48) as

$$C = \left(A_{ijk} + A_{ijkl} Q_l\right) x_i x_j x_k |^c. \tag{7.50}$$

In symmetric bifurcations the algorithm in section 7.3.4 can be supplemented with the following explicit forms. The vector y_j is computed from (7.9) as

$$\left(A_{ij} + A_{ijk} Q_k\right) y_j |^c = -A_i'. \tag{7.51}$$

Vector z_j in (7.22) results from

$$\left(A_{ij} + A_{ijk}Q_k\right) z_j|^c = -\left(A_{ij} + A_{ijkl}Q_l\right) x_j x_k|^c. \tag{7.52}$$

The stability coefficient in (7.26) is now

$$\tilde{V}_4 = \left[A_{ijkl}x_k x_l + 3\left(A_{ijk} + A_{ijkl}Q_l\right) z_k\right] x_i x_j|^c, \tag{7.53}$$

and the B coefficient, (7.14), becomes

$$B = \left(A_{ijk} + A_{ijkl}Q_l\right) x_i x_j y_k|^c. \tag{7.54}$$

The algorithm for **asymmetric** bifurcation presented in section 7.4.3 is now written in more explicit form in terms of the A_{ijk}, A_{ijkl} coefficients. The C and B coefficients are the same as in (7.50) and (7.54), but the A coefficient now becomes

$$A = \left(A_{ijk} + A_{ijkl}Q_l\right) x_i y_j y_k|^c. \tag{7.55}$$

The above equations simplify the computations in the case of a system with an energy form similar to (7.48). Frequent use of these equations is made in the examples presented in this chapter.

7.6 THEOREMS ABOUT POSTCRITICAL PATHS

Two theorems were presented by Thompson [21] regarding the geometry and stability of a primary path. These theorems explain what kind of response is permitted and what is denied along a path. The first theorem says the following:

Theorem 7.1. ''An initially stable (primary) equilibrium path rising monotonically with the loading parameter, cannot become unstable without intersecting a further distinct (secondary) equilibrium path'' [22, p. 62].

This means that a rising path cannot change its stability except if it crosses through a critical bifurcation state. Thus, the situation represented in Figure 7.5 cannot occur. It is very important to have the kind of certainty given by Theorem 7.1, because by looking at the geometry of the path and controlling the critical states, we know that changes in the stability will not occur.

Theorem 7.2. ''An initially stable equilibrium path rising with the loading parameter cannot approach an unstable equilibrium state from which the system would exhibit a finite dynamic snap without the approach of an equilibrium path (which may or may not be an extension of the original path) at values of the loading parameter less than that of the unstable state'' [22, p. 62].

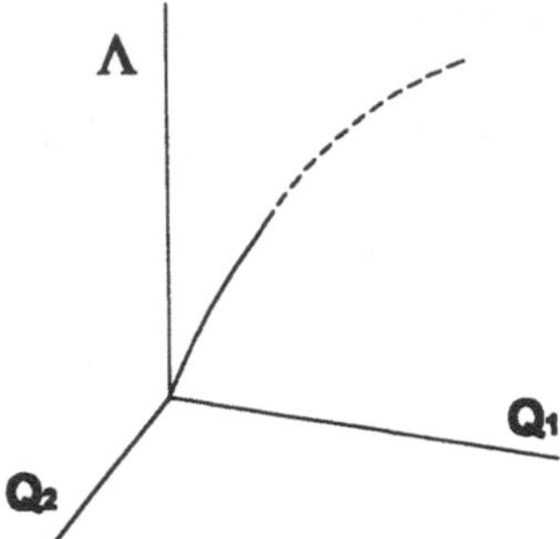

Figure 7.5 A situation that cannot occur.

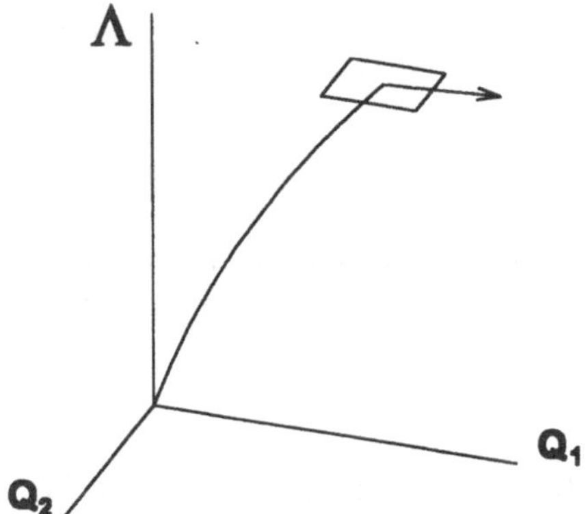

Figure 7.6 Another situation that cannot occur.

A situation not permitted by this second theorem is represented in Figure 7.6: A single path with no other equilibrium paths in its vicinity cannot have a sudden dynamic jump. Only at some critical states will dynamic jumps occur.

According to Theorem 7.2, finite dynamic snaps are only permitted at

(a) limit points;
(b) unstable symmetric bifurcations; and
(c) asymmetric bifurcations on the unstable branch.

A stable symmetric bifurcation, on the other hand, does not have a path with postcritical states at values lower than Λ^c and cannot have this kind of jump.

7.7 COLUMN UNDER AXIAL LOAD

This problem is identified as symmetric bifurcation and will be solved with the help of the results already obtained in section 5.6. Discretization of the column was made in the form

$$u = Q_1 L\left(\frac{x}{L}\right), \qquad w = Q_2 L \sin\left(\frac{n\pi x}{L}\right), \qquad \Lambda = \frac{P}{AE},$$

leading to the nonzero energy coefficients A'_1, A_{11}, A_{22}, A_{122}, and A_{2222}.

The fundamental or primary path was found to be

$$Q_j^F = \begin{Bmatrix} \Lambda \\ 0 \end{Bmatrix},$$

while the critical state is characterized by

$$P^c = AE\Lambda^c = -n^2\left(\frac{\pi}{L}\right)^2 EI,$$

$$x_j = \begin{Bmatrix} 0 \\ 1 \end{Bmatrix}.$$

This is a bifurcation problem, because $V_i' \, x_j = 0$.

In the postbuckling equilibrium path, we use Q_2 as a perturbation parameter, so that there is coincidence with the component of the eigenvector that was employed at the normalization, i.e.,

$$s = Q_2.$$

The matrix form of V_{ij}^c is now

$$V_{ij}^c = \begin{bmatrix} A_{11} & 0 \\ 0 & 0 \end{bmatrix}.$$

Let us compute vector y using $V_{ij} y_j = -V_i'$ and $y_2 = 0$:

$$\begin{bmatrix} A_{11} & 0 \\ 0 & 0 \end{bmatrix} \begin{Bmatrix} y_1 \\ 0 \end{Bmatrix} = -\begin{Bmatrix} A'_1 \\ 0 \end{Bmatrix}.$$

This leads to

$$y_j = \begin{Bmatrix} 1 \\ 0 \end{Bmatrix}.$$

This has the same form as the vector of displacements in the fundamental path.

Next, the vector required for the second-order displacement field is computed from $V_{ij} z_j = -V_{ijk} x_j x_k$ with $z_2 = 0$,

$$\begin{bmatrix} A_{11} & 0 \\ 0 & 0 \end{bmatrix} \begin{Bmatrix} z_1 \\ 0 \end{Bmatrix} = -\begin{Bmatrix} A_{122} \\ A_{2222} Q_2^c \end{Bmatrix},$$

from which

$$z_j = \begin{Bmatrix} -n^2\pi^2/2 \\ 0 \end{Bmatrix}.$$

The coefficients B and $\tilde{V}_4$ result in

$$B = A_{221} y_1 (x_2)^2 = n^2 \frac{\pi^2}{2} EA,$$

$$\tilde{V}_4 = A_{2222}(x_2)^4 + 3(A_{221} z_1 x_2^2) = \frac{3}{8} n^4 \pi^4 EA.$$

Finally, the second-order coefficients of the perturbation expansion are

$$\lambda^{(2)c} = -n^2 \left(\frac{\pi}{2}\right)^2,$$

$$q^{(2)c} = \begin{Bmatrix} -3/4\pi^2 n^2 \\ 0 \end{Bmatrix}.$$

The postbuckling path of the column for $n = 1$ can be written in explicit form as

$$P^c = EI \left(\frac{\pi}{L}\right)^2 \left[1 + \frac{1}{8} \frac{AL^2}{I} Q_2^2 + \cdots\right].$$

Under the assumptions of neglecting the membrane stiffness of the column, the critical load is the same, but the postcritical path changes to the form

$$P^c = EI \left(\frac{\pi}{L}\right)^2 \left[1 + \frac{\pi^2}{8} Q_2^2 + \cdots\right].$$

The axially loaded column is a structural system notorious for having an almost flat postbuckling behavior. Although the bifurcations stable at the postbuckling path

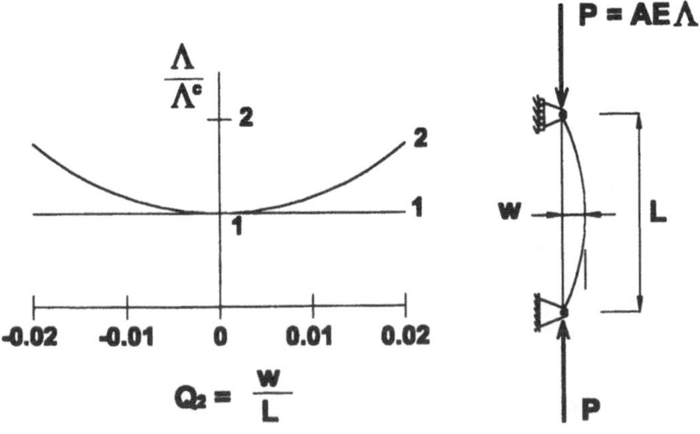

Figure 7.7 Postbuckling behavior of a strut (1) and a column (2).

are consequently stable, the load reserve of the column after buckling is very small. A flat plate, on the other hand, has a higher load reserve.

Example 7.2. *As an example, results for the postcritical path for a column with* $E = 10.5GPa$, $A = 7.5 \times 10^3 mm^2$, $I = 15 \times 10^6 mm^4$, *and* $L = 5000mm$ *are plotted in Figure 7.7. The column with membrane stiffness has a stronger postbuckling capacity than the simple strut, which displays an almost flat postbuckling behavior. The nature of the critical state in both cases is a symmetric bifurcation.*

7.8 TWO-BAR FRAME

The two-bar frame problem discussed earlier in Chapter 5 is revisited in this section to evaluate the postcritical behavior. But rather than looking at the same model of the frame, we consider here an approach in which membrane deformations are neglected. The bars of the frame become inextensible, and they deform by changes in curvature. This problem was solved using an analytical formulation by Koiter [15], and in [4, 22] using finite elements.

The kinematic relations of each bar are similar to what will be employed in the inextensible column on an elastic foundation. Similar to what we did in Chapter 5, the frame is discretized by means of finite elements with cubic interpolation of out-of-plane displacements. There are four unknowns in each individual element; thus each mode has two degrees of freedom associated with it.

For each element there is an end shortening due to the curvature of the mid-surface, which is a displacement in the direction of the axis of the element, with value

$$\Delta^e = L - \int \left[1 - \left(\frac{dw}{dx} \right)^2 \right]^{1/2} dx,$$

where the integral is carried out in the length of the element. Notice that in Chapter 6 the model employed included the end shortening as a degree of freedom at each end of the element.

The deflection at the load is given by the sum of the shortening of each element

$$\delta = \sum \Delta^e.$$

If each element of the frame is represented by just one finite element, then the number of unknowns is five: three rotations and the two displacements of the joint. The vertical displacement, however, is not independent, because it can be related to the other degree of freedom of the system. The horizontal displacement can be related to the rotation of the horizontal member. This has the consequence that only three degrees of freedom determine the total potential energy of the frame.

In the final form of the energy, the second-order coefficients are given by

$$A_{11} = 2\left(2 - \frac{1}{15}\lambda\right), \qquad A_{22} = 2\left(4 - \frac{1}{15}\lambda\right), \qquad A_{33} = 4,$$

$$A_{12} = 2 + \frac{1}{30}\lambda, \qquad A_{23} = 2.$$

The third-order coefficients result in

$$A_{222} = \frac{1}{25}\lambda, \qquad A_{122} = \frac{2}{5}\left(-3 + \frac{1}{30}\lambda\right), \qquad A_{112} = A_{113} = \frac{4}{5},$$

$$A_{123} = -\frac{1}{300}\lambda A, \qquad 223 = \frac{2}{5}\left(3 - \frac{1}{60}\lambda\right), \qquad A_{233} = A_{133} = \frac{1}{5}\left(-4 + \frac{1}{15}\lambda\right).$$

Finally, the fourth-order coefficients are

$$A_{1111} = \frac{2}{175}(1336 - 15\lambda), \qquad A_{2222} = \frac{48}{175}\left(\frac{313}{3} - \frac{4}{5}\lambda\right),$$

$$A_{3333} = \frac{8}{175}\left(334 - \frac{7}{5}\lambda\right), \qquad A_{1122} = \frac{34}{175}\left(2 - \frac{1}{15}\lambda\right),$$

$$A_{2233} = \frac{48}{175}\left(\frac{313}{3} - \frac{4}{5}\lambda\right), \qquad A_{1112} = \frac{6}{175}\left(250 + \frac{89}{120}\lambda\right),$$

$$A_{1113} = \frac{42}{175}\left(-1 + \frac{1}{60}\lambda\right), \qquad A_{2221} = \frac{1}{175}\left(229 - \frac{41}{10}\lambda\right),$$

$$A_{2223} = \frac{42}{175}\left(\frac{229}{7} + \frac{1}{12}\lambda\right), \qquad A_{3331} = -\frac{42}{175}, \qquad A_{3332} = \frac{350}{175}\left(\frac{5}{7} - \frac{1}{125}\lambda\right),$$

$$A_{1123} = -\frac{1}{25}, \qquad A_{1223} = \left(-2 + \frac{1}{1500}\lambda\right), \qquad A_{1233} = \frac{1}{25}.$$

For this problem, the linear primary path is trivial. The critical state is given by

$$\begin{bmatrix} A_{11} & A_{12} & 0 \\ 0 & A_{22} & A_{23} \\ 0 & 0 & A_{33} \end{bmatrix} \begin{Bmatrix} x_1 \\ x_2 \\ x_3 \end{Bmatrix} = \begin{Bmatrix} 0 \\ 0 \\ 0 \end{Bmatrix}$$

from which we get

$$96 - \frac{32}{5}\Lambda + \frac{1}{15}\Lambda^2 = 0,$$

and the critical load results in

$$\Lambda^c = 18.606.$$

A higher critical value is obtained for $\Lambda^c = 77.394$, but this will not be investigated in the following.

The eigenvector associated with the lowest critical load is

$$x_j = \begin{Bmatrix} 1 \\ -0.58 \\ 0.29 \end{Bmatrix}.$$

For the evaluation of the postcritical path we compute the coefficients

$$V_i' x_i \mid^c = 0,$$

$$C = A_{222}(x_2)^3 + 3(A_{122}x_2^2 + A_{112}x_2 + A_{113}x_3^3 + A_{223}x_2^2x_3 + A_{233}x_2x_3^2$$
$$+A_{133}x_3^2 + 2A_{123}x_2x_3) = -1.4821.$$

Since $C \neq 0$, we are in the presence of an asymmetric bifurcation. We continue the analysis with

$$B = A_{11}' + A_{22}'x_2^2 + 2A_{12}'x_2,$$

$$\lambda^{(1)c} = \frac{C}{2B} = 3.4181.$$

The finite element solution with only one element per member is an approximation in nonlinear problems such as this one, and to refine the results we need to include more elements in each bar, particularly in the vertical bar. The results already show

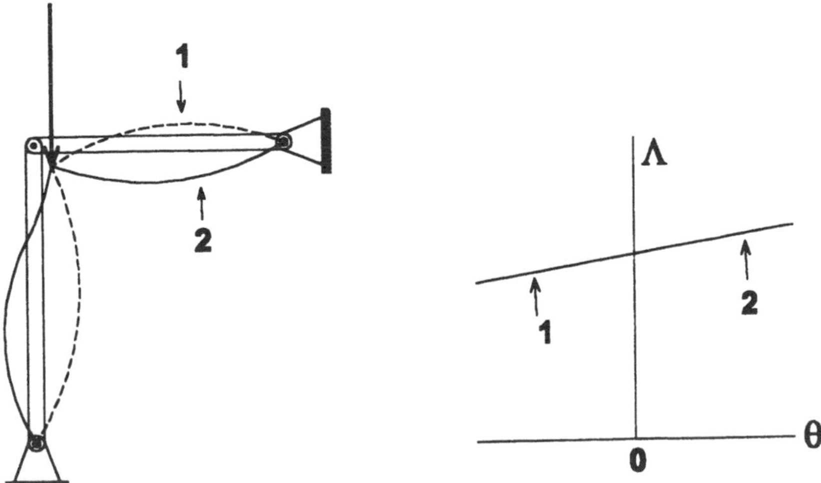

Figure 7.8 Postbuckling behavior of a two-bar frame.

that the nature of the bifurcation is asymmetric, a feature that is frequently found in frame structures but not so much in plates or shells.

A plot of the present results is shown in Figure 7.8, in which part of the path is stable and another part is unstable. Experiments for this problem were done by Roorda using a steel frame in [18], and the converged solution of this finite element model agrees quite well with the experiments.

7.9 COLUMN ON ELASTIC FOUNDATION

The influence of the elastic foundation on the postbuckling behavior of a column was first studied in [16]. Here we perform the analysis neglecting membrane stretching of the column, so that the internal energy is only due to bending of the strut

$$\chi_x = \frac{d^2w}{dx^2}\left[1 - \left(\frac{dw}{dx}\right)^2\right]^{-1/2}.$$

The total potential energy, up to fourth-order terms, results in the form

$$V = \frac{EI}{2}\int_0^L \left[\left(\frac{d^2w}{dx^2}\right)^2 + \left(\frac{dw}{dx}\right)^2\left(\frac{d^2w}{dx^2}\right)^2 + \cdots\right]dx + \frac{K}{2}\int_0^L w^2 dx - P\Delta.$$

The displacement of the point of application of the load is

$$\Delta = \int_0^L \left[\frac{1}{2}\left(\frac{dw}{dx}\right)^2 + \frac{1}{8}\left(\frac{dw}{dx}\right)^4 + \cdots\right].$$

The discretization of the problem can be made using a series solution in terms of sin functions. We restrict the analysis to only two terms to illustrate the way to obtain the solution

$$w(x) = Q_1 L \sin\left(\frac{\pi x}{L}\right) + Q_2 L \sin\left(\frac{2\pi x}{L}\right),$$

where Q_1 and Q_2 are dimensional degrees of freedom.

With the above degrees of freedom, the coefficients of the energy are

$$A_1 = A_2 = 0,$$

$$A_{11} = 2\left[(1+k) - \Lambda\right], \qquad A_{22} = 2\left[(16+k) - 4\Lambda\right],$$

$$A_{1111} = 6\pi^2\left(1 - \frac{3}{4}\Lambda\right), \qquad A_{2222} = 24\pi^2(16 - 3\Lambda),$$

$$A_{1122} = 4\pi^2(10 - 3\Lambda),$$

where the following notation has been used:

$$\Lambda = \frac{PL^2}{EI\pi^2}, \qquad k = \frac{KL^4}{EI\pi^4}.$$

Notice that the elastic foundation affects directly only quadratic terms. Quartic coefficients depend on Λ.

The fundamental path is trivial in the present case. The critical state satisfies the condition

$$\begin{bmatrix} A_{11} & 0 \\ 0 & A_{22} \end{bmatrix} \begin{Bmatrix} x_1 \\ x_2 \end{Bmatrix} = \begin{Bmatrix} 0 \\ 0 \end{Bmatrix},$$

and we notice that this is a diagonal form. From the determinant of V_{ij} we get two conditions:

$$\Lambda_1^c = 1 + k, \qquad x_j = \begin{Bmatrix} 1 \\ 0 \end{Bmatrix},$$

$$\Lambda_2^c = \frac{16 + k}{4}, \qquad x_j = \begin{Bmatrix} 0 \\ 1 \end{Bmatrix}.$$

The two modes are coincident when the stiffness takes the value $k = 4$, or $K = 4EI(\pi/L)^4$, that is, four times the Euler load of the column.

We can easily verify that both states satisfy the condition of symmetric bifurcation. Stability of each path depends on the value of k adopted.

The postbuckling path associated with Λ_1^c requires the evaluation of

$$B = A'_{11} x_1 x_1 = -2,$$

$$\tilde{V}_4 = A_{1111} (x_1)^4 = \frac{3}{2} \pi^2 (1 - 3k),$$

$$\lambda^{(2)c} = \frac{\pi^2}{4} (1 - 3k).$$

Thus, the critical path is unstable for values of $k \geq 1/3$.

Next, we investigate the behavior associated with the other critical state

$$B = V'_{ij} x_i x_j = A'_{22} x_2 x_2 = -8,$$

$$\tilde{V}_4 = A_{1111} (x_1)^4 = 24\pi^2 \left(4 - \frac{3}{4} k\right),$$

$$\lambda^{(2)c} = \pi^2 \left(4 - \frac{3}{4} k\right).$$

The second critical state has an unstable postcritical path for $k \geq 16/3$.

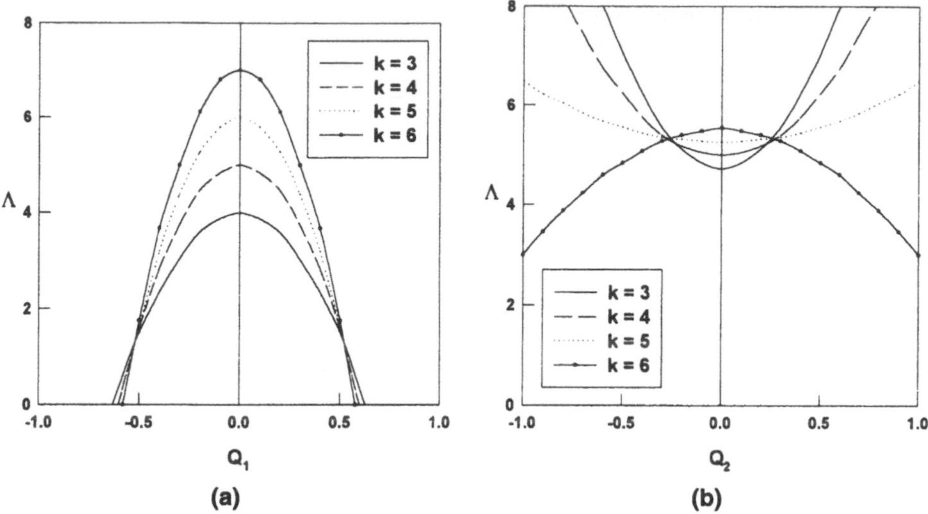

Figure 7.9 Column on elastic foundation for different values of k. (a) Displacement Q_1; (b) displacement Q_2.

Figure 7.9 shows the postcritical paths for several values of the parameter that control the stiffness of the elastic foundation. In the first case, $k = 3$, the lowest secondary path encountered by the structure is unstable. In the third example, $k = 5$, the lowest secondary path occurs for $n = 2$ and is stable, followed by an unstable path for $n = 1$. In the last case, for $k = 6$, the lowest secondary path is unstable and occurs for $n = 2$.

A more complete study requires the investigation including other modes for higher values of the stiffness of the elastic springs, k. As the value of k increases, the number of waves in the mode associated with the lowest buckling load also increases.

Finally, we observe that the lowest critical load in the present model is the same as in the column with axial deformation treated in Chapter 5. But there are significant differences between the two models in their postbuckling behavior.

7.10 CIRCULAR RING UNDER UNIFORM PRESSURE

The postbuckling behavior of a circular ring under uniform radial pressure can be obtained using the V-formulation and the results for critical states of Chapter 5. We shall not repeat here the information of section 5.12 and will proceed with the postbuckling path using the V-formulation. The generalized coordinates Q_1 and Q_2 are associated with the out-of-plane displacement, while the value of Q_3 is related to the in-plane displacement component.

To classify the critical state, with the eigenvector and load vector given by

$$x_i = \begin{Bmatrix} 0 \\ 1 \\ -1/n \end{Bmatrix}, \qquad V_i' = \begin{Bmatrix} -2 \\ 0 \\ 0 \end{Bmatrix},$$

we compute

$$V_i' x_i = 0,$$

which means that the critical state is a bifurcation. The coefficient C becomes

$$C = V_{ijk} x_i x_j x_k \mid^C = 0,$$

so that the bifurcation is symmetric.

To investigate the postbuckling path, we choose Q_2 as a perturbation parameter and remember that this degree of freedom describes out-of-plane displacements in the postbuckling path. The following coefficients result:

$$\lambda^{(1)c} = 0,$$

$$q_j^{(1)c} = x_j.$$

Next, we compute vector y_j from

$$[A_{ij} + Q_k^c \, A_{ijk}] \, y_j = -A_i'$$

with $y_2 = 0$. This yields

$$y_j = \begin{Bmatrix} 1 \\ 0 \\ 0 \end{Bmatrix}.$$

The other vector z required by the formulation is obtained from

$$\left[A_{ij} + A_{ijk} Q_k^c\right] z_j = -\left(A_{ijk} + A_{ijkl} Q_l^c\right) x_j x_k$$

with $z_2 = 0$. The solution yields

$$z_j = \frac{2\Gamma n^2 - n^4 - \Gamma^2}{2n^2} \begin{Bmatrix} 1 \\ 0 \\ 0 \end{Bmatrix}.$$

The stability coefficient was already evaluated as

$$\tilde{V}_4 = \frac{3}{4}\left(n - \frac{\Gamma}{n}\right)^4.$$

The coefficient B is now

$$\begin{aligned} B &= \left(V_{ijk}y_k + V'_{ij}\right)x_i x_j \\ &= A_{122}y_1x_2x_2 + 2A_{123}y_1x_2x_3 + A_{133}y_1x_3x_3 \\ &= \frac{\left(n^2 - \Gamma\right)^2}{n^2} > 0. \end{aligned}$$

The curvature of the postbuckling path results in

$$\lambda^{(2)c} = -\frac{\tilde{V}_4}{3B} = -\frac{1}{4}\left(n - \frac{\Gamma}{n}\right)^2.$$

The final results up to second-order perturbation coefficients are

$$\lambda = -\left[n^2\frac{I}{AR^2} + \frac{1}{8}\left(n - \frac{\Gamma}{n}\right)^2 Q_2^2\right],$$

where Γ is either 1 or 0, depending on the ring theory employed.

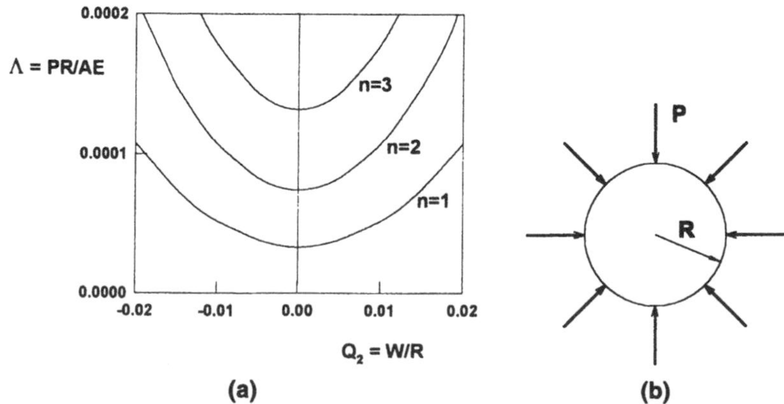

Figure 7.10 The ring under radial pressure. (a) Postbuckling paths for different mode shapes; (b) geometry and load.

Example 7.3. *Results for one case are presented in Figure 7.10, with* $I/AR^2 = 8.33 \times 10^{-6}$; $E = 200GPa$; *and* $AE/R = 120GPa.m$. *As the mode number increases, that is, the number of waves in the buckling mode increases, the critical load is also increased.*

The postbuckling path is stable, and the path increases with the load. However, the curvature of the postbuckling path is not affected so much by the mode considered in the first few modes, at least for the case in which $\Gamma = 1$.

7.11 RECTANGULAR ORTHOTROPIC PLATE

We consider in this section the buckling behavior of a rectangular plate, loaded by in-plane loads in one or two directions. The material of the plate is assumed to be orthotropic, made of a fiber-reinforced composite. The information about the critical state was provided in Chapter 5.

The coefficient C results in this case as

$$C = V_{ijk}x_i x_j x_k \,|^c = (A_{555} + Q_\ell A_{555\ell})\,(x_5)^3 \,|^c = 0. \tag{7.56}$$

This means that this plate problem displays a symmetric bifurcation behavior.

The vector y is obtained as

$$y_1 = -\left(\frac{A_1' + A_{12}y_2}{A_{11}}\right), \qquad y_2 = \left(\frac{A_{12}A_1' - A_{11}A_2'}{A_{11}A_{22} - A_{12}^2}\right), \qquad y_3 = y_4 = y_5 = 0, \tag{7.57}$$

where the last component was set to zero.

The vector z is solved and results in

$$z_1 = -\left(\frac{A_{155} + A_{12}z_2}{A_{11}}\right), \qquad z_2 = \left(\frac{A_{12}A_{155} - A_{11}A_{255}}{A_{11}A_{22} - A_{12}^2}\right),$$

$$z_3 = -\left(\frac{A_{355} + A_{34}z_4}{A_{33}}\right), \qquad z_4 = \left(\frac{A_{34}A_{355} - A_{33}A_{455}}{A_{33}A_{44} - A_{34}^2}\right), \tag{7.58}$$

where the last component was set to zero.

The following coefficients are computed:

$$B = A_{255}y_2x_5x_5 + A_{155}y_1x_5x_5,$$

$$\tilde{V}_4 = A_{5555}\,(x_5)^4 + 3(A_{551}z_1 + A_{552}z_2 + A_{553}z_3 + A_{554}z_4).$$

The coefficient $\lambda^{(2)c}$, which represents the curvature of the postbuckling path, results in

$$\lambda^{(2)c} = \frac{A_{5555}\,(x_5)^4 + 3(A_{551}z_1 + A_{552}z_2 + A_{553}z_3 + A_{554}z_4)}{3\,(A_{255}y_2x_5x_5 + A_{155}y_1x_5x_5)}.$$

The explicit form becomes

$$\lambda^{(2)c} = \frac{\pi^2}{80}\left[5\left(a^4n^4K_{22} + b^4m^4K_{11}\right) + 2b^2a^2n^2m^2(2K_{66} - 3K_{12})\right] \times \left[b^2a^2\left(f_1b^2m^2 + f_2n^2a^2\right)\right]^{-1}. \tag{7.59}$$

Examples of the use of the plate equations are left as problems at the end of the chapter.

7.12 FINITE ELEMENTS IN POSTCRITICAL PROBLEMS

The examples in this chapter have been simplified in such a way so that Ritz solutions can be obtained with a few degrees of freedom. Thus, for a given structure, there are functions that approximate the displacements and satisfy the boundary conditions. Furthermore, the approximations are such that both the fundamental path and the secondary paths can be modeled with just a few degrees of freedom.

Whenever the fundamental path is trivial, i.e., the displacements $Q_i^F = 0$, the problem becomes simpler because only the displacements along the secondary path need to be represented. But if the fundamental path is nontrivial, then it is more complex to find appropriate displacement fields for a Ritz solution. Further complications arise whenever the boundary conditions are not simply supported or when the loads are not uniform. One can say that most practical situations found in engineering projects have such complexity that simple Ritz functions cannot be found for them.

A convenient alternative in complex problems is to carry out the analysis using finite elements. The same formulation developed in this and previous chapters can be used for finite element analysis; however, the notation should change to matrix rather than index notation. For the present V formulation, this has been done in [8], and the notation is assumed to be similar to classical references in finite elements [24, 25].

For a given problem, the displacement field u_i is denoted by a vector $\mathbf{u}$, while the degrees of freedom Q_i are represented by a vector $\mathbf{a}$. The interpolation of u_i in terms of Q_i within an element is carried out as follows:

$$\mathbf{u} = \mathbf{\Phi}\,\mathbf{a},$$

where $\mathbf{\Phi}$ is a matrix containing the interpolation functions of one element.

Following standard finite element notation, the strains are related to nodal unknowns as

$$\varepsilon = [B_0 + B_1\,(\mathbf{a})]\ \mathbf{a}$$

in which the differential operator B may be expressed as a constant part B_0 (independent of $\mathbf{a}$) and a linear operator B_1, which is linear in $\mathbf{a}$.

The constitutive equations for elastic material are

$$\sigma = C\,\varepsilon.$$

Following the total Lagrangian formulation, the total potential energy can be written in the form

$$V = \frac{1}{2}\int \sigma^T\ \varepsilon dv - \mathbf{a}^T \int \Phi^T \mathbf{b}\, dv - \mathbf{a}^T \int \Phi^T \mathbf{p}\, dS.$$

Here σ is the second Piola–Kirchhoff stress tensor; ε is the Green–Lagrange strain tensor; $\mathbf{b}$ is the vector of body forces; and $\mathbf{p}$ is the vector of surface forces acting on the boundary of the structure.

To carry out the analysis of bifurcation the required derivatives are

$$V_i \qquad V_{ij} \qquad V_{ijk} \qquad V_{ijkl}$$

$$V_i' \qquad V_{ij}' \qquad V_i'' \qquad V_{ijk}' \qquad V_{ij}'' \qquad V_i'''$$

However, for the specialized system the number of derivatives reduces to five:

$$V_i \qquad V_{ij} \qquad V_{ijk} \qquad V_{ijkl} \qquad V_i'$$

Details about the computations of such derivatives are given in [8]. The arrays V_{ijk} and V_{ijkl} are organized in such a way that they are never required in isolation but enter into the formulation multiplied by vectors. Thus, matrices D_1 and D_2 are defined such that

$$D_1 = V_{ijk} q_k,$$

$$D_2 = V_{ijkl} q_k q_l.$$

7.12.1 Shells of Revolution

Shells have several features in their behavior that are different from other structures. The fundamental path may involve both membrane and bending stress resultants coupled to satisfy equilibrium. The geometry of the shell may be a function of the space coordinates. Membrane and bending actions may be again coupled in the secondary path.

For shells of revolution, the geometry is illustrated in Figure 7.11.

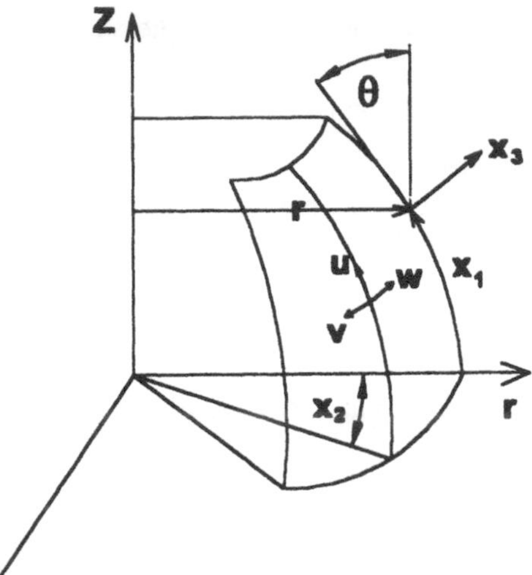

Figure 7.11 Shell of revolution. Reproduced from [9] with permission of John Wiley.

The displacements in this case may be interpolated taking advantage of the axisymmetric nature of the geometry. In this case we write

$$\begin{Bmatrix} u \\ v \\ w \end{Bmatrix} = \sum \begin{bmatrix} \cos n\theta & 0 & 0 \\ 0 & \sin n\theta & 0 \\ 0 & 0 & \cos n\theta \end{bmatrix} \begin{bmatrix} \Phi & 0 & 0 \\ 0 & \Phi & 0 \\ 0 & 0 & \Psi \end{bmatrix} \begin{Bmatrix} q_u^n \\ q_v^n \\ q_w^n \end{Bmatrix},$$

where n is the number of harmonics taken into account in the analysis; Φ are the interpolation functions for the membrane components; and Ψ is the interpolation for the displacement component w.

Further details of the finite element formulation are given in [9], with quintic interpolation for w and cubic for u and v.

Example 7.4. *An ellipsoidal head under external pressure is shown in Figure* 7.12. *This is a thin shell, with thickness $h = 1$, dimension $a = 750$, modulus $E = 10^6$, and* $\nu = 0.3$.

The problem has been studied with 18 elements along the meridian, and the perturbation approximation includes first- and second-order terms. Because the load is axisymmetric, the fundamental path also has axisymmetric displacements and is linear until a bifurcation state is found, with membrane stresses in most of the shell and a small area affected by bending action. The critical mode is shown in Figure 7.12.b, *with axisymmetric shape and large displacements at the apex.*

Example 7.5. *Theoretical and experimental investigations have been performed for shallow spherical domes under external pressure. The geometry of the shell is shown*

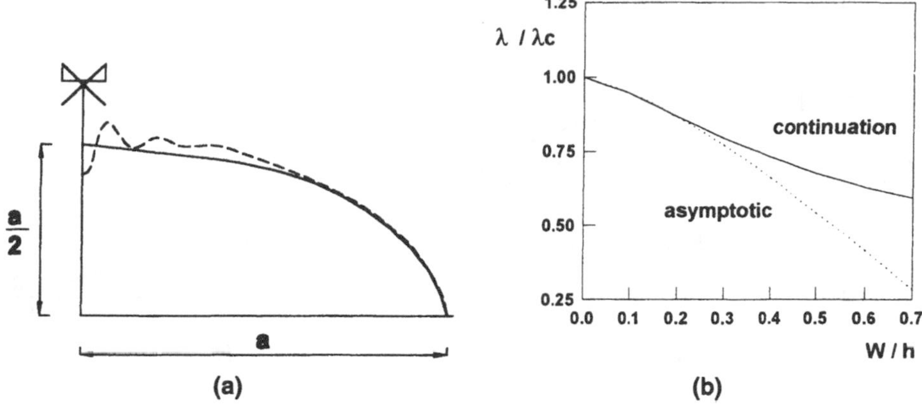

Figure 7.12 Ellipsoidal head under pressure (a) geometry; (b) postcritical path. Reproduced from [9] with permission of John Wiley.

in Figure 7.13, *and the specific data (*$a = 21$, $R = 100$, $H = 2.23$, $h = 0.1$, $E = 10^4$, $\nu = 0.3$*) is taken from* [7], *for which*

$$\left[3\left(1-\nu^2\right)\right]^{1/4}\left(\frac{H}{h}\right)^{1/2} = 6.$$

The classical critical load calculated from a linear fundamental path is smaller (92.8%) *than the equivalent computation for a complete sphere. But if nonlinearity of the fundamental path is taken into account, a lower value* (78.04%) *of the classical buckling load of the complete sphere is obtained. Buckling occurs in a nonaxisymmetric mode, with 7 waves in the circumferential direction* ($n = 7$). *The perturbation parameter chosen to follow the secondary path is the degree of freedom with the largest value in the critical mode.*

Both the critical load and the postbuckling path are in agreement with values presented in [7]. *The postbuckling path is a highly nonlinear path shown in Figure* 7.13. *It is initially unstable and displays a stiffening behavior associated with the value of* $\lambda^{(4)} > 0$.

Example 7.6. *The annular plate with axial load is shown in Figure* 7.15, *in which the ends are prevented from rotation. The data in this problem is* $a = 20$, $b = 12.4$, $h = 1$, $E = 10^4$, $\nu = 0.3$. *Contrary to what is found in the circular plate, the fundamental path in the annular plate does not have a constant stress field. The resultant radial stress has significant variations, from zero at the inner boundary to large values inside the plate.*

The ratio b/a *dominates the behavior. For small values of* b/a, *the lowest critical load has an axisymmetric mode* ($n = 0$). *The number of circumferential waves increases with* b/a. *The results in Figure* 7.14 (*from* [9]) *were obtained for* $b/a = 0.62$, *leading to a secondary path, which is essentially quadratic.*

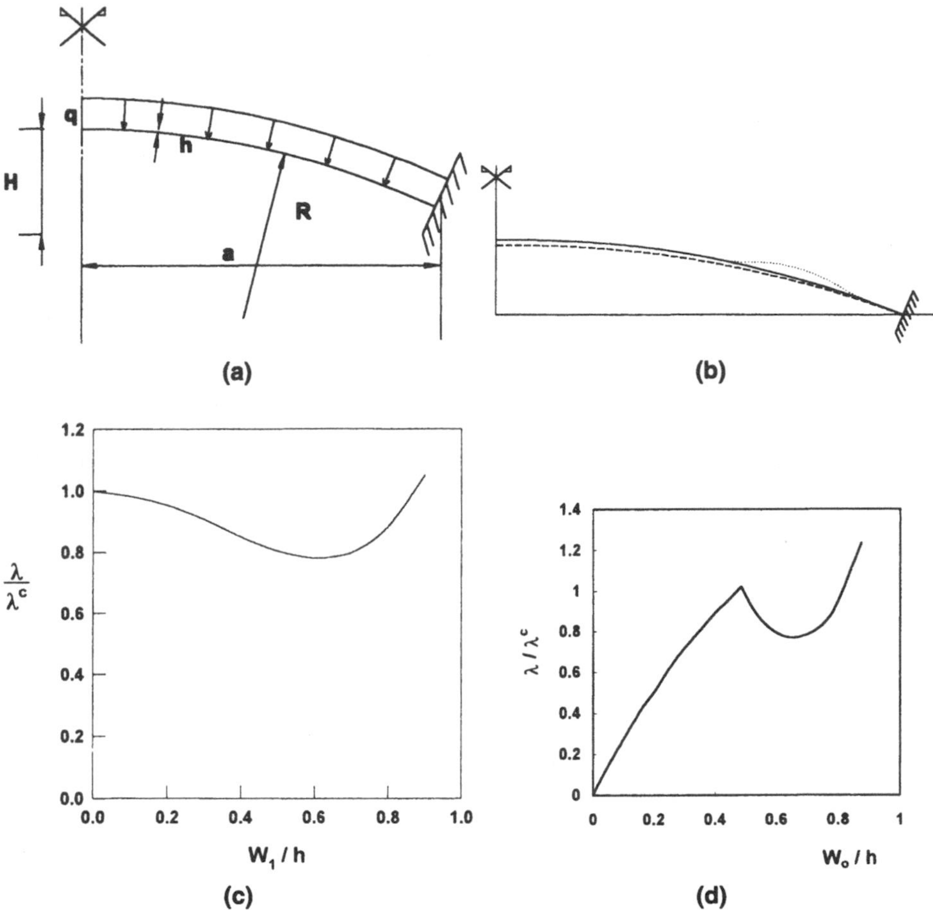

Figure 7.13 Shallow spherical dome under pressure (a) geometry; (b) the dashed line indicates a deformed shape at the critical point (enlarged by a factor of 10; the dotted line shows the critical mode at the harmonic $n = 7$); (c) postcritical path; (d) postcritical path. Reproduced from [9] with permission of John Wiley.

7.13 FINAL REMARKS

Bifurcation points have been identified in this chapter, and the spatial form of the three cases studied are presented in Figure 7.15. In symmetric bifurcations, the tangent to the postcritical path and the eigenvector have coincident directions. This is true for stable as well as unstable symmetric bifurcations. However, the tangent and the eigenvector have different directions in the asymmetric bifurcation. In this case, the eigenvector is parallel to the plane of displacements, while the tangent to the new path is inclined.

Koiter concluded in the early 1960s that the first stages of postbuckling behavior are governed completely by whether the bifurcation state itself is stable or unstable [14]. We have seen that the same coefficient $\tilde{V}_4$ employed in Chapter 5 to investigate

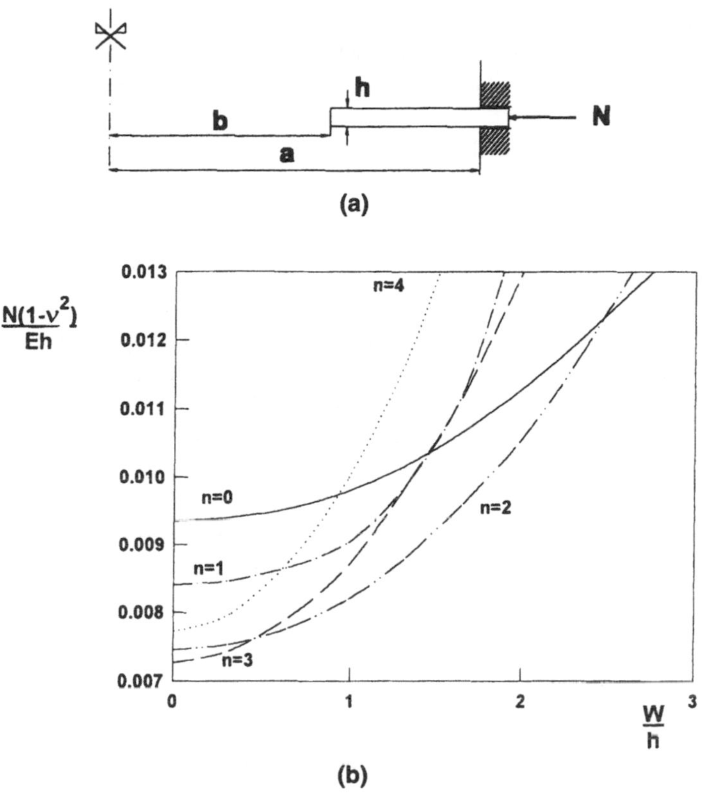

Figure 7.14 Annular plate (a) geometry; (b) asymptotic postcritical paths. Reproduced from [9] with permission of John Wiley.

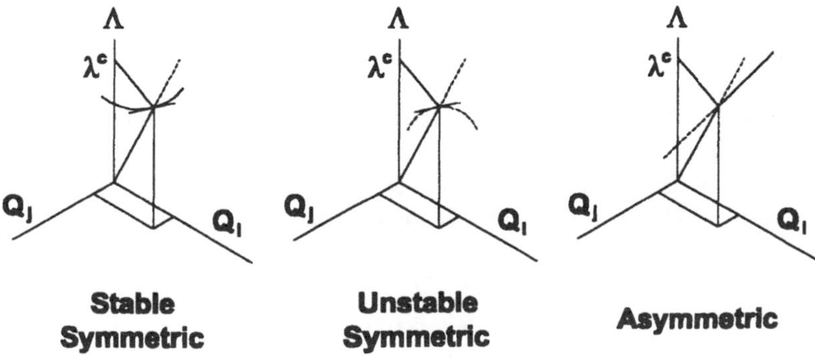

Figure 7.15 Types of bifurcation.

the stability of a critical state is also present in this chapter to investigate postbifurcation behavior.

A second comment can be made regarding the use of the contraction mechanism in bifurcation analysis. The aim of the contraction mechanism has been twofold:

- first, to add a condition, so that the rth system of equations has a unique solution; and
- second, because in the contraction mechanism we require that the sum of the terms associated with the rth derivatives is orthogonal to x_i; this is a way to assure that the $(r+1)$th perturbation equation has a solution.

The problem of symmetry in bifurcation states is a very interesting one and has been discussed by several authors [12]. Many structural shapes are designed with geometric symmetry, such as shells of revolution, cylinders, spheres, etc. At the critical state the structure changes its shape and develops a new displaced pattern with less symmetry than the original shape. Still, there is some symmetry in this new shape. But symmetry now arises at the level of the equilibrium path, in which the displaced shape has a mirror image that is also possible to achieve by the system. Thus, symmetry in terms of actual geometry is lost, but there is a new symmetry in terms of the equilibrium path.

7.14 PROBLEMS

Review questions. (*a*) Explain the meaning of the condition $V_i' x_i = 0$. (*b*) What are the analytical differences between symmetric and asymmetric bifurcations? (*c*) Can a symmetric bifurcation have $\tilde{V}^4 > 0$ and $\lambda^{(2)c} < 0$? (*d*) What is the consequence of $q_j^{(1)c} \neq x_j^c$ in asymmetric bifurcations? (*e*) Is symmetric bifurcation a special case of asymmetric bifurcation? (*f*) Is it possible to have an equilibrium state along a path rising with the load parameter, without any path in its vicinity, and still change its stability from stable to unstable? (*g*) In what cases is it possible to have a dynamic jump from an equilibrium position?

Problem 7.1 (Theory). Show that $y_j|^c$ is the slope of the fundamental path at the critical state.

Problem 7.2 (Theory). Obtain an explicit form for the second-order coefficients in asymmetric bifurcation.

Problem 7.3 (Two-degrees-of-freedom system). For a given problem, the energy is written as in (7.48), with the values of the coefficients given by

$$A_1' = A_2' = -1,$$

$$A_{11} = A_{22} = 90.16, \qquad A_{12} = 71.92,$$

$$A_{111} = A_{222} = -2402.4, \qquad A_{112} = A_{122} = 0.4,$$

$$A_{1111} = A_{2222} = 24040, \qquad A_{1122} = 12024, \qquad A_{1112} = A_{2221} = -12026.$$

A bifurcation point has been detected at the lowest critical load, obtained in Problem 5.2 in Chapter 5. Identify the nature of the bifurcation and calculate the perturbation coefficients of the secondary path.

Problem 7.4 (Theory). Show that the curvature of the secondary path in symmetric bifurcation can be obtained by specialization of the asymmetric curvature.

Problem 7.5 (Trifurcation and bifurcations) [1, 11]. A simple mechanical model with two degrees of freedom is shown in Figure 7.16. The drawing shows the unloaded configuration, the deflected shape in the fundamental path, and the postcritical equilibrium configuration. The shortening of each spring is Q_2, while Q_1 is the rotation of one of the members. The total potential energy becomes

$$V = \frac{1}{2} k Q_2^2 + 2C Q_1^2 - P\Delta$$

with

$$\Delta = (L - Q_2) \cos(Q_1) .$$

The $\cos(Q_1)$ function should be expanded in power series of Q_1. Consider three cases of stiffness of the rotational spring: (a) $C = kL^2/32$; (b) $C = 15kL^2/128$; (c) $C = 10kL^2/128$. Find the critical state and the postcritical paths emerging from it. Explain the results that you obtain.

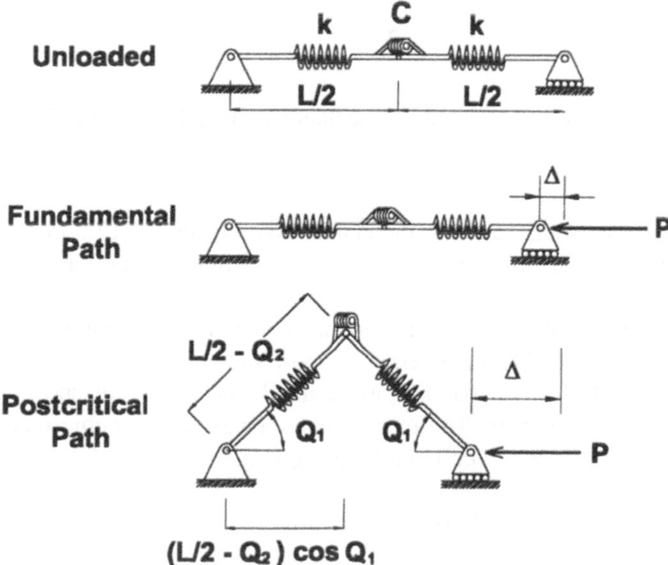

Figure 7.16 Problem 7.5: Two-degrees-of-freedom mechanical model, showing the geometry in the unloaded configuration, deflections in the fundamental path, and deflections in the postbuckling path.

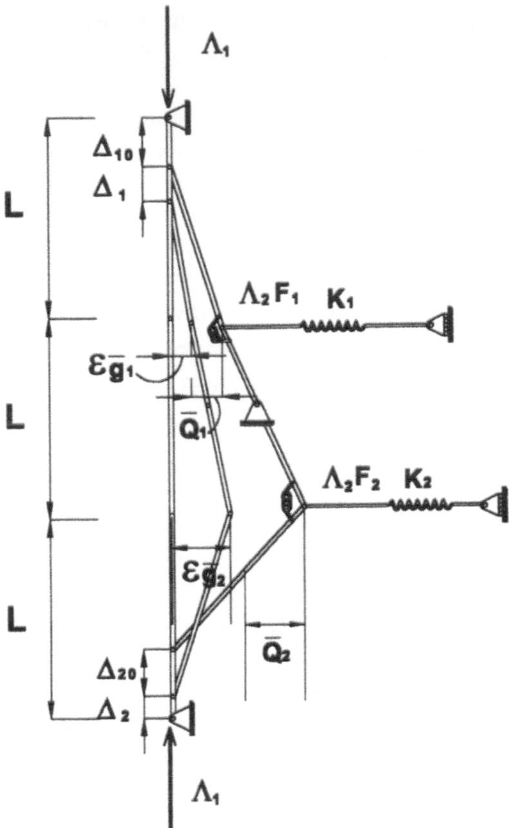

Figure 7.17 Two-degrees-of-freedom mechanical model, Problem 7.6, showing the geometry and loads.

Problem 7.6 (Nonlinearity in the constitutive equations) [2]. In some cases the unstable behavior arises from nonlinear elastic constitutive relations rather than from kinematic nonlinearity alone.

Consider the two-degrees-of-freedom mechanical model of Figure 7.17, with three rigid bars. The model has an internal support that allows rotation and horizontal displacement of the central bar and extensional and rotational springs attached to the joints between members. The extensional springs have a cubic relation with the displacements, so that the force in each spring is given by

$$F_1 = k_{11} L Q_1 + k_{12} L^2 (Q_1)^2 + k_{13} L^3 (Q_1)^3 ,$$

$$F_2 = k_{21} L Q_2 + k_{22} L^2 (Q_2)^2 + k_{23} L^3 (Q_2)^3 .$$

The displacements are normalized with respect to the length L of the columns, i.e., $Q_1 = \bar{Q}_1 / L$. Use $k_{11} = k_{21} = C/L^2$, $k_{12} = k_{22} = -8.5$, and $k_{13} = k_{23} = 25$.

The total potential energy of the system is

$$V = V_{k1} + V_{k2} + V_{C1} + V_{C2} + V_{\Lambda 1} + V_{\Lambda 2},$$

where

$$V_{k1} = k_{11}L^2 (Q_1)^2 + k_{12}L^3 (Q_1)^3 + k_{13}L^4 (Q_1)^4,$$

$$V_{k2} = k_{21}L^2 (Q_2)^2 + k_{22}L^3 (Q_2)^3 + k_{23}L^4 (Q_2)^4,$$

$$V_{C1} = \frac{1}{2}C\left[(4Q_1^2 - 4Q_1Q_2 + Q_2^2) + \frac{1}{3}(4Q_1^4 - 8Q_1^3Q_2 + 9Q_1^2Q_2^2 - 15Q_1Q_2^3 + Q_2^4)\right],$$

$$V_{C2} = \frac{1}{2}C\left[(4Q_2^2 - 4Q_1Q_2 + Q_1^2) + \frac{1}{3}(4Q_2^4 - 8Q_2^3Q_1 + 9Q_2^2Q_1^2 - 15Q_2Q_1^3 + Q_1^4)\right],$$

$$V_{\Lambda 1} = -\Lambda_1 (\Delta_1 + \Delta_2),$$

$$V_{\Lambda 2} = -\Lambda_2 L (f_1 Q_1 + f_2 Q_2),$$

$$(\Delta_1 + \Delta_2)/L = (Q_1^2 - Q_1Q_2 + Q_2^2) - \frac{1}{4}\left(Q_1^4 - 2Q_1^3Q_2 + 3Q_1^2Q_2^2 - 2Q_1Q_2^3 + Q_2^4\right).$$

Consider $\varepsilon = 0$ and find the bifurcation states using the V-formulation and $f_1 = f_2 = 1$.

Problem 7.7 (Mechanical system). A mechanical system is shown in Figure 7.18. It is formed by four rigid bars and a spring BC at the center. All bars have the same length L. At the unloaded situation, $P = 0$, the angle between the bars is zero $Q_1 = 0$. Obtain the postcritical states and identify the type of bifurcation.

Problem 7.8 (Cantilever column). A vertical steel column with one end clamped and the other end free is loaded at the free edge by a vertical load P. It is assumed that the column is inextensional, and the out-of-plane displacement is approximated using a cubic polynomial. Find the critical state and postcritical path of the column.

Problem 7.9 (Rectangular plate). Find the critical load and the postcritical path for the isotropic rectangular plate under biaxial loading.

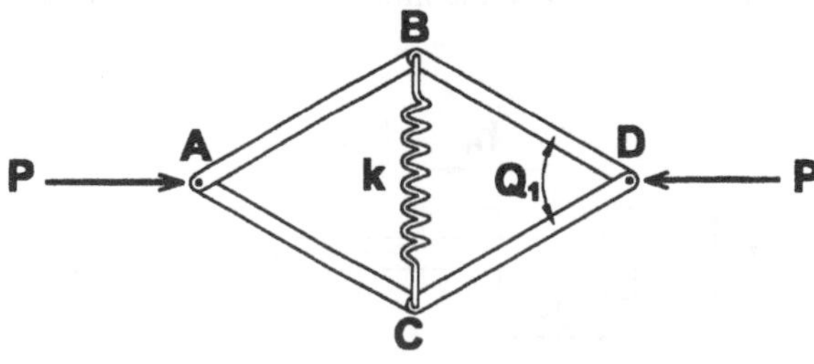

Figure 7.18 Mechanical system of Problem 7.7.

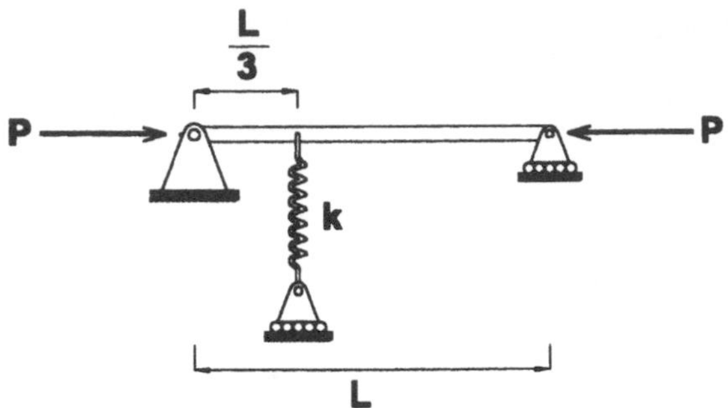

Figure 7.19 Column with partial bracing, Problem 7.10.

Problem 7.10 (Column with partial bracing). A column (assumed inextensional) of total length L simply supported at the ends is shown in Figure 7.19. A bracing element (modeled as an extensional spring) is located at $L/3$ of one of the supports and has the nonlinear law

$$F = k_1 w + k_2 w^2 + k_3 w^3,$$

where F is the force on the spring. Consider $L = 5m$; $EI = 157,500$; $k_1 = 150,000$; $k_2 = -1.5 \times 10^6$; $k_3 = -2 \times 10^7$. Find under what conditions the structure has unstable postbuckling behavior. Use trigonometric functions and three terms in the approximation of out-of-plane displacements.

Problem 7.11 (Rectangular plate). Use the data of Example 5.8 to plot the postbuckling path of a rectangular plate under uniaxial load.

Problem 7.12 (Rectangular plate). Draw the interaction diagram of critical loads for the above problem. An interaction diagram has f_1 on one of the axes and f_2 on the other and contains the load combinations that lead to buckling.

Problem 7.13 (Column on elastic foundation). Explain what would happen if you tried to solve the postbuckling problem of a column on an elastic foundation using exactly the same formulation as in section 5.7.

Problem 7.14 (Stiffened plate). A plate with stiffeners (illustrated in Figure 7.20) can be modeled using what is called a "smeared-out" approach, in which the constitutive properties are modified to include the influence of the stiffeners. Such an approach

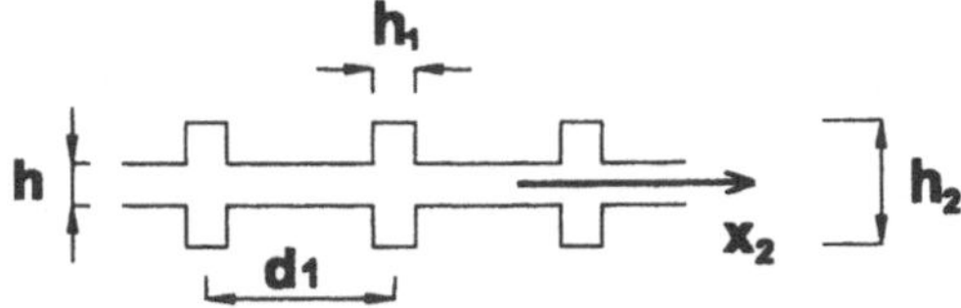

Figure 7.20 Stiffened plate, Problem 7.14.

would not be adequate for a stress analysis, but it is a reasonable approximation in problems that depend on global structural properties, such as buckling or vibrations.

Consider the constitutive relations

$$N_{11} = A_{11}\varepsilon_{11} + A_{12}\varepsilon_{22}, \qquad N_{22} = A_{12}\varepsilon_{11} + A_{22}\varepsilon_{22}, \qquad N_{12} = 2A_{66}\varepsilon_{12},$$

$$M_{11} = D_{11}\chi_{11} + D_{12}\chi_{22}, \qquad M_{22} = D_{12}\chi_{11} + D_{22}\chi_{22}, \qquad M_{12} = 2D_{66}\chi_{12},$$

with

$$A_{11} = K + \frac{EA_1}{d_1}, \qquad A_{22} = K + \frac{EA_2}{d_2}, \qquad A_{12} = \nu K, \qquad A_{66} = (1-\nu)\,K,$$

$$D_{11} = D + \frac{EI_1}{d_1}, \qquad D_{22} = D + \frac{EI_2}{d_2},$$

$$D_{12} = \nu D, \qquad D_{66} = (1-\nu)\,D + \frac{G}{2}\left(\frac{J_1}{d_1} + \frac{J_2}{d_2}\right),$$

$$K = \frac{Eh}{1-\nu^2}, \qquad D = \frac{Eh^3}{12\left(1-\nu^2\right)},$$

where it has been assumed that there are stiffeners in the two directions x_1 and x_2. For stiffeners in the x_1 direction, A_1 is the area of the stiffener; I_1 is the moment of inertia of the stiffener cross section; J_1 the torsional constant; and d_1 is the distance between the center lines of stiffeners. Obtain the equations that define the postcritical path.

Problem 7.15 (Frame). The structure in Figure 7.21 is formed by two rigid members (AB and BC) and two flexible members (AC and CD). The initial length

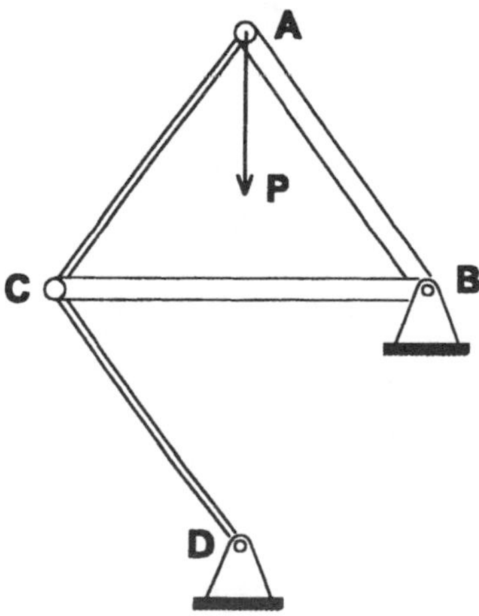

Figure 7.21 Frame of Problem 7.15.

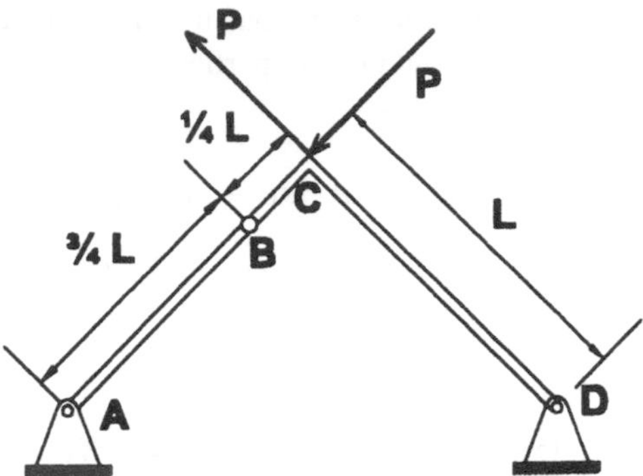

Figure 7.22 Frame with internal hinge of Problem 7.16.

of all the bars is the same. The flexible bars have the same material and cross section and can have stretching and bending. Find the critical and postcritical states.

Problem 7.16 (Plane frame). The frame in Figure 7.22 has an internal hinge in one of the members. The member AB is deformable, but the two other members BC and CD are rigid. $L = 600mm$, $E = 200GPa$, $A = 1.6 \times 25.4mm^2$. Use the formulation for trusses in Appendix B. Find the postbuckling path and compare it with the postbuckling path of a column with length $3L/4$. The values of EI are constant.

Problem 7.17 (Plane frame). A frame problem shown in Figure 2.13 was studied in Problem 5.16 to obtain the critical state. Obtain the postcritical path of the frame.

Problem 7.18 (Plane truss). Consider again Problem 5.16. What is the maximum postcritical displacement and load that can have member BC?

Problem 7.19 (Truss). The structure in Figure 7.23 is modeled as a truss with pin joints at A, B, and C. The members are standard steel pipes with nominal diameters $= 152.4mm$, wall thickness $= 7.112mm$, $L_1 = 9144mm$, $L_2 = 7620mm$, $A = 3600mm^2$, $I = 1.17 \times 10^7 mm^4$, $E = 200GPa$. A preliminary analysis based on linear

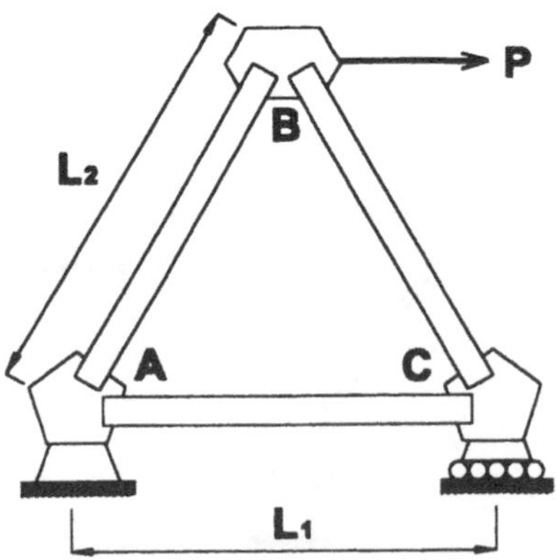

Figure 7.23 Truss of Problem 7.19.

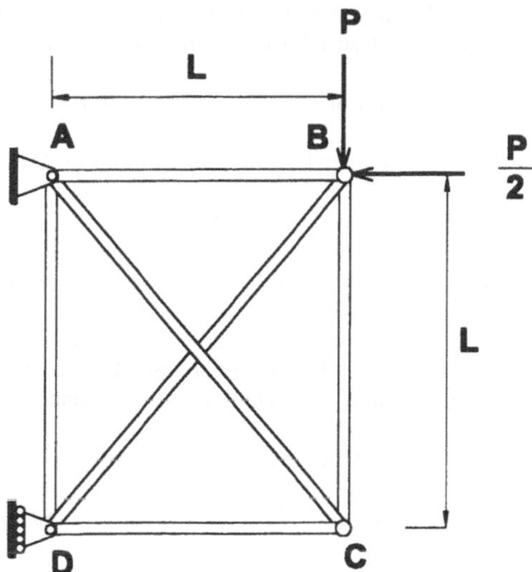

Figure 7.24 Truss of Problem 7.20.

theory shows that the force at AB is tension $0.83P$, the force at BC is compression $0.83P$, and the force at AC is tension $0.5P$. Evaluate the critical load and postbuckling path of the truss. *Hint:* you can use the degree of freedom for out-of-plane displacement of a bar only for the member with the highest compression.

Problem 7.20 (Truss). The truss in Figure 7.24 is formed by six members. Evaluate critical and postcritical states. *Hint:* you can use the degree of freedom for out-of-plane displacement of a bar only for the member with the highest compression.

7.15 BIBLIOGRAPHY

[1] Bellini, P. X., and Sritalapat, P., Post-buckling behavior illustrated by two DOF model, *J. Engrg. Mech., ASCE*, 114(2), 314–327, 1988.

[2] Brewer, A. T., and Godoy, L. A., Dynamic buckling of discrete structural systems under combined step and static loads, *Internat. J. Non-Linear Dynamics*, 9, 249–264, 1996.

[3] Budianski, B., Theory of buckling and post buckling behavior of elastic structures, in *Advances in Applied Mechanics*, Academic Press, New York, 14, 1–65, 1974.

[4] Casciaro, R., Di Carlo, A., and Pignataro, M., A finite element technique for bifurcation analysis, *Proc. XIV IUTAM Congress*, Delft, 1976.

[5] Donnell, L. H., *Beams, Plates, and Shells*, McGraw-Hill, New York, 1976.

[6] El Naschie, M. S., *Stress, Stability and Chaos in Structural Engineering: A Variational Approach*, McGraw-Hill, London, 1990.

[7] Fitch, J. R., and Budianski, B., Buckling and postbuckling behavior of spherical caps under axisymmetric load, *AIAA J.*, 8(4), 686–693, 1970.

[8] Flores, F. G., and Godoy, L. A., Elastic postbuckling analysis via finite element and perturbation techniques: Part 1, Formulation, *Internat. J. Numer. Methods Engrg.*, 33, 1775–1794, 1992.

[9] Flores, F. G., and Godoy, L. A., Elastic postbuckling analysis via finite element and perturbation techniques: Part 2, Application to shells of revolution, *Internat. J. Numer. Methods Engrg.*, 36, 331–354, 1993.

[10] Gibson, R. F., *Principles of Composite Material Mechanics*, McGraw-Hill, New York, 1994.
[11] Godoy, L. A., and Banchio, E. G., General trifurcation analysis in elastic stability, *J. Mech. Struc. Machines*, 27(3), 253–274, 1999.
[12] Hunt, G. W., Hidden (a)symmetries of elastic and plastic bifurcation, *Appl. Mech. Rev.*, 39(8), 1165–1186, 1986.
[13] Koiter, W. T., *On the Stability of Elastic Equilibrium*, Ph.D. thesis, Delft Institute of Technology, Delft, 1945.
[14] Koiter, W. T., Elastic stability and post-buckling behavior, in *Non-linear Problems*, R. Langer, Ed., Wisconsin University Press, Madison, 1963, 257–275.
[15] Koiter, W. T., Post-buckling of a simple two-bar frame, in *Progress in Applied Mechanics, The Folke Odqvist Volume*, Almqvist & Wiksell, Stockholm, 1967.
[16] Lekkerkerker, J. G., On the stability of an elastically supported beam subjected to its smallest buckling load, *Proc. Konink. Nederl. Akad. Wetensch.*, Ser. B, 65, 190, 1962.
[17] Riks, E., An incremental approach to the solution of snapping and buckling problems, *Internat. J. Solids Structures*, 15, 529–551, 1979.
[18] Roorda, J., *Instability of Imperfect Elastic Structures*, Ph.D. thesis, University of London, London, 1965.
[19] Sewell, M. J., On the branching of equilibrium paths, *Proc. Roy. Soc. London Ser. A*, 315, 499, 1970.
[20] Thompson, J. M. T., Basic principles in the general theory of elastic stability, *J. Mech. Phys. Solids*, 11, 13–20, 1963.
[21] Thompson, J. M. T., Basic theorems of elastic stability, *Internat. J. Engrg. Sci.*, 8, 307, 1970.
[22] Thompson, J. M. T., and Hunt, G. W., *A General Theory of Elastic Stability*, John Wiley, London, 1973, pp. 61–65.
[23] Walker, A. C., Segal, Y., and McCall, S., The buckling of thin-walled ring-stiffened steel shells, in *Buckling of Shells*, E. Ramm, Ed., Springer-Verlag, Berlin, 1982.
[24] Zienkiewicz, O. C., and Taylor, R., *The Finite Element Method*, Fourth edition, vol. 1, McGraw-Hill, London, 1989.
[25] Zienkiewicz, O. C., and Taylor, R., *The Finite Element Method*, Fourth edition, vol. 2, McGraw-Hill, London, 1991.

CHAPTER

EIGHT

EQUILIBRIUM OF SYSTEMS WITH IMPERFECTIONS

8.1 INTRODUCTION

Critical states and postcritical equilibrium paths have been characterized in Chapters 4 to 7. In Chapters 8 to 11 we investigate changes in the equilibrium and stability behavior as a consequence of modifications of the structure. In other words, thc studies concentrate on modifications of the system and the associated modifications in the response, and this is part of what is known as sensitivity analysis.

Concern about modifications in the response as a consequencc of changes in the design are very old. For example, the statute of the Strasbourg stonecutters (dated 1459) reads in Article 10:

> If a master mason has agreed to build a work and has made a drawing of the work as it is to be executed, he must not change this original design. But he must carry out the work according to thc plan that he has presented to the lords, towns, or villages in such a way that the work will not be diminished or lessened in value. [8, p. 71]

In modern times, the existence of imperfections and changes in a design are acknowledged in industry to such an extent that precise tolerance specifications are usually established to assure quality of the products or structures. For example, tolerances for fiberglass-reinforced structural shapes are established in [3] concerning the following geometric aspects

- cross-sectional dimensions, such as tubular wall thickness and line dimensions, with allowable deviations from nominal dimensions for rods, bars, and shapes;
- wall thickness eccentricity in round and square tubes;
- mean wall thickness in round and square tubes;

- straightness, with allowable deviations from straight;
- flatness;
- twist, with allowable deviations from straight in angles;
- angularity, with allowable deviations from straight in angles;
- length;
- squareness of end cut, with allowable deviation from straight in angles.

Guidelines, *ASTM* standards, and codes of practice for specific structures vary according to what is considered to have a potentially adverse effect on the structural integrity and normal function.

But tolerances are specified at a design stage, while other imperfections and defects may develop during the service life of the structure. An overview of structural imperfections may be found in [10], with special emphasis on the stress redistributions associated with geometrical imperfections and changes in the cross section of structural members.

The analytical evaluation of sensitivity of nonlinear systems is a relatively new field in mechanics, in which variations or derivatives of state fields are found due to variations in the system parameters. In linear problems, sensitivity analysis may detect only marginal changes in the response, but in nonlinear problems (as in buckling and postbuckling) there are qualitative and quantitative changes of great importance to the behavior and performance of the system.

There are two main cases of modifications of the structure:

- *Changes induced by imperfections and damage suffered by the structure.* These are undesired modifications and may affect the geometry of the structure, the loads, the boundary conditions, the material, etc. Typically they are associated with tolerance specifications, quality control, accidents, etc. The influence of such changes will be denoted as "imperfection sensitivity," and is studied in Chapters 8, 10, and 14.
- *Changes introduced by the engineer at a design stage.* Such changes should be reflected by design parameters, which may be, again the geometry, a dimension of the structure, or a property of the material. Design changes are modifications voluntarily introduced by an engineer after a basic design was produced. Typically they arise as a consequence of questions such as, "What if we reduce the thickness of the component in this area?" or "What if we have an opening to introduce a crossing pipe?" or "What if we substitute the material with a different one?" This is the subject of Chapter 11.

In section 8.2 we explain what is modified when a geometric imperfection, a load imperfection, or a design parameter are considered. A second parameter is introduced in section 8.3 to show how the energy functional is modified and what new terms arise in conjunction with imperfection or design parameters. Geometric imperfections are considered, and a model based on initial strains is introduced to represent deviations from an idealized geometry. Load imperfections are usually simpler and are also discussed. Some records from different industries are discussed to illustrate examples of practical imperfections.

In section 8.4 we consider systems governed by a limit-point type of behavior and show that changes in the system modify quantitatively the value of the critical load but do not affect the nature of the response; i.e., the system still shows a limit point. In systems governed by bifurcations (section 8.5) there are two very different possibilities: either the modification destroys the bifurcation, so that the system does not display branching from a primary path to a secondary path, or else the modification preserves the nature of the problem (we still have a bifurcation) and changes the quantitative aspects, such as values of critical loads and postcritical equilibrium paths.

8.2 THEORY VERSUS EXPERIMENTS IN BIFURCATION BUCKLING

8.2.1 On Disagreements between Theory and Experiments in Shells

The buckling pressure p^c of a spherical shell under uniform external pressure was found by Zoelly in 1915 [22], i.e.,

$$p^c = \frac{2E}{\sqrt{3\left(1-\nu^2\right)}}\left(\frac{h}{r}\right)^2. \tag{8.1}$$

This is a theoretical result and can be reproduced using the bifurcation analysis presented. But experimental results lead to different numbers.

Many experiments have been conducted on thin-walled spheres and spherical caps in laboratory conditions (see, for example, Chapter 4 in [9]). A summary of years of experiments was presented in [13] and is schematically shown in Figure 8.1. The vertical axis is the ratio between experimental pressure and that given by (8.1), while the horizontal axis contains the ratio between the radius of the sphere and the thickness.

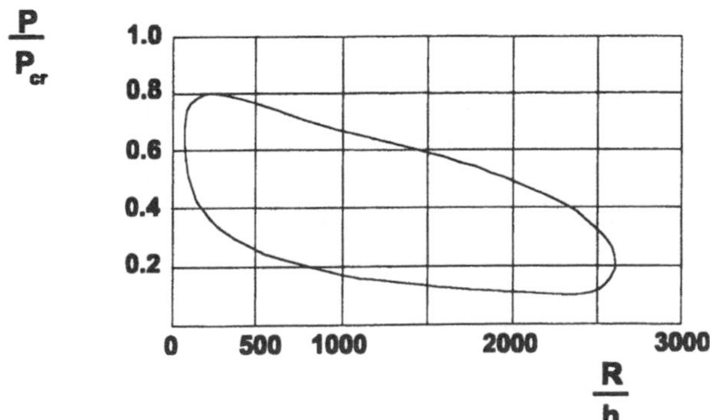

Figure 8.1 Experimental results for spherical shells under external pressure.

As in any comparison of theory versus experiments, we expect to have good agreement; however, this is not the case in Figure 8.1. The data come from many laboratories, where tests were performed on shells made of different materials (steel, araldite, copper, etc.), geometries (ratios between radius and thickness), and load systems, and there is a systematic lack of agreement. The experiments lead to values of buckling loads lower than the theory. Furthermore, there seems to be a pattern in the differences between theory and experiments: Thinner shells have lower ratios of experimental buckling loads with respect to theory than thicker ones.

What is worse, many other thin shells exhibited a similar disagreement: the cylinder under axial load, cylindrical panels, etc. The situation was even more confusing, since other structural types showed good agreement (plates, columns, etc.).

All these differences made it difficult to employ such unreliable theoretical results in design. For years, industries relied almost exclusively on experimental measurements to carry out design, and the theoretical results for perfect shells were only used for normalization of tests or as a reference value to apply high safety coefficients.

The stimulus to save the gap between theory and experiments was big, and years of research attempted to explain why one could not reproduce the high buckling loads given by the theory in laboratory conditions. Testing techniques were improved, and tests were done more carefully, but still there was an unexplained gap.

8.2.2 Experimental Buckling and Theoretical Postbuckling Paths

First, we notice that differences between computations and lab tests were important only in problems governed by bifurcation behavior. Sensitivity in arches is not very high, as will be clear from the examples in this chapter.

To understand the differences between bifurcation results from analytical models and experimental tests, we have to consider the postbuckling path of the bifurcation analysis.

- Problems in which the postbuckling path is ascending (i.e., stable symmetric bifurcation points) show good agreement between theory and experiments.
- Cases in which the secondary path develops at values of load lower than the bifurcation load (unstable symmetric bifurcation and asymmetric bifurcation) are found to be in disagreement with lab results. Here we remember that descending paths are unstable, so that cases in which there are unstable equilibrium states close to and below the critical bifurcation load display disagreement with tests.

A possible explanation is that if for some reason the system jumps from the stable primary path to an unstable secondary path, then the maximum load would be governed by the postbuckling path, not by the bifurcation load. This is illustrated in Figure 8.2.a and was discussed by several authors for some time. If this explanation was correct, then the actual load that the system can take is the minimum found along the secondary path. However, if there is no reserve along the postbuckling path, i.e., the minimum is zero, then the structure would be unable to take load, and this is contrary to what is found in experiments.

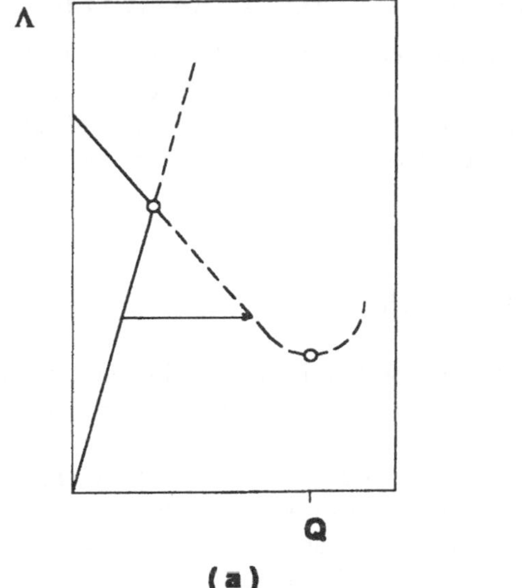

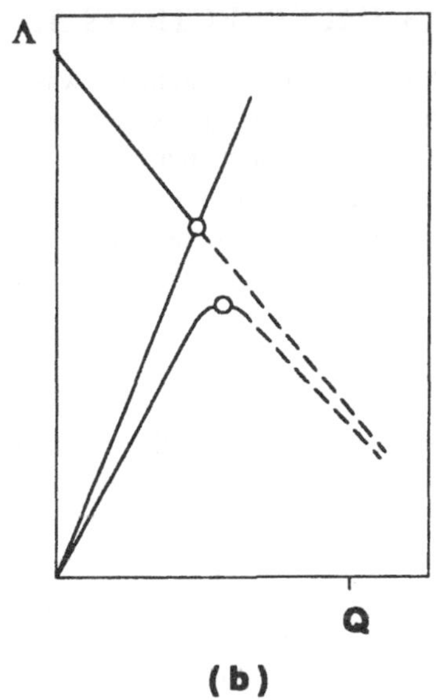

Figure 8.2 Possible explanations for discrepancies between theory and experiments: (a) jumps between primary and secondary equilibrium paths; (b) new equilibrium paths are generated as a consequence of small imperfections.

The above explanation received a great deal of attention in the 1960s for axially loaded cylinders. The original computations of [21] employed two degrees of freedom (using Fourier series terms) to model the postbuckling path, and the lowest point along this path was $\Lambda^{lower} = 0.34\Lambda^c$. But this result depends on the number of generalized coordinates employed in the analysis: Several authors increased the set of generalized coordinates and found that with 11 terms the value of $\Lambda^{lower} = 0.108\Lambda^c$ [1]. Finally, it was found that as the number of terms increases, the value of Λ^{lower} tends to zero. An interesting discussion of such results is presented in [9], according to which ''this put an end to the practice of representing the post-buckling behavior of isotropic cylinders by a single post-buckling curve'' (p. 112). Thus, a far more refined analysis is necessary to accurately predict the postbuckling path and extract conclusions about design loads of structures from them.

A second explanation is that some neglected effects become important when the secondary path is unstable. Such effects can be small geometric imperfections, small changes in the load, differences in boundary conditions, material inhomogeneities, small environmental effects, etc. The answer is that some of these small effects, which can be grouped as small imperfections, may have a large consequence on the maximum load that the system can attain, as illustrated in Figure 8.2.b. Some bifurcation buckling problems are extremely sensitive to the influence of even small changes in the geometry and load and generate families of new equilibrium paths for each imperfection considered. This will be highlighted by means of examples of bifurcations with small imperfections.

8.3 IMPERFECTIONS AS A SECOND CONTROL PARAMETER

A single control parameter was considered in the developments of Chapters 1 to 7 of this book, although a system has multiple response parameters (also called generalized coordinates, or displacement and rotations). We now take into account the influence of a second control parameter and investigate sensitivity of the stability problem with respect to this second parameter.

The energy V is written in the form

$$V = V\,[Q_j, \Lambda, \xi], \tag{8.2}$$

where (Q_j, Λ, ξ) define an energy surface. This surface is obtained by independent variations of the generalized coordinates Q_j, the load parameter Λ, and a new parameter ξ.

Out of many problems on the stability of a structure, which may be studied by means of a new parameter ξ, we shall consider here only two: the influence of imperfections and the influence of modifications in parameters of design. Thus we restrict our attention to problems in which the load Λ is the main control parameter, and ξ is a secondary control parameter that modifies the response. In general, the second control parameter ξ does not necessarily produce equilibrium paths in the physical sense.

An imperfection can be modeled by some space distribution and a scalar amplitude ξ, so that the new energy has the general form

$$V\,[Q_j, \Lambda, \xi] = V^{perf}\,[Q_j, \Lambda, 0] + \; V^{imp}\,[Q_j, \Lambda, \xi]. \tag{8.3}$$

In general, linear and quadratic terms in ξ may arise. We study two types of imperfections in the following sections: geometric and load imperfections.

8.3.1 Geometric Imperfections

The term "geometric" refers here to the position of the midsurface, the volume, or the boundary conditions of a thin-walled or slender structure. In general, the stiffness matrix does not exhibit an explicit dependence on geometric variables. The latter define the limits of the volume integrals, the coordinates of the midsurface, or are included in the kinematic boundary conditions.

In many engineering problems, the shape of the structure has significant deviations from an ideal, or as-designed geometry. The most common situations of this kind occur during the construction or fabrication of a structure, because of the difficulty of building something with the exact shape specified in the design. Another source of imperfection can be damage after the structure is completed.

To understand the structural significance of geometric imperfections, one must remember that smooth shells and thin-walled structures carry loads by means of a membrane mechanism, and only a marginal part of the load is equilibrated by bending action. But in zones of high changes in geometric curvature (as would occur due to local changes in the position of the midsurface), a local redistribution of stresses induces membrane as well as bending stresses. Furthermore, deviations in the geometry of the order of the thickness of a shell may induce additional membrane stresses of the same order as those acting in the perfect shell. This is the subject of [10].

There are two main possibilities to model a geometric imperfection:

- First, the midsurface may be redefined to include the imperfect shape. This is possible if the surface is represented by some discretization technique such as finite differences or finite elements. If we seek to obtain analytical or Ritz solutions, then the definition of a complex geometry makes it more difficult to obtain a good approximate solution.
- A second possibility is to model a geometric imperfection as an initial displacement field measured from the perfect geometry. This is explored in the following.

To include the geometric imperfection in V we must start again from the continuous formulation. The imperfection can be included in V by means of a simple concept of initial strains. In this chapter we show how to compute V in systems with imperfections and then employ the modified V, which already accounts for imperfections. These models may be found in [6, 10] and several others.

The coefficients present in a specific example depend on whether the structure is composed of rigid links (in which displacements fields include trigonometric functions), struts (which have no axial strains), or deformable structures, such as thin-walled structures (which exhibit both membrane strains and changes in curvature). Reference [4] presents an excellent account of rigid-link structures, and we concentrate here on deformable thin-walled structures.

Example 8.1 (Imperfect column under axial load). *To illustrate the use of an initial strain model, let us consider the midsurface of a slender column, as depicted in Figure* 8.3.

Let us assume that the deviations from the "ideal" or "perfect" geometry involve only displacements in the out-of-plane component w. These "initial displacements" define the position of the midsurface before the main load Λ is applied and are denoted by w^0. The displacements measured with respect to the perfect geometry, w^P, are

$$u^P = u,$$

$$w^P = w^0 + w. \tag{8.4}$$

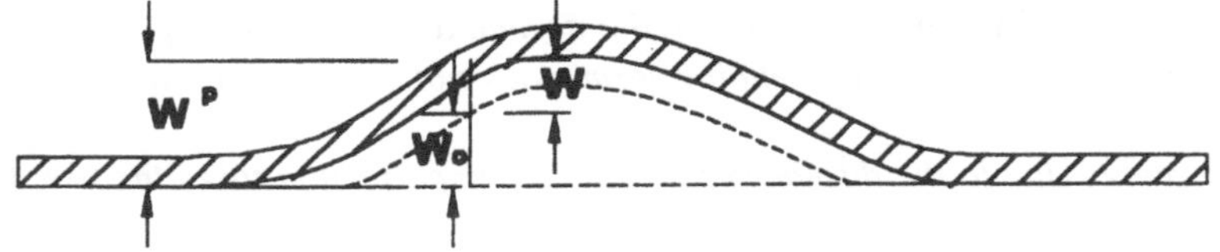

Figure 8.3 Geometric imperfection as initial displacement.

Following Donnell,

the net strains occurring under load can then be calculated as those which would be produced by a lateral deflection of $\left(w^0 + w\right)$ minus those which would be produced by a lateral deflection w^0 alone. [6, p. 349]

The strains due to w^P are

$$\varepsilon_x^P\left(w^0+w\right) = \frac{du}{dx} + \frac{1}{2}\left[\frac{d\left(w^0+w\right)}{dx}\frac{d\left(w^0+w\right)}{dx}\right]$$

$$= \frac{du}{dx} + \frac{1}{2}\left[\left(\frac{dw}{dx}\right)^2 + \left(\frac{dw^0}{dx}\right)^2 + 2\frac{dw^0}{dx}\frac{dw}{dx}\right].$$

The strains that would be produced by an initial lateral deflection w^0 alone are

$$\varepsilon_x^0\left(w^0\right) = \frac{1}{2}\left(\frac{dw^0}{dx}\right)^2.$$

Finally, the net strains due to the load result in

$$\varepsilon_x^P\left(w^0+w\right) - \varepsilon_x^0\left(w^0\right) = \left[\frac{du}{dx} + \frac{1}{2}\left(\frac{dw}{dx}\right)^2\right] + \left(\frac{dw^0}{dx}\frac{dw}{dx}\right). \tag{8.5}$$

The term in brackets is the usual strain in a perfect column. The new strain term is a function of the imperfection (or initial deflection w^0) coupled with the real displacement w

$$\bar{\varepsilon}_x = \left(\frac{dw^0}{dx}\frac{dw}{dx}\right).$$

The new term in the stress resultant N_x associated with $\bar{\varepsilon}_x$ becomes

$$\bar{N}_x = AE\left(\frac{dw^0}{dx}\frac{dw}{dx}\right).$$

It may be easily shown that, under the assumption of linear dependence on the out-of-plane displacement, the bending term does not lead to any new contribution in ξ.

The energy due to the displacements and the imperfection takes the new form

$$V\left(u, w, \Lambda, w^0\right) = V\left(u, w, \Lambda, 0\right) + V^{imp}\left(0, w, \Lambda, w^0\right),$$

where the term $V(u, w, \Lambda, 0)$ is the same energy computed for the perfect system, while the contribution V^{imp} is due to the imperfection in the geometry of the column. This latter contribution (see Problem 8.1) takes the form

$$V^{imp} = \frac{1}{2}AE\int_0^L \left[\frac{dw^0}{dx}\frac{dw}{dx}\frac{du}{dx} + \frac{dw^0}{dx}\left(\frac{dw}{dx}\right)^3\right] dx$$

$$+\frac{1}{2}AE\int_0^L \left(\frac{dw^0}{dx}\right)^2\left(\frac{dw}{dx}\right)^2 dx. \tag{8.6}$$

Notice that the first line in (8.6) is a linear function of w^0, while the second line depends on the square of w^0.

Let us consider the following approximation for the displacements and the imperfection

$$u = Q_1\frac{x}{L}, \qquad w = Q_2 \sin\frac{\pi x}{L},$$

$$w^0 = \xi \sin\frac{\pi x}{L},$$

in which ξ is the amplitude of the imperfection at the center of the column. Then (8.6) becomes

$$V^{imp}(Q_1, Q_2, \xi) = \frac{1}{2}AE\left[\left(\frac{\pi^2}{L^2}Q_1Q_2 + \frac{3}{8}\frac{\pi^4}{L^3}Q_2^3\right)\xi + \left(\frac{3}{8}\frac{\pi^4}{L^3}Q_2^2\right)\xi^2\right]. \tag{8.7}$$

If the amplitude of the imperfection is very small, i.e., $\xi \ll 1$, then the quadratic term in ξ can be neglected in comparison with the linear term. An important conclusion from this study is that for geometric imperfections the new energy contributions do not depend on Λ, but they are nonlinear in Q_j and in ξ.

8.3.2 Load Imperfections

Load imperfections can be included directly in the analysis without having to employ a model of initial strains as in the previous section. The usual way to model a load imperfection is by a new small load.

Example 8.2 (Column with eccentric axial load). *An example of load imperfection in an axially loaded column is an off-set e in the load P (Figure 8.4.b). This induces a new statically equivalent moment M with value*

$$M = P\,e.$$

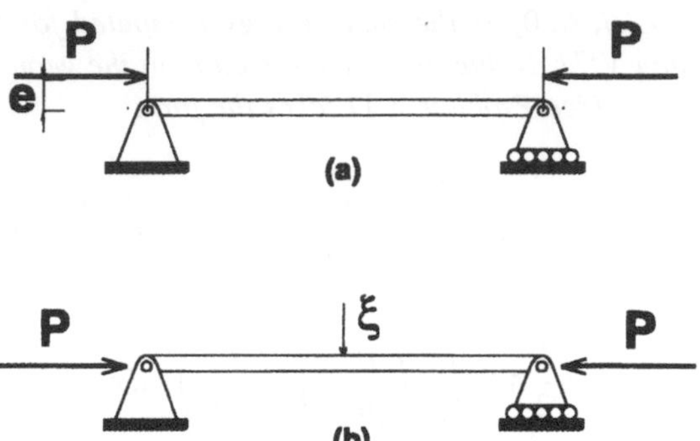

Figure 8.4 Various forms of load imperfections in columns: (a) load eccentricity; (b) small lateral load.

If the load is increased by means of a parameter Λ, *and the imperfection is controlled by a parameter* $\xi = e/L$, *then we get*

$$M(\Lambda, \xi) = (\Lambda P)(\xi L) = \Lambda \, \xi \, (P \, L).$$

A new load potential term is given by the product of the moment and the rotation at both ends of the column

$$V^{imp} = -2M \frac{dw}{dx}\Big|_{x=0} .$$

For a simply supported column under eccentric axial load, let us assume

$$w = Q_2 L \sin \frac{\pi x}{L}.$$

Then

$$V^{imp}(Q_2, \Lambda, \xi) = -(2\pi P \, L) \Lambda \xi \, Q_2. \tag{8.8}$$

Because of the presence of a load imperfection, the new term that has to be added to the energy contains both the load and the imperfection parameter and is linear in Q_j. *This does not occur in geometric imperfections.*

Example 8.3 (Circular arch with an eccentric point load). *A circular arch was studied in several chapters of this book, adopting the assumption of constant axial strain (see Example* 2.5 *in Chapter* 2). *Under the same assumptions, let us investigate the modifications in the load potential to include the influence of a load imperfection shown in Figure* 8.4.b.

A small eccentricity e in the point of application of the load induces a moment similar to what was seen in the column problem, and the load potential is given by

$$V^{imp} = -\Lambda (P \, e) \frac{dw}{R d\theta}\Big|_{\theta=0}. \tag{8.9}$$

The displacement w is given in terms of nondimensional degrees of freedom Q_1 and Q_2 as

$$w = Q_1 R \cos\left(\frac{\pi\theta}{2\theta_0}\right) + Q_2 R \sin\left(\frac{\pi\theta}{\theta_0}\right).$$

Then the load potential becomes, for $\xi = e/R$,

$$V(Q_1, Q_2, \Lambda, \xi) = V(Q_1, Q_2, \Lambda) + V^{imp}(Q_1, Q_2, \Lambda, \xi),$$

where the new term in the energy is given by

$$V^{imp}(Q_1, Q_2, \Lambda, \xi) = -\left(\frac{\pi}{\theta_0} PR\right) \Lambda \xi Q_2 = B_2 \Lambda \xi Q_2.$$

Again, we find that in the arch the new term in the energy due to the load imperfection is liner in Q_j and contains Λ. For the computations of the arch, the actual energy was normalized in Chapter 2 by multiplication with the factor $\frac{1}{EAR\theta_0}$.

8.3.3 Some Records of Structural Imperfections

It is very important to collect data regarding the imperfections that are usually found in real thin-walled structures. However, only one attempt has been made to create a data bank of imperfections recorded in one industry, and this is the aerospace industry. Very thin shells in astronautic applications buckle with imperfection-sensitive behavior. Researchers at the universities of Delft and Haifa recorded systematic information about geometric imperfections in metal cylindrical shells, and these constitute the basis of the International Initial Imperfection Data Bank [2]. To compare data from different sources the information is represented in terms of a double Fourier series

$$\frac{\xi}{h} = \sum_m \sum_n \sin\left(\frac{n\pi x}{L}\right)\left[C_{kl}\cos\left(\frac{my}{r}\right) + D_{kl}\sin\left(\frac{my}{r}\right)\right], \tag{8.10}$$

where C_{kl} and D_{kl} are the amplitudes in the cosine and sine components, and the summation extends to about 30 sine waves in the circumferential direction (summation in n), and about 5 half waves in the longitudinal direction (summation in m). All deviations are measured with respect to the so-called best-fit cylinder.

The use of Fourier series makes it easier to compare data from different shells. It is also possible to correlate imperfections with different fabrication techniques. Several shells were considered in [2], and different characteristics were found for shells made by machining and shells assembled from panels.

Imperfections have also been measured in the offshore industry. In compliant offshore structures there are large metal cylindrical shells that form part of the legs of the platforms, and these are subject to both initial imperfections and damage during their service life. Measurements of imperfections have been reported, for example, in [5]. Those shells were formed by an assembly of panels welded at three joints.

They had ring stiffeners located closer than the radius apart and stringer stiffeners in some parts. Such large tubular members are employed to provide buoyancy. The imperfections in this case show a clear influence of the number of panels used in the fabrication. More localized imperfections occur in the vicinity of the welds, some of which are clearly visible, leading to cusps toward the inside of the shell.

Cooling towers are a third type of construction in which there are records of imperfections. These are very large reinforced concrete structures, in the form of a thin-walled hyperboloid of revolution. Dimensions of such shells can be 150m in height, with thicknesses of the order of 0.25m. Records of imperfect towers became common after the collapse of several shells in England in 1965 and in 1973. Imperfections in this case cannot be described simply in geometric terms but should also include cracks in the concrete whenever they are present. A further complication arises if imperfections in the geometry grow with time. Measurements of imperfections and cracking have been reported, for example, in [7, 14]. In 1984 the Central Electricity Generating Board of Great Britain (*CEGB*) operated 139 large reinforced concrete cooling towers, of which 74 were 114m high. Following the failure of a cooling tower at Fiddlers Ferry, the *CEGB* investigated the imperfections in the towers and produced perhaps the most important assessments available in this field. Out of a number of towers surveyed, most appeared to have imperfections in the geometry and cracks. The cause of deformed shapes in cooling towers has been the subject of some recent debate [12, 11]. An example of an imperfection in a cooling tower is shown in Figure 8.5.

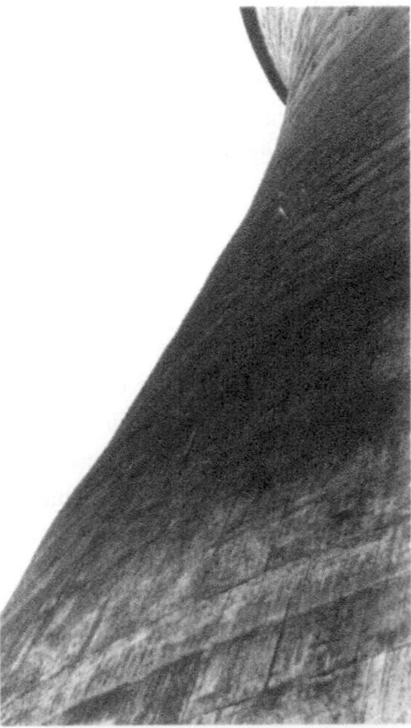

Figure 8.5 Geometric imperfection in a shell.

Large steel spherical shells are used in the nuclear industry and as gas containers. The spheres are assembled by welding carefully controlled panels; however, the fabrication process introduces small imperfections and, furthermore, damage can occur during the final construction of the shell [19].

This short review shows that the type of imperfection is associated with the fabrication process and varies from one industry to another. But all structural components have one thing in common: the very presence of imperfections. Because they are unavoidable, we must make sure that we understand well how they influence the buckling behavior of the structure, and this is the subject of the following sections.

8.4 EQUILIBRIUM AND STABILITY OF SYSTEMS WITH IMPERFECTIONS

It is assumed that an imperfection can be modeled by means of a single imperfection parameter ξ. Such imperfections can be associated with deviations from the as-designed geometry, with changes in the constitutive material, in the load system, or with other defects. In the limit, for $\xi = 0$, the imperfect system should reproduce the behavior of the perfect one.

The total potential energy results now in the form

$$V = V\left[Q_i, \Lambda, \xi\right] = V\left[Q_i, \Lambda, 0\right] + V^{imp}\left[Q_i, \Lambda, \xi\right]. \tag{8.11}$$

In general, V^{imp} may include linear and nonlinear terms in ξ. Similar to what we did with the load potential, for which we distinguished between the general system and the specialized system, we do the same for the imperfection potential. In a specialized system the imperfection potential is linear in ξ, while other cases are treated as a general system. Let us concentrate on the specialized system; in this case one can write

$$V\left[Q_i, \Lambda, \xi\right] = V\left[Q_i, \Lambda, 0\right] + \xi \frac{\partial V}{\partial \xi} \Big|_{\xi=0} . \tag{8.12}$$

If the derivative is denoted as

$$\frac{\partial V}{\partial \xi} \Big|_{\xi=0} = V^*, \tag{8.13}$$

then

$$V\left[Q_i, \Lambda, \xi\right] = V\left[Q_i, \Lambda, 0\right] + \xi V^*. \tag{8.14}$$

Notice that V^* is the part of the total potential energy that is due to the imperfection, for a unit value of imperfection amplitude ξ.

The equilibrium condition is

$$V_i = V_i\left[Q_i, \Lambda, 0\right] + \xi V_i^*. \tag{8.15}$$

If we choose to carry out a perturbation analysis, then an appropriate perturbation parameter should contain a major component in the eigenvector of the problem. For example, $s = Q_1$, if the eigenvector of the perfect system is normalized with $x_1 = 1$. The first-order perturbation equation is

$$\frac{d}{ds}(V_i) = 0 = V_{ij}\dot{Q}_j + V_i'\dot{\Lambda} + \xi V_{ij}^*\dot{Q}_j + \xi V_i^{*\prime}\dot{\Lambda}$$

or else

$$\left[V_{ij} + \xi V_{ij}^*\right]\dot{Q}_j + \left(V_i' + \xi V_i^{*\prime}\right)\dot{\Lambda} = 0. \tag{8.16}$$

The second-order perturbation equation is

$$\left[V_{ij} + \xi V_{ij}^*\right]\ddot{Q}_j + \left(V_i' + \xi V_i^{*\prime}\right)\ddot{\Lambda}$$
$$= -\left[V_{ijk} + \xi V_{ijk}^*\right]\dot{Q}_j\dot{Q}_k - 2\left[V_{ij}' + \xi V_{ij}^{*\prime}\right]\dot{Q}_j\dot{\Lambda}.$$

Solution of the above system leads to the approximation

$$Q_j(s) = Q_j^0 + s\dot{Q}_j + \frac{1}{2}s^2\ddot{Q}_j,$$

$$\Lambda(s) = \Lambda^0 + s\dot{\Lambda} + \frac{1}{2}s^2\ddot{\Lambda}.$$

The fundamental path becomes nonlinear even in cases in which the associated perfect system ($\xi = 0$) has a linear path. The only case in which an imperfection does not affect the system is when

$$V_i^* x_i^c = 0; \tag{8.17}$$

i.e., the imperfection does not have a component in the eigenvector of the perfect system.

As before, we assume that the path emerging from the origin is initially stable, until a critical point is reached. Critical stability of the system with imperfections can be investigated as follows: The second derivatives are

$$V_{ij} = V_{ij}\left[Q_i, 0, 0\right] + \Lambda V_{ij}'\left[Q_i, 0, 0\right] + \xi V_{ij}^*\left[Q_i, \Lambda, 0\right]. \tag{8.18}$$

For the perfect system $\xi = 0$ and

$$\left[V_{ij}\left[Q_i, 0, 0\right] + \Lambda^c V_{ij}'\left[Q_i, 0, 0\right]\right] x_j^c = 0.$$

For the imperfect system, we shall denote critical states by M, not C. Then the condition of critical state becomes

$$\left[V_{ij}\left[Q_i, 0, 0\right] + \Lambda^M V_{ij}'\left[Q_i, 0, 0\right] + \xi V_{ij}^*\left[Q_i, \Lambda^M, 0\right]\right] x_j^M = 0.$$

Notice that, in general, $\Lambda^M \neq \Lambda^c$, $x_j^M \neq x_j^c$, and $Q_i^M \neq Q_i^c$.

The computation of a nonlinear equilibrium path can also be done with a standard computer package for geometrically nonlinear analysis of structures. Each time we modify ξ there is a new nonlinear equilibrium path. We start by setting ξ to a given value and compute the equilibrium path in terms of increments of Λ similar to what we do in the perfect structure.

Example 8.4. *Consider an imperfection vector with the form of the eigenvector*

$$V_i^* = x_i^c.$$

Then

$$V\left[Q_i, \Lambda, \xi\right] = V\left[Q_i, \Lambda, 0\right] + \Lambda V_i'\left[Q_i, 0, 0\right] + \xi x_i^c.$$

First-order perturbation is

$$V_{ij}\dot{Q}_j + V_i'\dot{\Lambda} + \xi\, \dot{x}_i^c = 0.$$

If we choose $x_1 = 1$ and $s = Q_1$, then

$$\dot{x}_i^c = \begin{Bmatrix} 1 \\ 0 \\ 0 \\ \cdots \\ 0 \end{Bmatrix} \qquad \textit{and} \qquad \ddot{x}_i^c = 0.$$

The second-order perturbation equation is

$$V_{ij}\ddot{Q}_j + V_i'\ddot{\Lambda} = -V_{ijk}\dot{Q}_j\dot{Q}_k - 2V_{ij}'\dot{Q}_j\dot{\Lambda}.$$

The terms that contain ξ do not have Λ, so that the condition of critical state reduces to

$$\left\{V_{ij}\left[Q_i, 0, 0\right] + \xi V_{ij}^*\left[Q_i, 0, 0\right] + \Lambda^M V_{ij}'\left[Q_i, 0, 0\right]\right\} x_j^M = 0.$$

The imperfection has the effect of changing the stiffness matrix of the structure but does not affect the load-geometry matrix $V_{ij}'\left[Q_i, 0, 0\right]$.

8.5 NONLINEAR ANALYSIS OF LIMIT POINT SYSTEMS WITH IMPERFECTIONS

Let us consider several examples of nonlinear analysis for a circular arch under a central point load.

Example 8.5 (Arch with geometric deviation). *In this example the geometry of the circular arch is modified so that the distance between supports remains the same, but the radius R, the central angle θ_0, and the elevation H are modified. The resulting arch is still a circular arch, for which we write*

$$H = H_0(1 + \xi), \tag{8.19}$$

where H_0 is the elevation of the arch geometry investigated, and ξ describes the imperfection. Notice that the imperfection has the same form of the eigenvector for limit point. Other relevant geometric parameters are

$$R = \frac{H^2 + L^2}{2H}, \qquad \cos(\theta_0) = 1 - 2\frac{H^2}{H^2 + L^2}.$$

A detailed expression of the total potential energy is given in Problem 8.3. *An important feature is that the shape of the imperfection is the same as the shape of the buckling mode. In this case, if we compute derivatives with respect to ξ, we find that*

$$\frac{\partial^2 V}{\partial Q_i \partial \xi} \mid^c = \dot{V}_i x_i \mid^c \neq 0;$$

that is, the imperfection vector $\dot{V}_i$ is not orthogonal to the eigenvector at the limit point.

We compute the equilibrium path as a quadratic approximation (the exact path is cubic) using $L = 9.345$, $H_0 = 2.4$, and $A = 0.00774$. Figure 8.6.a *shows that as H increases, there is also an increase in the maximum load at the limit point.*

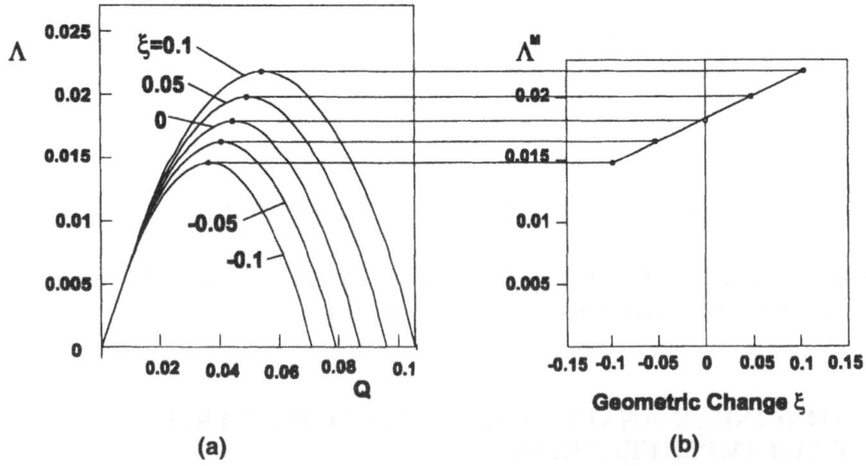

Figure 8.6 An arch with geometric imperfection (a) nonlinear equilibrium paths; (b) imperfection sensitivity.

Instead of looking at the complete path, it is interesting to concentrate just on the values of the limit load (the maximum of each curve in Figure 8.6.a) *as a function of the amplitude of the imperfection,* ξ. *This is plotted using circles in Figure* 8.6.b *and is known as an imperfection sensitivity diagram. The limit load changes almost linearly with increasing amplitude of imperfection.*

Example 8.6 (Circular arch with load imperfection). *We continue investigating the behavior of a circular arch under a point load, and we take into account the influence of a small eccentricity in the point of application of the load. This produces a load imperfection and can be modeled as described in Example* 8.3. *Notice that the primary equilibrium path with the load centered at the apex has displacements only in* Q_1, *with* $Q_2 = 0$. *With an imperfection the fundamental path is no longer in the* $\Lambda - Q_1$ *plane, and displacements* Q_2 *occur from the beginning of the loading process.*

The equilibrium equations, up to quadratic terms, become

$$V_1 = A_1\Lambda + A_{11}Q_1 + \frac{1}{2}A_{111}Q_1^3 + \frac{1}{2}A_{122}Q_2^2 = 0,$$

$$V_2 = B_2\Lambda\xi + A_{22}Q_2 + A_{122}Q_1Q_2 = 0,$$

where

$$B_2 = -\pi\frac{e}{R\theta_0}.$$

The A coefficients were given in Example 2.5 (*Chapter* 2), *and the* B_2 *coefficient was given in Example* 8.3.

An exact solution of the two quadratic equations leads to a set of values as a function of ξ. *The geometry of the arch considered is given by* $R = 19.4$, $I = 0.000249$, $A = 0.00774$, *and* $\theta_0 = 0.5026 rad$. *As a reference eccentricity we adopted* $e = R/50$. *In Figure* 8.7 *we have plotted the equilibrium paths projected on the* $\Lambda - Q_1$ *plane for different values of* ξ *ranging from* 0 *to* 1.

The load path changes slightly, and the maximum load is reduced by the influence of the imperfection. The maximum load is still reached in the form of a limit point, but it occurs for values of Q_1 *larger than in the perfect system.*

In the present example, the load imperfection cannot increase the maximum load, and for all values of ξ, *positive and negative, yields a reduction in the maximum load.*

We anticipate that in this problem the following product is zero:

$$\dot{V}_i x_i \mid^c = 0.$$

Although it is possible to compute the nonlinear equilibrium path each time the system is modified, either due to imperfections or due to a change in the design, such computations are expensive. Thus, one can attempt to compute the sensitivity curve having evaluated only once the nonlinear equilibrium path of the perfect system. This is the subject of Chapter 10.

We computed sets of nonlinear equilibrium paths for one example, i.e., the circular arch, and found that even when imperfections or design changes are included, the

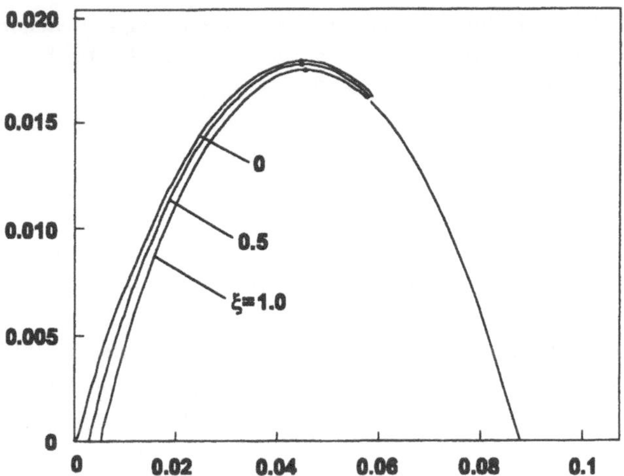

Figure 8.7 Equilibrium paths for load imperfection in an arch.

nature of the behavior is the same. This means that the path still has a limit point, although the actual load and displacements at this limit state are different. Such conclusion is general for the limit point and can be obtained for other structures and imperfections. This is very important, because then we know what we are looking for: a family of curves all showing limit points.

8.6 NONLINEAR ANALYSIS OF BIFURCATION SYSTEMS WITH IMPERFECTIONS

In this section we consider one example of each type of bifurcation (stable symmetric, unstable symmetric, and asymmetric bifurcation) discussed in Chapter 7 and examine the influence of small imperfections as generators of new nonlinear equilibrium paths.

Example 8.7 (The axially loaded column with geometric imperfection). *An imperfection in the geometry of an axially loaded column was discussed in Example* 8.2, *and the expression for the energy was formulated there. The coefficients of V with two degrees of freedom (axial shortening Q_1 and lateral deflection Q_2) are given in explicit form in Example* 8.2 *and section* 5.6.

Equilibrium of the imperfect column takes the form

$$V_1 = A_1'\Lambda + A_{11}Q_1 + \frac{1}{2}A_{122}Q_2^2 + \frac{1}{2}B_{12}Q_2 = 0, \tag{8.20}$$

$$V_2 = A_{22}Q_2 + A_{122}Q_1Q_2 + \frac{1}{6}A_{2222}Q_2^3 + \frac{1}{2}B_{12}Q_1 + \frac{1}{2}B_{222}Q_2^2\xi + C_{22}Q_2\xi^2 = 0, \tag{8.21}$$

where

$$B_{12} = \frac{\pi^2}{L^2} AE, \qquad B_{222} = \frac{9}{8}\frac{\pi^4}{L^3} AE, \qquad C_{22} = \frac{3}{8}\frac{\pi^4}{L^3} AE.$$

Analytical solution of the two equations is achieved as follows: The value of Q_1 *can be obtained from* (8.21) *and substituted in* (8.20) *to get*

$$\Lambda = \frac{P}{AE}$$

$$= -\frac{\pi^2}{2}\left(\xi Q_2 + \frac{1}{2} Q_2^2\right) + \frac{\pi^2 \left(9 Q_2^2 \xi + 6\xi^2 Q_2 + 3 Q_2^3 + 8 \frac{I}{AL^2} Q_2\right)}{8\,(\xi + Q_2)}. \tag{8.22}$$

Unlike the perfect column, the imperfect column does not have a primary path with zero lateral displacements. The path emerges from zero load with $Q_2 \neq 0$ *and is nonlinear with the load. Results for a specific case (a thin composite layer) have been computed and are plotted in Figure* 8.8 *for several values of the amplitude of imperfection, ranging from* $0 < \xi < h/3$. *The data assumed is* $b = 0.1$, $h = 0.003$, $E = 3858.6$, $A = 3 \times 10^{-5}$, *and* $I = 2.25 \times 10^{-11}$.

The primary path is bent from the onset of the loading process and displays a decrease in the stiffness. For small values of ξ *the path approaches the perfect case and departs most significantly in the vicinity of the bifurcation state. But there is no bifurcation in the imperfect path: The equilibrium states are uniquely defined along the single fundamental path. The stiffness increases as this path gets closer to the perfect postbuckling path, so that the column with imperfections shows no loss of stability.*

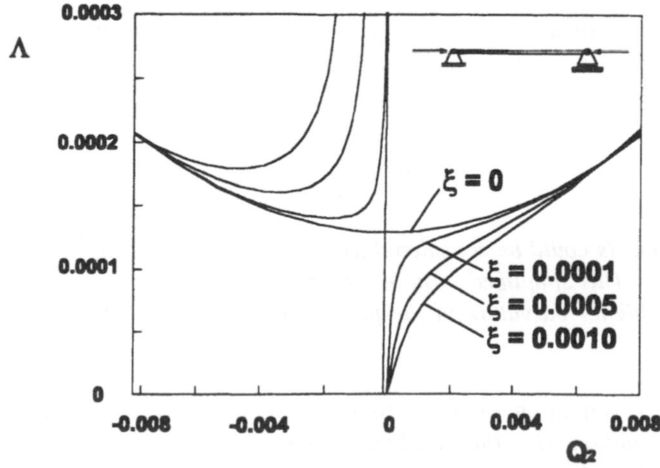

Figure 8.8 Equilibrium paths of a column with geometric imperfections.

What is also surprising in the behavior of the column is the presence of a new equilibrium path for each imperfection amplitude, which does not cross the fundamental path and does not start from the origin. Thus, there is no way to reach this path if we increase the load or the displacement from the origin of the graph. This is known as **complementary path**, and we notice that it attempts to follow the other branch of the postcritical path in the perfect system and the unstable part of the primary path for loads higher than the bifurcation load. If we investigate stability of such a path we find that some of the states are stable (those close to the stable secondary path of the perfect system) and some are unstable (close to the primary path in its unstable part). The combination of the two paths, the fundamental and the complementary paths, approaches the primary and secondary paths of the perfect system in a bifurcation. The only way to observe part of the complementary path is to load the system at loads higher than the minimum of the path, and force a displacement in the opposite direction as it deflects, to reach the complementary path. Once there, the system will stay in equilibrium.

Example 8.8 (Column, simplified analysis). *The nonlinear imperfect path of the column may be simplified, because the postbuckling path of the perfect structure was shown to be fairly flat. The imperfect behavior can be linearized so that it becomes asymptotic to the horizontal postbuckling path. One can start from* (8.22)*, and eliminate nonlinear terms in* Q_2 *and in* ξ*. Proof of this linearization is the subject of Problem* 9.6*. The result is*

$$Q_2 + \xi - \frac{\Lambda^c}{\Lambda} Q_2 = 0. \tag{8.23}$$

Equation (8.23) *can also be written as*

$$\Lambda = \Lambda^c \frac{Q_2}{Q_2 + \xi}, \tag{8.24}$$

which is the approximate version of (8.22)*. Notice that the dependence between* Λ *and* Q_2 *is nonlinear. Finally, the relation between the displacements and the amplitude of the imperfection is often written as*

$$Q_2 = \frac{\Lambda}{\Lambda^c - \Lambda} \xi. \tag{8.25}$$

The displacements could be considered as an amplification of the imperfection as the load increases. This simplified form of the path is adequate for the column; however, systems with higher curvature in the postbuckling path cannot be simplified in this way.

Example 8.9 (Column on an elastic foundation). *Our second case is the column on an elastic foundation. The stiffness of the springs is given as before (section* 7.9*) in the form* $k = K\,(L/\pi)^4\,/EI$*; the load parameter is* $\Lambda = PL/\left(EIL^2\right)$*; and two degrees of freedom are included in the analysis, i.e.,*

$$w = Q_1 L \sin\left(\frac{\pi x}{L}\right) + Q_2 L \sin\left(\frac{2\pi x}{L}\right).$$

We investigate here the system with $k = 2$ for which the lower critical state in the perfect system is an unstable symmetric bifurcation.

To make the system imperfect, a small load P_2 acting transverse to the column is included. This leads to a new term in the total potential energy of the system given by

$$\bar{V} = -\xi Q_1, \tag{8.26}$$

where

$$\xi = \frac{4}{\pi^4}\frac{L^2}{EI}P_2, \tag{8.27}$$

and the energy V has been normalized through multiplication by a factor $4L/\left(\pi^4 EI\right)$.

The equilibrium conditions are

$$V_1 = 2(1 + k + \Lambda) Q_1 + \pi^2\left(1 - \frac{3}{4}\Lambda\right) Q_1^3 + 2(10 - 3\Lambda) Q_1 Q_2^2 - \xi = 0,$$

$$V_2 = 2(16 + k - 4\Lambda) Q_2 + 4\pi^2 (16 - 3\Lambda) Q_2^3 + 2(10 - 3\Lambda) Q_2 Q_1^2 = 0.$$

The solution of the fundamental path is

$$Q_2 = 0,$$

$$\Lambda = 4\frac{\pi^2 Q_1^3 + 2Q_1(1 + k) - \xi}{Q_1\left(8 + 3\pi^2 Q_1^2\right)}.$$

The plots for different values of imperfection amplitude are shown in Figure 8.9. *Again, we find a nonlinear fundamental path, but now this has a maximum and then the load drops for increasing displacements. The behavior is similar to a limit point. For displacements with opposite signs, there are complementary equilibrium paths. As the imperfection is small, the fundamental and complementary paths approach the solution of the perfect case.*

For each imperfection amplitude there is a clear maximum starting from the unloaded state, and this is the maximum load that the imperfect column can take. This value decreases as the imperfection increases.

Example 8.10 (The two-bar frame). *Our final example of nonlinear behavior of a system with imperfections is the asymmetric bifurcation in a two-bar frame. This problem was investigated originally by Roorda* [15], *and was previously studied in Chapters 5 and 7. The imperfection is chosen as an eccentricity of the load, and the additional term in the energy is*

$$V^{imp} = -\Lambda \xi L Q_2.$$

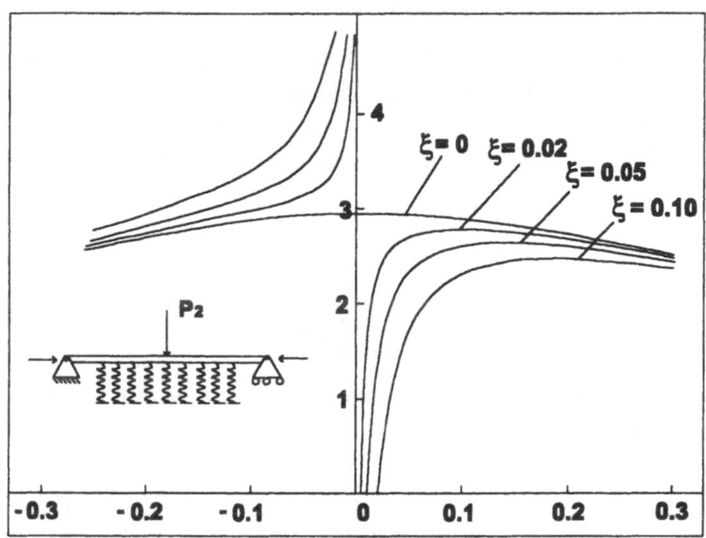

Figure 8.9 Column on an elastic foundation. Curve (1) $\xi = 0$, (2) $\xi = 0.02$, (3) $\xi = 0.05$, (4) $\xi = 0.10$.

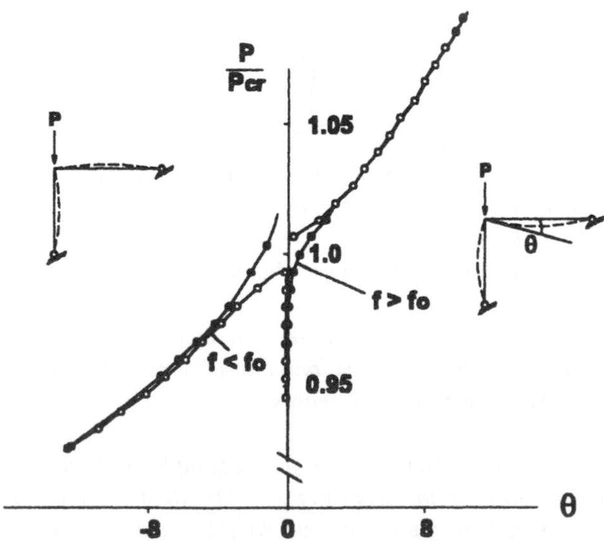

Figure 8.10 Two-bar frame. Reproduced with permission of the American Society of Civil Engineers from a work by Roorda [15].

Rather than showing theoretical results, for this case we reproduce in Figure 8.10 *experimental values obtained by Roorda over 30 years ago* [15]. *There is an equilibrium path with a maximum on the side of the unstable postbuckling path, while for the stable postbuckling part the fundamental path of the imperfect system is always ascending. One should worry about the sensitivity to imperfections that pushes the system toward the unstable side. As imperfections are unknown a priori, one should assume the worst case in the analysis and expect the system to be sensitive to imperfections.*

8.7 THE SOUTHWELL PLOT

In 1932, Southwell presented an interesting method of extracting buckling loads in imperfect columns [20]. In this section we discuss the basics of the method and the application in understanding experiments.

To illustrate the Southwell plot we can start from (8.22), the load versus lateral displacement Q_1 of an imperfect column under axial load. We proceed to linearize the equation, neglecting terms with ξ^2 and with Q_1^2, and get, after some algebraic manipulation, the following relation (see Problem 8.10):

$$(\xi + Q_2) - \Lambda^c \left(\frac{Q_2}{\Lambda}\right) = 0. \tag{8.28}$$

The above relation can be plotted with the displacement Q_2 as the abscissa and Q_2/Λ in the vertical axis, as shown in Figure 8.11. Following the linearized derivation of the equations leading to (8.28), this should be a straight line in which the slope gives the critical load, and the horizontal value at zero load is the amplitude of the imperfection.

We can now test a structure with unknown imperfections and obtain data for load and displacements at successive load steps. The data can be represented in a Southwell plot, and the best straight line fitted to the data available, as shown in Figure 8.11 from experiments of [17]. The slope is the buckling load of the real structure, while the value of displacement at the origin is the imperfection in the form of an equivalent displacement. Notice that a load imperfection would also be represented in the Southwell plot as a geometric imperfection.

Some limitations of the Southwell plot have been shown, for example, in [17]. First, the plot is useful for small deflections of the system; however, in real experiments the results have to be measured for large values of displacements in order to have observable data. Second, the plot assumes an imperfect equilibrium path with an asymptotic value at the critical load level, while real experimental data departs from the asymptotic value. This means that the experimental values do not lie on a straight line (as shown in Figure 8.11), not because of matters of accuracy, but because the relation in the Southwell plot should not be linear in most cases. Thus, a warning is made in [17]:

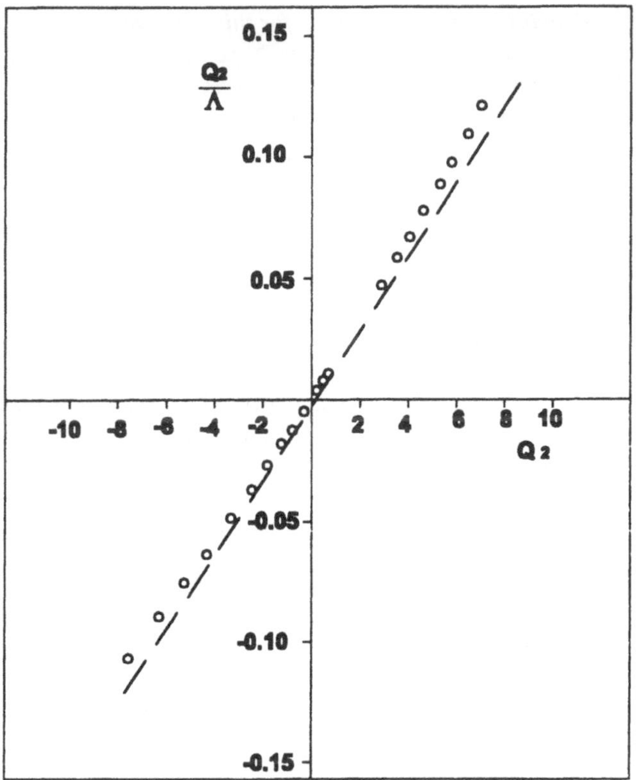

Figure 8.11 Southwell plot for a Warren truss. Reprinted with permission of the American Society of Civil Engineers from a work by Roorda [17].

If the Southwell plot is applied to structures other than the column, great care must be taken in the interpretation of the results. Correct interpretation of the Southwell plot for such structures depends upon a knowledge of the post-buckling behavior of the idealized structure.

8.8 PROBLEMS

Review questions. (*a*) What is the difference between design sensitivity and imperfection sensitivity? (*b*) Give examples of geometric and load imperfections. (*c*) How is an imperfection modeled via an initial strain? (*d*) What is a complementary path? (*e*) Explain ways of observing experimentally a complementary path. (*f*) Describe the Southwell plot. (*g*) When does a Southwell plot yield a good approximation of the critical load?

Problem 8.1 (Theory). Show how to obtain (8.6).

Problem 8.2 (Circular ring). For a circular ring with radial pressure directed to the center, consider a geometric imperfection with a shape given by the lowest buckling mode. Obtain the energy for the imperfect system.

Problem 8.3 (Circular arch). The circular arch of Example 2.5 has a shape imperfection in the form of the mode for the limit point. Find the energy for this case.

Problem 8.4 (Plane frame). A steel frame is formed by two bars of equal length $L = 4m$ and properties given by $E = 200GPa$, $I = 2.9 \times 10^{-4} m^4$, and $A = 0.0193 m^2$. The central angle between the two members is 130°. The members are pin supported at the ends, and there is a rigid connection between them. A load P is applied at the connection between members, with direction at equal angles of the members. Consider one finite element for each bar of the frame. Find the equilibrium paths for eccentric load in which the eccentricity is $1/50$ of the length L. You may neglect in-plane deformations.

Problem 8.5 (Plane frame). Repeat Problem 8.4 for the frame with a geometric imperfection in the form of the fundamental path.

Problem 8.6 (Plane frame). This is the same as Problem 8.4, but with an imperfection in the form of a load applied to the center of one of the members in a direction perpendicular to it.

Problem 8.7 (Column with geometric imperfection). For the imperfect column with a geometric imperfection, compute the fundamental and complementary paths. Use the following data for a steel column: $E = 200GPa$, $I = 2.9 \times 10^{-4} m^4$, $A = 0.0193 m^2$, and $L = 4m$.

Problem 8.8 (Column with load imperfection). Repeat Problem 8.7 for a load imperfection. Consider a small load acting at the center of the column and perpendicular to it.

Problem 8.9 (Column with load imperfection). Plot the nonlinear fundamental path and the complementary path for the column under axial load with a load imperfection. Use data as in Figure 8.8. Consider an eccentricity of the load P, and imperfections $\xi = 0.0$, $\xi = 0.0001$, $\xi = 0.0005$, $\xi = 0.0010$.

Problem 8.10 (Southwell plot). Show that (8.28) can be obtained from (8.22) by linearization.

Problem 8.11 (Southwell plot). Discuss what shape is expected from a Southwell plot for (*a*) stable symmetric bifurcation; (*b*) unstable symmetric bifurcation; (*c*) asymmetric bifurcation. Use [17] to further explore this question in more detail.

Problem 8.12 (Circular ring). The total potential energy of an elastic ring with an imperfection in the geometry was found in Problem 8.2. Obtain the equilibrium paths of the imperfect ring. The radius is $200mm$; the cross section is rectangular with $5mm$ height and $30.5mm$ width. The material is duraluminum, and according to the manufacturer the modulus of elasticity is $E = 70GPa$.

Problem 8.13 (Column). A simply supported column of length L has an axial load P. The column is assumed to have bending and extensional strains. Consider the influence of a small load p uniformly distributed along the length. Compute the equilibrium paths for different values of p. Because this is a stable symmetric bifurcation, sensitivity analysis is not important, but the path itself is relevant. Use the following data: $L = 3000mm$; $E = 17,245 N/mm^2$; $A = 819.35 mm^2$; $I = 2.78 \times 10^5 mm^4$.

Problem 8.14 (Shallow shell). Formulate the terms due to a geometric imperfection for a shallow shell in Cartesian coordinates. Consider the membrane strains

$$\varepsilon_{ij} = \frac{1}{2}\left(\frac{\partial u_i}{\partial x_j} + \frac{\partial u_j}{\partial x_i}\right) + \frac{1}{2}\left(\frac{\partial u_3}{\partial x_i}\frac{\partial u_3}{\partial x_j}\right), \qquad \chi_{ij} = \frac{\partial^2 u_3}{\partial x_i \partial x_j}$$

for $i, j = 1, 2$. Consider initial displacements only in u_3.

Problem 8.15 (Column with nonlinear bracing). In Problem 7.10 we computed postcritical behavior of a column with a single bracing element located at $L/3$ of one of the supports. Evaluate the equilibrium paths including the influence of a small eccentricity in the load.

Problem 8.16 (Experiment). Use a displacement-controlled machine for testing aluminum commercially available cans (used for soft drinks). Carry out the tests on 10 cans under axial compression, and read the maximum load reached by the system before buckling. Plot the load for each experiment in a single graph and discuss the imperfection sensitivity of the results.

Problem 8.17 (Rectangular plate). Include the influence of a geometric imperfection with the shape of the lowest critical mode in an isotropic aluminum plate with rectangular shape under biaxial compression. The sides of the plate are simply supported and have dimensions $610mm \times 460mm$, thickness $h = 0.5mm$, $E = 70GPa$, and $\nu = 0.3$.

8.9 BIBLIOGRAPHY

[1] Almroth, B. O., Post-buckling behavior of axially compressed circular cylinders, *AIAA J.*, 1, 630–633, 1963.

[2] Arbocz, J., Shell stability analysis: Theory and practice, in *Collapse, The Buckling of Structures in Theory and Practice*, J. M. T. Thompson and G. W. Hunt, Eds., Cambridge University Press, Cambridge, UK, 1983.

[3] Creative Pultrusions, *Design Guide*, Creative Pultrusions Inc., AlumBank, PA, 1990.

[4] Croll, J. G. A., and Walker, A. C., *Elements of Structural Stability*, Macmillan, London, 1972.

[5] Dwight, J. B., Imperfection levels in large stiffened tubulars, in *Buckling of Shells in Off-shore Structures*, J. E. Harding et al., Eds., Granada Publishing, London, 1982, 393–412.

[6] Donnell, L. H., *Beams, Plates, and Shells*, McGraw-Hill, New York, 1976.

[7] Ellinas, C. P., Croll, J. G. A., and Kemp, K. O., Cooling towers with circumferential imperfections, *J. Structural Division, ASCE*, 106, 2405–2423, 1980.

[8] Erlande-Brandenburg, A., *Cathedrals and Castles: Building in the Middle Ages*, Thames and Hudson, London, 1993.

[9] Esslinger, M., and Geier, B., *Post-Buckling Behavior of Structures*, Springer-Verlag, Wien, 1975.

[10] Godoy, L. A., *Thin-Walled Structures with Structural Imperfections: Analysis and Behavior*, Pergamon Press, Oxford, 1996.

[11] Godoy, L. A., Discussion of "Cause of deformed shapes in cooling towers", *J. Struct. Engrg., ASCE*, 122, 220–221, 1996.

[12] Jullien, J. F., Aflack, W., and L'Huby, Y., Cause of deformed shapes in cooling towers, *J. Struct. Engrg., ASCE*, 120(5), 1471–1488, 1994.

[13] Kollar, L. and Dulacska, E., *Buckling of Shells for Engineers*, Wiley, Chichester, UK, 1984.

[14] Pope, R. A., Grubb, K. P., and Blackhall, J. D., Structural deficiencies of natural draught cooling towers at U.K. power stations, Part 2: Surveying and structural appraisal, *Proc. Inst. Civil Engineers, Structures and Buildings*, 104, 11–23, 1994.

[15] Roorda, J., Stability of structures with small imperfections, *J. Engrg. Mech. Div., ASCE*, 91, 87–105, 1965.

[16] Roorda, J., The buckling behavior of imperfect structural systems, *J. Mech. Phys. Solids*, 13, 267, 1965.

[17] Roorda, J., Some thoughts on the Southwell plot, *J. Engrg. Mech. Div., ASCE*, 93, 37–48, 1967.

[18] Roorda, J., On the buckling of symmetric structural systems with first and second order imperfections, *Internat. J. Solids Structures*, 4, 1137, 1968.

[19] Sanchez Sarmiento, G., *Analisis Tensional de Recipientes de Presion Esfericos*, Ph.D. thesis, Balzeiro Institute, Bariloche, Argentina, 1993.

[20] Southwell, R. V., On the analysis of experimental observations in problems of elastic stability, *Proc. Roy. Soc. London Ser. A*, 135, 601, 1932.

[21] von Karman, T., and Tsien, H. S., The buckling of thin cylindrical shells under axial compression, *J. Aeronautical Sci.*, 8, 303–312, 1941.

[22] Zoelly, R., *Uber ein Knickungsproblem an der Kugelschale*, Ph.D. thesis, Zurich, 1915 (in German).

CHAPTER

NINE

INFLUENCE OF NONLINEAR MATERIAL BEHAVIOR ON BUCKLING STATES

9.1 INTRODUCTION

In the developments of Chapters 1 to 8 it was assumed that the constitutive material of the structure is elastic until elastic buckling occurs. In many cases this is true, especially in very thin walled members, which tend to buckle elastically from a configuration with small displacements. But in less optimized structures the critical state is reached at loads for which the material may have plasticity or some other form of failure.

The term **plastic buckling** is usually employed when buckling occurs ''at loads for which some or all of the structural material has been stressed beyond its proportional limit'' [2]. However, there are different phenomena that fall under the definition of plastic buckling. Reviews of the literature in this field were written by Sewell in 1972 [20], Hutchinson in 1974 [15], and Bushnell in 1981 [2].

To understand the different scenario that is present when plasticity occurs, it is useful to recall the nonlinear behavior of elastic structures.

- *Perfect structural systems*. Three stages were found during the loading process: first, there is a primary or fundamental equilibrium path, either linear or nonlinear; second, there is a critical state, which could be stable or unstable; and, third, there is a secondary or postcritical equilibrium path, which is nonlinear.
- *Imperfect structural systems*. Whenever small imperfections are included in the analysis, then the primary equilibrium path is nonlinear, as seen in Chapter 8. This equilibrium path follows the perfect primary path for small loads, it follows the postcritical path for large displacements, and it has the most significant differences with the path in the perfect structure for loads close to the elastic critical load.

To include the influence of plasticity in the analysis, we shall consider three possibilities:

- There are structural systems for which the material has plastic deformations along the fundamental path and well below the elastic critical load. This is illustrated in Example 9.1. The problem is dominated by the material behavior, and the role of elastic instability is simply to induce deflections larger than in a linear fundamental path, due to the influence of imperfections and geometric nonlinearity.
- If plastic deformations occur in the secondary equilibrium path, then the structure first experiences elastic instability and then plasticity. The problem is now dominated by elastic instability, and plasticity limits the postcritical load reserve of the structure. As the occurrence of plasticity is far away from the critical state, then imperfections may be marginally important and serve only to refine the analysis, because perfect and imperfect equilibrium paths tend to get close away from the critical load. This is shown in Example 9.2.
- Finally, there are structural systems for which plasticity and elastic instability both occur at approximately the same load level. There may be a strong interaction between both effects, and the phenomenon should be considered as **elastoplastic instability**. In an imperfect system and close to an elastic critical load, there are significant deflections and high stress levels, and this couples with plasticity as the stresses reach the yield surface for the material. This form of buckling is sensitive to imperfections. Example 9.3 shows this case.

Different approaches to consider the influence of plasticity on elastic instability are discussed in section 9.2. The study is then restricted for systems with stable post-buckling behavior in section 9.3. A column is taken as theme structure, and different types of behavior are illustrated for different column lengths. Section 9.4 is about the influence of plasticity in systems with unstable critical states. This includes unstable symmetric bifurcation, asymmetric bifurcation, and limit-point behavior. Two further examples are discussed in sections 9.5 and 9.6: They are a circular plate, for which two-dimensional yield surfaces are necessary, and a thin-walled composite column with local buckling, for which first-ply failure is identified. Section 9.7 contains a brief discussion of buckling in the design of structures.

9.2 PLASTICITY AS A CONSTRAINT IN ELASTIC BUCKLING

Several approaches can be followed to account for the influence of plasticity in the buckling process.

9.2.1 Computational Approach Including Nonlinear Material and Geometric Models

The analysis of instability under elastoplastic conditions requires a model with both nonlinear kinematic relations and nonlinear material behavior. According to [2], ''accurate prediction of critical loads corresponding to either mode [nonlinear collapse and bifurcation buckling] requires a simultaneous accounting for moderately large deflections and nonlinear, irreversible, path-dependent material behavior.''

Imperfections should also be included, and the solution can be obtained by means of step-by-step algorithms. Because of the material behavior with plasticity, the total potential energy cannot be used in the formulation, and virtual work may be an alternative.

A further warning is often made in the literature: ''The simultaneous presence of moderately large deflections and non-linear path dependent material behavior leads to a multitude of possibilities for misusing sophisticated state of the art computer programs created to explore plastic buckling.'' This is further termed as a ''frequent mismatch between the computer program and the user'' [2].

Problems that are frequently encountered in computational elastoplastic instability are the following:

- The solution fails to converge, mainly due to the choice of a large load increment.
- The solution may converge to the wrong equilibrium state, due to a coarse mesh, an inadequate model of the material behavior, or the algorithm cannot change to a new path because it is always attracted by another path.
- Excessive computer time due to the choice of a very small load increment.

Such analysis is general, but even for present day computational facilities it is expensive. A computational approach can only be done for some special structures, and in the final stages of design. Details about computational strategies for coupled material and geometric nonlinearity are beyond the scope of this book and may be seen, for example, in [4].

9.2.2 Material Behavior as a Constraint

In many buckling problems we acknowledge that plasticity does not occur until large deformations are reached. Thus, one could also treat separately the nonlinear kinematic problem from the nonlinear material problem and use plastic as a constraint of the elastic instability problem. The total potential energy can still be used to evaluate the elastic instability problem, and plasticity (or material failure) can be considered as a constraint in the analysis.

This approach is attractive because the complete formulation for elastic instability can still be employed and can help as a guide to the expected behavior. In this chapter we follow this second approach; i.e., plasticity is considered as a constraint of the elastic instability problem.

Equilibrium of an elastic path was seen to satisfy the condition

$$\boxed{V_i[Q_j, \Lambda, \xi] = 0 \qquad \Longrightarrow \qquad \text{equilibrium}} \tag{9.1}$$

In the following, it is assumed that the material is elastic until a yield surface is reached. We shall not attempt to explain here details of the formulation of plasticity theory, and this is found in a number of good books [19, 11, 17, 8].

Let the components of the stress tensor, for a given point in the structure, in principal coordinates be $[\sigma_I, \sigma_{II}, \sigma_{III}]$. A yield surface is defined by a condition of the form

$$\boxed{\mathcal{F}\left(\sigma_I, \sigma_{II}, \sigma_{III}, C_y\right) = 0 \qquad \Longrightarrow \qquad \text{yield condition}} \tag{9.2}$$

where C_y is some experimental parameter (or set of parameters) to identify plasticity in the material. In one-dimensional problems, C_y may be the yield stress of the material in a uniaxial test, but in more complex (two- or three-dimensional) stress states we have a more elaborate definition to specify when plasticity develops. There are several criteria useful for yield surfaces in metal structures, such as the criteria of Tresca, von Mises, etc. Other materials do not have plasticity, but they have some form of failure that may be associated with it. This is the case of composite materials, for which the criteria (Tsai-Hill, Tsai-Wu) identify first-ply failure or last-ply failure [21].

Two important aspects should be mentioned here:

- Elastic instability is a global phenomenon, in the sense that it is a property of the complete structure, not of its parts. Plasticity, on the other hand, is local: It occurs for a fiber at some point of the thickness in a thin-walled structure.
- The yield surface is written in terms of principal stresses; however, the energy V is described by generalized coordinates (displacements or rotation).

It is no longer sufficient to have the energy in terms of Q_j, and one must also look at the mechanics of the problem to compute the stresses. This can be done as follows: From the definition of the displacement field one can compute strains using kinematic relations and stresses using constitutive equations. Thus, the principal stresses result in

$$\sigma_I = \sigma_I(Q_j, \Lambda), \qquad \sigma_{II} = \sigma_{II}(Q_j, \Lambda), \qquad \sigma_{III} = \sigma_{III}(Q_j, \Lambda). \tag{9.3}$$

Such relations for principal stresses can next be introduced in the yield surface $\mathcal{F}$, leading to

$$\mathcal{F}(Q_j, \Lambda, C_y) \neq 0. \tag{9.4}$$

One can then solve the equilibrium equations and verify if the yield condition is reached at each load level using (9.2). For example, in a two-dimensional problem in which the criterion of Tresca is employed as yield surface, the stress state may be a point A in Figure 9.1.a.

But one can also evaluate how far from the yield condition is the current stress state associated with Q_j. This is indicated by the distance $\overline{AA'}$ in the space of principal stresses. Similar information can be plotted in the equilibrium space by defining

$$\Lambda^P = \frac{\overline{OA'}}{\overline{OA}} \Lambda\left(Q_j^A\right), \tag{9.5}$$

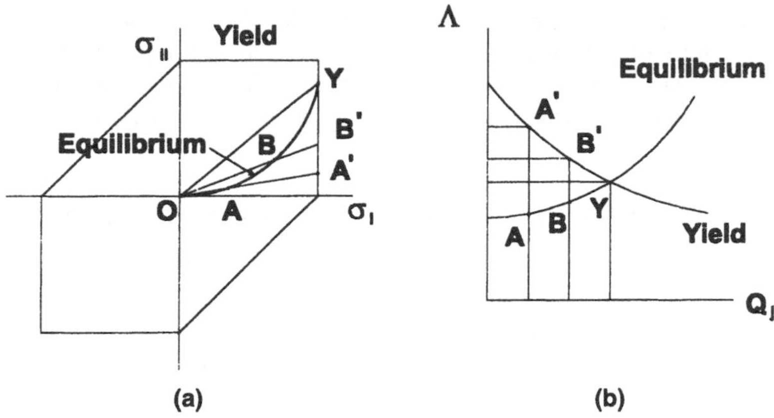

Figure 9.1 (a) Space of principal stresses; (b) space of equilibrium path.

where Λ^P is the load required to reach yield for a given displacement amplitude. In general, $\Lambda^P \neq \Lambda^A$, so that a curve different from the equilibrium path is generated. This locus of Λ^P for different values of Q_j is plotted in the load-displacement space of Figure 9.1.b. Clearly, this is not an equilibrium path and only gives indication about how far a given equilibrium state is from yield. For $Q_j = 0$ the value of Λ^P is maximum.

If we increase the displacement by an increment in Q_j, the new stress field is along a different line, because the kinematic equations are nonlinear. This would lead to point B in Figure 9.1.a. Now the yield surface is closer. Finally, there is a displacement field Q_j^Y for which the stresses are on the yield surface, and the new load is $\Lambda^P = \Lambda^Y$. Notice that for this point both the equilibrium and the yield paths are coincident, and the structure will have **first yield** at a fiber.

The curve $OABY$ in Figure 9.1.a is the equilibrium locus in the space of principal stresses, while the curve $OA'B'Y$ is the yield curve in the equilibrium space.

In summary, there are several ways to identify first yield in buckling problems using plasticity as a constraint:

- to follow the equilibrium path and consider if the yield criteria is satisfied at each point,
- to draw the yield curve in the equilibrium space and look at the intersection with the equilibrium path,
- to draw the equilibrium path in the space of principal stresses and identify the intersection with the yield surface.

Notice the following points.

- The yield surface in the space of principal stresses is a property of the material; it does not change for each structure.
- The yield locus in the equilibrium space (as defined here) is not a property of the material but also depends on the equilibrium path to which it is associated.

The above discussion referred to the first yield of a fiber, but it could also be extended to **first full plasticity** of a section. This leads to a new curve in the load-displacement space, which limits states along the equilibrium path for which the first section becomes plastic. This type of analysis is presented, for example, in [7].

9.3 SYSTEMS WITH STABLE POSTBUCKLING BEHAVIOR

In this section we consider buckling problems in which the critical state itself is stable, i.e., stable symmetric bifurcations. The theme structure is a column with simple supports under axial load. The perfect geometry was studied in section 7.8, and the equilibrium paths for the imperfect column were derived in Example 8.7.

9.3.1 Perfect Systems with Plasticity in the Secondary Path

First, consider the case in which plasticity occurs along the secondary path. There is no need to include imperfections in a first approximation, so that the secondary path is

$$P(s) = AE\left(\frac{\pi}{L}\right)^2\left[1 + \frac{1}{2}\lambda^{(2)}s^2\right], \tag{9.6}$$

$$Q_1^s(s) = Q_1^c + \frac{1}{2}q_1^{(2)}s^2, \tag{9.7}$$

where $s = Q_2$, the out-of-plane displacement, and

$$\lambda^{(2)} = \frac{A}{I}\left(\frac{L}{2}\right)^2, \qquad Q_1^c = \frac{P^c}{AE}, \qquad q_1^{(2)} = \frac{3}{4}\pi^2.$$

For a given value of P, one can obtain the associated postbuckling displacements Q_j. The displacement field results in

$$u = xQ_1^s, \qquad w = Q_2^s L\sin\left(\frac{\pi x}{L}\right). \tag{9.8}$$

The strains result in

$$\varepsilon = \frac{du}{dx} + \frac{1}{2}\left(\frac{dw}{dx}\right)^2 = \frac{I}{A}\left(\frac{\pi}{L}\right)^2 + s\pi\cos\left(\frac{\pi x}{L}\right) + s^2\frac{3}{8}\pi^2,$$

$$\chi = \frac{d^2w}{dx^2} = -s\frac{\pi^2}{L}\sin\left(\frac{\pi x}{L}\right). \tag{9.9}$$

The stress resultants are

$$N_x = AE\varepsilon, \qquad M_x = EI\chi.$$

At the central section one has $x = L/2$, and the maximum stress is

$$\sigma = \frac{N_x}{A} + \frac{M_x}{W_e} = E\left[\frac{I}{A}\left(\frac{\pi}{L}\right)^2 - \frac{\pi^2 I}{W_e L}s + s^2\frac{3}{8}\pi^2\right], \tag{9.10}$$

where W_e is the elastic modulus of the cross section, $W_e = I/c$ and c is the distance from the neutral axis to the extreme fibers. The maximum stress σ in the central section is a function of the perturbation parameter of the secondary path s.

How far the stresses are from first plasticity is given by

$$\rho = \frac{\sigma}{\sigma_y}. \tag{9.11}$$

The load required to produce σ is Ps, but it has been identified here by the perturbation parameter s. Thus,

$$P^y = \frac{P^s}{\rho}. \tag{9.12}$$

For each value of Q_j there is a load P^s, a ratio ρ, and a load necessary to reach plasticity P^y. The value of P^y results in

$$P^y = A\left(\frac{\pi}{L}\right)^2 \frac{\left[1 + \frac{1}{2}\frac{A}{I}\left(\frac{L}{2}\right)^2 s^2\right]}{\left[\frac{I}{A}\left(\frac{\pi}{L}\right)^2 - \frac{\pi^2 I}{W_e L}s + s^2\frac{3}{8}\pi^2\right]}\sigma_y. \tag{9.13}$$

9.3.2 Imperfect Systems

Problems with initial imperfections were modeled in Chapter 8, leading to a nonlinear fundamental equilibrium path. The procedure of analysis with plasticity as a constraint is very similar to what was shown in the previous section; however, the equilibrium path in (9.6) should be defined in a different way to account for the nonlinear fundamental path. For example, if the column has a geometric imperfection w_0, then we assume

$$w_0 = \xi L \sin\left(\frac{\pi x}{L}\right), \tag{9.14}$$

so that ξ represents a deviation measured as a fraction of the length L of the column. From Example 8.7 we get the path as

$$\Lambda = \frac{P}{AE} \tag{9.15}$$

$$= \frac{\pi^2}{2}\left[-\left(\xi Q_2 + \frac{1}{2}Q_2^2\right) + \frac{\left(9Q_2^2\xi + 6\xi^2 Q_2 + 3Q_2^3 + 8\frac{I}{AL^2}Q_2\right)}{4(\xi + Q_2)}\right].$$

The procedure now follows toward the evaluation of σ, ρ, and P^y.

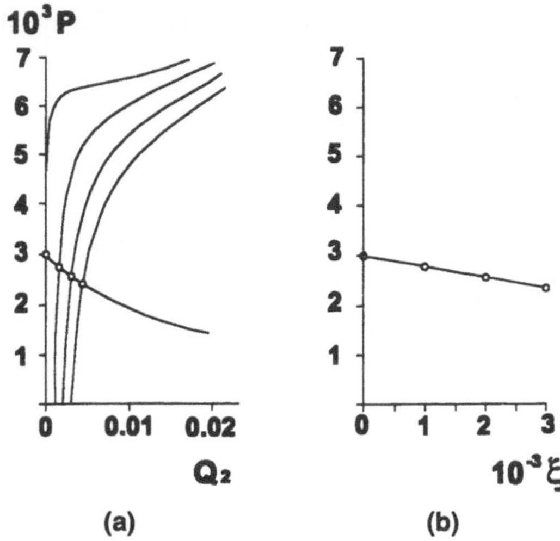

Figure 9.2 Example of a short column, including first yield (a) equilibrium paths; (b) imperfection sensitivity.

The influence of plasticity in systems with stable critical states is further discussed for several examples of columns with different slenderness ratio but the same stress-strain curve.

Example 9.1. *Consider a steel column with section* $W200 \times 52$, *i.e.,* $h = 21.45$, $b = 52$, $d = 206$, $A = 6650$, $I = 52.7 \times 10^6$, $W_e = 511650$, $E = 200$, $\sigma_y = 0.45$, *all values in mm and kN. It is assumed that buckling occurs on the strong axis and in the form of global buckling (not local buckling of flanges). The behavior depends on the length L of the column, as will be shown by the computations. First, we consider* $L = 4000$, *a short column, and geometric imperfections for which the maximum amplitude is* ξ.

Several imperfections have been introduced in the analysis, namely, $\xi = 0.0001L$, $\xi = 0.001L$, $\xi = 0.002L$, *and* $\xi = 0.003L$. *The imperfect paths are plotted in Figure* 9.2, *and the curve of first yield intersects the imperfect paths well below the elastic critical load. For this short column, plasticity occurs before elastic buckling. Figure* 9.2.b *shows the sensitivity of the first yield locus, and the variation is linear with* ξ. *Imperfection sensitivity is moderate.*

First plasticity initiates a material nonlinear behavior, with unloading on part of the structure. The structure will eventually reach a maximum load. To evaluate such a load, a reduced modulus approach has been followed, in which a critical load is determined with the same Euler formula but with a different value of the modulus of elasticity.

Example 9.2. *For the same cross section of Example* 9.1 *and the same material properties, consider a long column with* $L = 9000mm$. *The elastic buckling load in this case is reduced so that plasticity is only possible for equilibrium states in the*

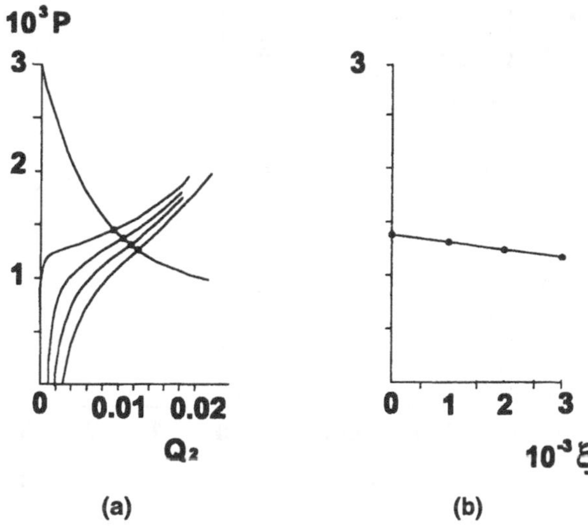

Figure 9.3 Example of a long column, including first yield (a) equilibrium paths; (b) imperfection sensitivity.

postcritical path. This is shown in Figure 9.3.a. *Sensitivity of this problem is again linear, as shown in Figure* 9.3.b.

Example 9.3. *An intermediate length column with* $L = 6000mm$ *has a very different behavior from Examples* 9.1 *and* 9.2. *Both plasticity and elastic instability occur for approximately the same load level, as shown in Figure* 9.4.a. *Even for very small imperfections* $\xi = 0.0001L$ *there is a drop from the maximum load of the perfect structure. This drop is due to the high imperfection sensitivity in the elastic range, which makes a difference between the perfect and imperfect columns. The imperfection sensitivity of first yield is in Figure* 9.4.b: *It is high, with a sharp change for very small imperfections.*

From the above examples we conclude that the influence of imperfections is a secondary effect in the evaluation of first fiber plasticity of either short or long columns, but it is of primary importance whenever plasticity and elastic bifurcation occur for nearly the same loads. Such systems are said to have sensitivity of first yield to imperfections.

There have been many attempts to derive a simplified formula similar to the Euler buckling load but including plasticity (for a review, see, for example, [20]). Of course, this is of utmost practical importance, since the ultimate load of the column is given by a condition of material failure. The most accepted model is based on the concept of **tangent modulus**, and the load at which the structure buckles plastically is given by

$$P^c = \left(\frac{\pi}{L}\right)^2 I \; E_t,$$

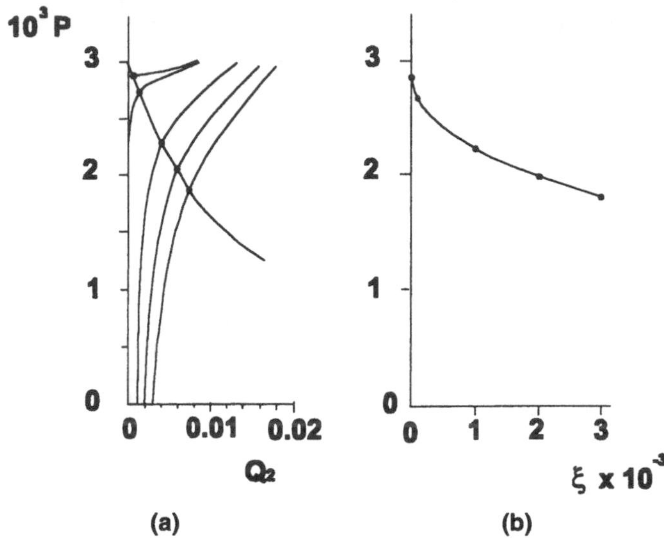

Figure 9.4 Example of an intermediate length column, including first yield (a) equilibrium paths; (b) imperfection sensitivity.

where E_t is the tangent modulus. The tangent modulus is computed as the slope of the stress-strain diagram in the plastic range, so that it is a function of the stress level. Design recommendations are usually given using some form of tangent modulus approach, and excellent reviews of design information are given in [10, 9].

9.4 SYSTEMS WITH UNSTABLE POSTBUCKLING BEHAVIOR

A column on an elastic foundation was seen in Chapter 7 to have unstable postbuckling behavior for some relations between the stiffness of the column and the foundation.

The displacements are modeled as $w = Q_1 L \sin(\pi x/L) + Q_2 L \sin(2\pi x/L)$, and there is an imperfection with amplitude ξ. The energy has the form

$$V = \frac{\pi^2 EI}{4L}\left[(1 + k - \Lambda)Q_1^2 + (16 + k - 4\Lambda)Q_2^2 + \left(\frac{\pi}{2}\right)^2\left(1 - \frac{3}{4}\Lambda\right)Q_1^4 \right.$$
$$\left. + \pi^2(16 - 3\Lambda)Q_2^4 + \pi^2(10 - 3\Lambda)Q_1^2 Q_2^2 - \xi Q_1\right].$$

The general procedure follows the same as in the column problem of stable postbuckling behavior in the last section. Details of the calculations are omitted, and results are next shown to illustrate the behavior.

Example 9.4 (Column on elastic foundation). *Identify first yield for a steel column resting on an elastic foundation, simply supported at the ends and loaded axially.*

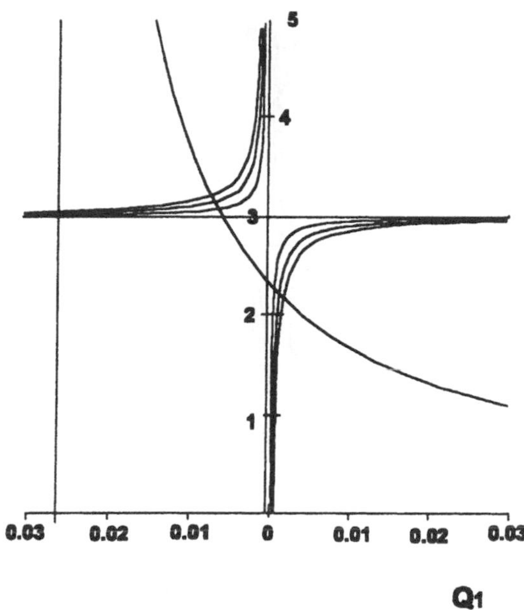

Figure 9.5 Example of column on elastic foundation including first yield, $k = 2$, $L = 9000$.

The cross section is a $W200 \times 52$, *with* $A = 6650$, $d = 206$, $I = 52.7 \times 10^6$, $L = 9000$, $E = 200$ *(all units are mm and kN). Let us consider a load normalized as* $P = \Lambda P^{Euler}$, *where* $P^{Euler} = EI(\pi/L)^2$. *The stiffness of the foundation is* $K = k(\pi/L)^2 P^{Euler}$, *and we adopt* $k = 2$. *The yield stress of the material is* $\sigma_y = 0.45$.

The results are plotted in Figure 9.5 *for imperfection amplitudes* $\xi = 0$, 0.001, 0.002, *and* 0.003. *The plot is very similar to Example* 9.1, *mainly because plastic buckling of the structure occurs well before the elastic buckling load.*

Example 9.5 (Column on elastic foundation). *We repeat Example* 9.4 *for* $k = 3$, $L = 12000$. *The results have been plotted in Figure* 9.6 *and show that the material behavior is important only for large deflections of the structure due to elastic buckling.*

Example 9.6 (Column on elastic foundation). *The same data of Example* 9.4 *is again studied, but now for* $k = 2$, $L = 12000$. *Results are shown in Figure* 9.7. *This time the phenomenon of elastic instability and first yield of the material occur at approximately the same load level, so that the problem is imperfection sensitive. Both nonlinear geometric and nonlinear material behavior are important in this case. The results show that not only the length of the column, but also the relative stiffness between the column and the foundation are relevant to determine when first yield occurs.*

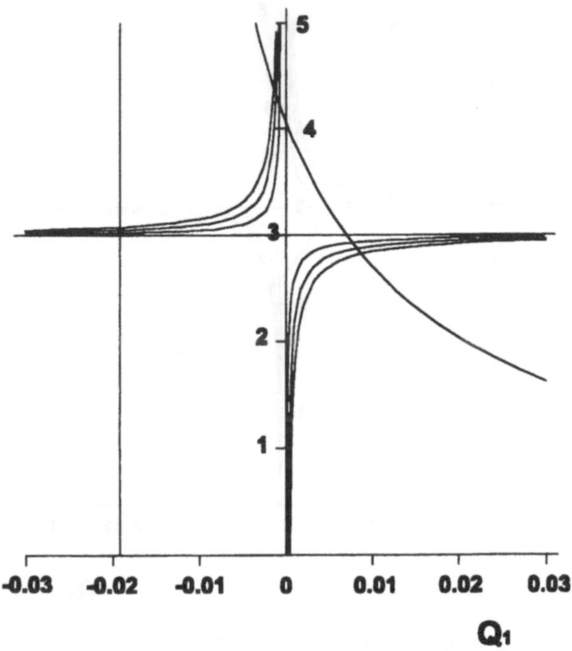

Figure 9.6 Example of column on elastic foundation including first yield, $k = 3$, $L = 12000$.

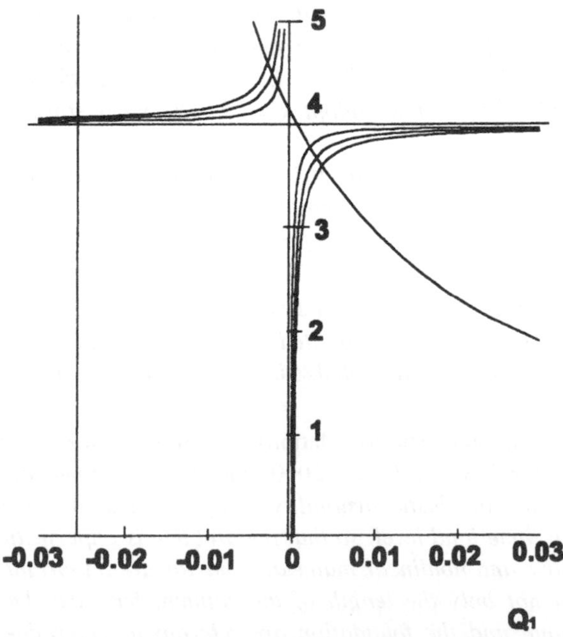

Figure 9.7 Example of column on elastic foundation including first yield, $k = 2$, $L = 12000$.

9.5 CIRCULAR PLATE

A circular plate under radial load and simply supported at the circular boundary has been studied in several chapters of this book, using a simplified two-degrees-of-freedom model. The displacements are assumed as

$$w(r) = Q_1 \cos\left(\frac{\pi r}{2R}\right), \qquad u(r) = \frac{r}{R} Q_2, \tag{9.16}$$

where R is the radius of the plate and r is the radial coordinate. It was shown in Example 2.5 that the energy results in the form

$$\begin{aligned} V[Q_1, Q_2, \Lambda] = \pi \Bigg\{ & -2R\Lambda Q_2 + \frac{1}{2}\left[\frac{1}{2}\pi^2 (1.191 + \nu) \frac{D}{R^2} Q_1^2 + 2C(1+\nu) Q_2^2\right] \\ & + \frac{1}{6}\left[3\frac{\pi^2 + 4}{8R} C(1+\nu)\right] Q_1^2 Q_2 \\ & + \frac{1}{24}\left[\frac{3}{8}\left(\frac{\pi}{4}\right)^2 \left(3\pi^2 + 16\right) \frac{C}{R^2}\right] Q_1^4 + \pi \Lambda \xi e Q_1 \Bigg\}. \end{aligned} \tag{9.17}$$

A new term is now added with respect to the energy in Chapter 2, namely, the presence of an eccentricity e in the load Λ, for which the energy term is

$$V = 2\pi R \Lambda e \left(\frac{dw}{dr}\Big|_{r=R}\right) = \pi^2 R \Lambda \xi Q_1,$$

where $\xi = e/R$. If we perform a perturbation analysis using the displacement Q_1 as perturbation parameter, then the equilibrium path is defined as

$$\begin{aligned} Q_1(s) &= s, \\ Q_2(s) &= Q_2^E + s\,\dot{Q}_2 + \frac{1}{2} s^2 \ddot{Q}_2, \\ \Lambda &= \Lambda^E + s\,\dot{\Lambda} + \frac{1}{2} s^2 \ddot{\Lambda}. \end{aligned} \tag{9.18}$$

With the results for the equilibrium path, one obtains the displacement field using (9.16). The strains are

$$\begin{aligned} \varepsilon_r &= \frac{\partial u}{\partial r} + \frac{1}{2}\left(\frac{\partial w}{\partial r}\right)^2, \qquad & \varepsilon_\theta &= \frac{u}{r}, \\ \chi_r &= \frac{\partial^2 w}{\partial r^2}, \qquad & \chi_\theta &= \frac{1}{r}\frac{\partial w}{\partial r}. \end{aligned} \tag{9.19}$$

The stress and moment resultants can be obtained from the constitutive equations for a linear, homogeneous, and elastic material

$$N_r = \frac{Eh}{1-\nu^2}(\varepsilon_r + \nu\varepsilon_\theta), \qquad N_\theta = \frac{Eh}{1-\nu^2}(\varepsilon_\theta + \nu\varepsilon_r),$$

$$M_r = \frac{Eh^3}{12(1-\nu^2)}(\chi_r + \nu\chi_\theta), \qquad M_\theta = \frac{Eh^3}{12(1-\nu^2)}(\chi_\theta + \nu\chi_r). \tag{9.20}$$

The stresses are next computed in the form

$$\sigma_r = \frac{N_r}{A} \pm \frac{M_r}{W_e}, \qquad \sigma_\theta = \frac{N_\theta}{A} \pm \frac{M_\theta}{W_e}.$$

These are principal stresses, and to verify how far they are from yield it is necessary to employ a yield criterion. For example, for von Mises criterion the effective stresses are

$$\sigma_{effective} = \frac{1}{\sqrt{2}}\left[(\sigma_r - \sigma_\theta)^2 + (\sigma_r)^2 + (\sigma_\theta)^2\right].$$

The distance from the actual stresses to the yield stresses is given by

$$\rho = \frac{\sigma_{effective}}{\sigma_y}.$$

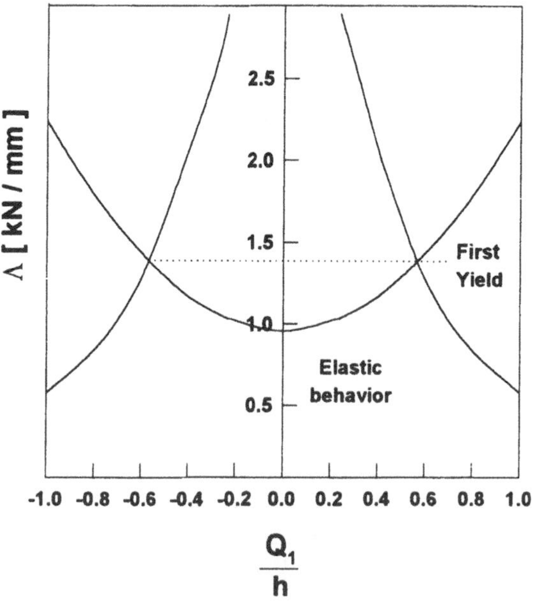

Figure 9.8 Circular plate including first yield.

The ratio in terms of the radial loads Λ is given by

$$\Lambda^y = \frac{\Lambda}{\rho},$$

where Λ^y is the load necessary to reach plasticity for a displacement $Q_j(s)$. It is then possible to plot the value of the load Λ^y versus the displacement in the same graph with the equilibrium equation of the problem. This is the plasticity constraint of the buckling problem.

Example 9.7. *A plate has $R = 100mm$, $h = R/20$, and is made with steel with $E = 200GPa$ and $\nu = 0.3$. Results for the perfect plate are plotted in Figure 9.8. For $\sigma_y = 0.001E$, and von Mises yield criterion, the plate first buckled elastically and then reached first yield. Notice that the dimensions were chosen so that the plate is not so thin, and even for this case elastic buckling dominates the behavior.*

9.6 FIRST-PLY FAILURE OF A COMPOSITE COLUMN

The theme structure in this section is an I-section column made with a composite fiber-reinforced polymer, as reported in [1]. Thin-walled columns may also have local buckling of flanges, and cases in which material yield or failure occur in the postbuckling region are discussed in this section.

The first step in the analysis is the computation of the equilibrium paths using elastic constitutive properties. The material has been modeled using classical lamination theory. The generalized coordinates are based on a Ritz approximation including the axial displacement q_1, the rotation of the flanges q_3, and the lateral deflection of the column q_2. Along the fundamental path only axial displacements are present.

The postbuckling equilibrium paths of such a column have been evaluated using elastic studies leading to a stable symmetric bifurcation. The postbuckling path is obtained using perturbations with a few degrees of freedom to account for rotation of the cross section and shortening of the column.

However, how far the structure is from the maximum load it can attain is not given by such eigenvalue calculations. A failure criteria is needed to compute more realistic maximum loads and thus guide and improve the material performance. The most frequently used criteria for first-ply failure in composites are extensions of similar yield criteria for isotropic materials and include the maximum stress, the maximum strain, and the quadratic criteria. The extension of the Hill criterion to predict the onset of failure in an orthotropic lamina was suggested by Tsai, and the resulting equation is often referred to as the Tsai–Hill criterion [21, 16], i.e.,

$$\frac{\sigma_1^2}{S_L^2} - \frac{\sigma_1\sigma_2}{S_L^2} + \frac{\sigma_2^2}{S_T^2} + \frac{\tau_{12}^2}{S_{LT}^2} = 1. \tag{9.21}$$

In (9.21) the stresses in a lamina are σ_1, σ_2, and τ_{12}, and the strengths of a unidirectional fabric ply should be determined from tests: the longitudinal compressive strength

S_L, the transverse compressive strength S_T, and the longitudinal shear strength S_{LT}. In order to avoid failure, the left-hand side of (9.21) must be less than 1, and failure is predicted if the result is ≥ 1. The failure surface generated by this equation is an ellipsoid. Such an ellipsoid may be nonsymmetric because tension and compression failures in composites do not have the same values.

For laminated composites, the procedure to identify first yield is more complex than in structures made with isotropic materials. From the generalized coordinates,

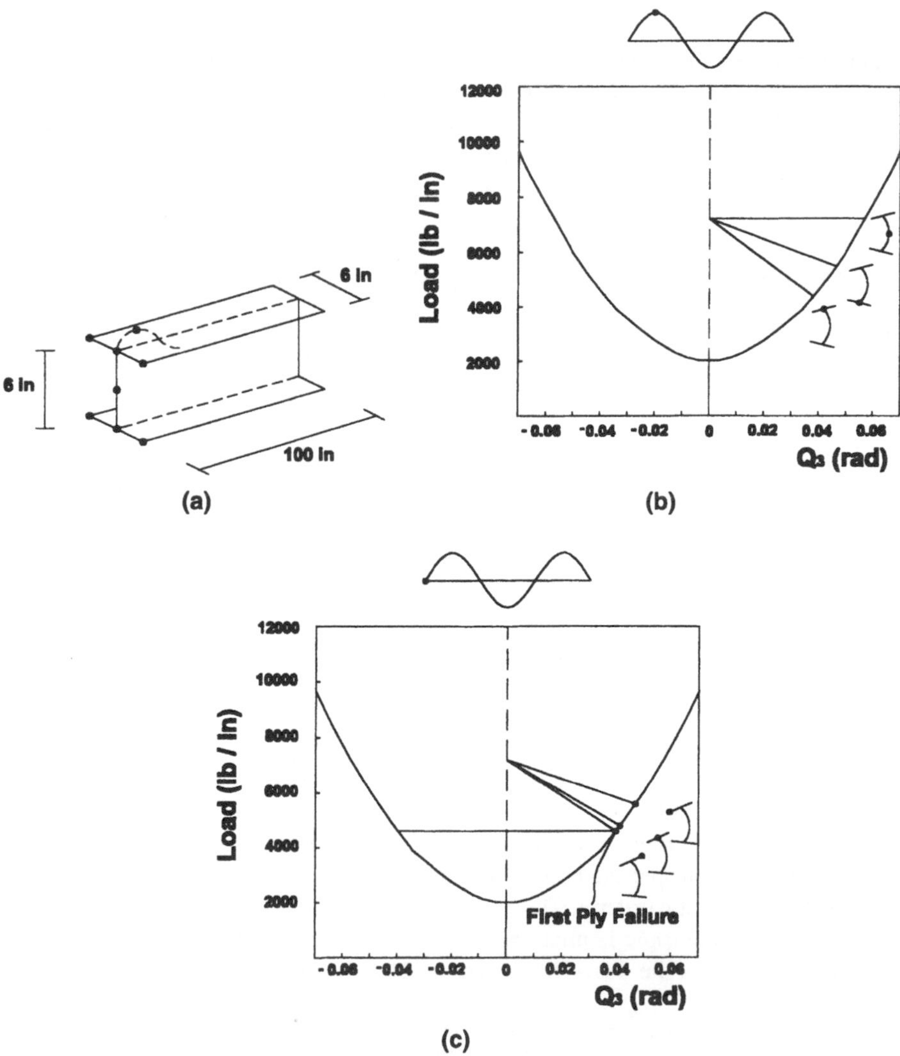

Figure 9.9 First-ply failure of a composite column in the postbuckling range. (a) location of points selected to identify stresses, (b) failure in the web, (c) failure in the flanges.

one computes the displacement fields and strains and curvatures, followed by the stress resultants of the laminate. The new aspect is that with the stress resultants, the stresses should be computed at each lamina and in the principal axis. Such stresses are compared using the failure criterion to determine how far they are from the fist-ply failure surface.

Example 9.8. *Consider a column with dimensions* $6 \times 6 \times 1/4in$ *(*$15.24 \times 15.24 \times 0.635mm$*). The specific composite studied has 7 layers with a fiber direction angle of* ± 45 *and volume fraction* 0.30 *(the volume ratio between fibers and polymer). For a length* $L = 2540mm$*, local buckling occurs with a wave number of* 12 *(there are* 12 *waves in the deflected shape of the flanges along the length of the column).*

The experimental data for the example is adopted as $S_L^2 = 9860MPa$*,* $S_T^2 = 171.1MPa$*, and* $S_{LT} = 9.3MPa$*. The stresses have been identified in different locations, as indicated in Figure* 9.9.a*, because we do not know a priori at what location first-ply-failure will start.*

The failure path has been calculated by identifying which layer of the composite has the highest stress and computing the load needed to reach failure for the first time. Thus, when the postbuckling path of the column reaches the same load level of this failure path, the ply will fail.

In the I*-section composite column, failure can occur in the web or in the flanges. Therefore, the failure path was calculated for different locations of the web, as well as for different locations of the flanges. The load for which the web fails is represented in Figure* 9.9.b *by the intersection of failure path and postbuckling path. The same procedure for the flanges is shown in Figure* 9.9.c*. From the above figure, one can identify which point of the column fails first, including its failure load and displacement. And from the analysis, the layer of the composite that failed can also be known.*

In this case failure was identified at a load $\Lambda = 2.2\Lambda^c$ *and occurs on the edge of the top flange. Of course the structure still has considerable load reserve after first failure, and collapse is still far away. A more detailed analysis of the collapse process requires the use of computer models.*

9.7 BUCKLING IN THE CONTEXT OF STRUCTURAL DESIGN

Buckling studies are very important to the engineer because they are part of the requirements in design. The studies in this book emphasize the development of a general mathematical and mechanical framework to model instability, the understanding of the complex mechanics of behavior of geometrically nonlinear structures, and the types of nonlinear behavior that can be expected in certain cases. However, there is a gap between the theory of elastic stability and the needs of the practical engineer, who uses design guides and recommendations as part of his or her everyday work.

Design is an intellectual activity, in which alternatives are considered in order to choose one based on some criteria adopted. More specifically, design implies the generation of possible alternatives of action, their exploration, and comparison, in order to choose the most desirable one. There are practical limitations in the work of the designer, and these also have to be taken into account. Design cannot be considered as a general field, because the needs of an aeronautical and a mechanical engineer are not identical; they are also different from those of the civil engineer and the architect; and those in turn differ from the work of the offshore engineer and the ship designer. Differences are found in materials, slenderness of the structure, loads, failure criteria, and a number of other factors. Thus, it is not practical to have the same design guidance for any structure, and more specialized codes have been developed to account for the specific features frequently found in a certain field. The theory, on the other hand, is the same for all applications.

This book does not cover the practical aspects of buckling design in specific contexts, and this can be found in design books such as [10, 5, 9] and several others. Buckling is usually considered as a constraint in a design; i.e., the dimensions and material of the structure are given by other considerations, and the designer verifies that the structure and its components are stable and have enough margin of safety with respect to buckling loads. Most regulatory bodies accept two classes of design: design by code and design by analysis. In a design by code the engineer follows a procedure specified in the code, including tables and graphs prepared by a number of experts in the field, who know the theory. In a design by analysis the engineer employs the best tools available to him or her to carry out the analysis, including computer programs, solutions from manuals, etc.

In the context of civil engineering, in which columns and frames are the most common structural types, codes of practice and design guidelines are based on three different philosophies of safety: the allowable stress design (ASD), the plastic design (PD), and the load and resistance factor design (LRFD). The ASD approach employs service loads and design considers that the stresses should not exceed certain allowable values (i.e., first yield should not be reached in the structure). The other two philosophies are concerned with the margin of safety with respect to an ultimate criteria of the structure. In the LRFD there are two modification factors affecting both the loads and the nominal strength of the material. Thus, the LRFD uses load factors and combinations to increase the service loads, while a second factor is used to affect the resistance of the material. For columns, this latter factor is 0.85.

As an illustrative example, consider the American Institute of Steel Construction (AISC) design curves for columns, following ASD and LRFD approaches (the plastic design leads to a curve similar to the LRFD approach in this case). Figure 9.10 shows curves of design loads versus slenderness factor for I-section columns, where the slenderness factor λ is defined as

$$\lambda = \frac{KL}{\pi}\sqrt{\frac{P_y}{EI}}. \tag{9.22}$$

K is the effective length factor and depends on the boundary conditions of the column. For simply supported columns, the values of K is 1. Design loads in the ASD approach only reach values of $P = 0.6P_y$, but in the other approaches they reach $P = 1.0P_y$.

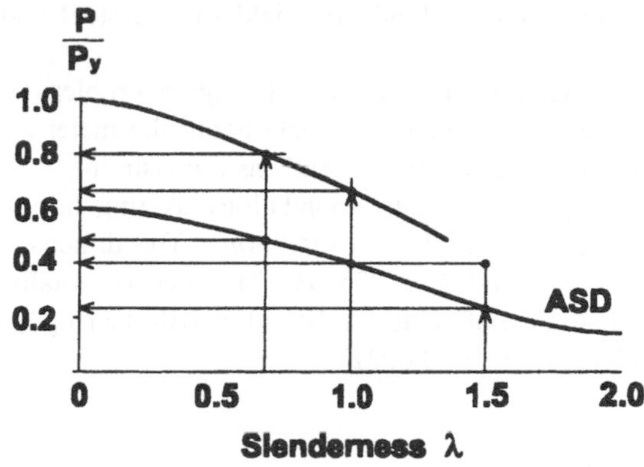

Figure 9.10 Design curves for steel columns, AISC.

Example 9.9. *Consider the columns studied in Examples* 9.1 *to* 9.3, *which have different slenderness ratios. The short column* (*Example* 9.1) *has a slenderness* $\lambda = 0.68$, *and the design loads according to each criterion are* $P/P_y = 0.48$ ($P = 1436$) *in the ASD, and* $P/P_y = 0.8$ ($P = 2393$) *in the LRFD. The column with intermediate length* (*Problem* 9.3) *has slenderness* $\lambda = 1.02$, *and design loads* $P/P_y = 0.4$ *and* 0.65, *respectively. In the long column of Problem* 9.2, *the slenderness is* 1.52 *and the loads for design are* $P/P_y = 0.23$ *and* 0.4. *The long column falls within the region identified as elastic buckling, which in the studies of Example* 9.2 *showed to have elastic buckling before the occurrence of plasticity. The shorter column is in the region of plastic buckling, again in agreement with the nonlinear studies of Example* 9.1. *The intermediate column falls within a transition region in the design curves.*

9.8 PROBLEMS

Review questions. (*a*) What is a yield surface? (*b*) Discuss why imperfections may be important in elastoplastic buckling. (*c*) Discuss the difference between first yield and collapse of a structure. (*d*) Identify the differences between ASD and LRFD approaches to design.

Problem 9.1 (Circular arch). An arch with cross section as a circular pipe is part of a structure. The dimensions are $E = 200$, $A = 9419.34$, $I = 1.16 \times 10^8$, $R = 3500$, $\theta_0 = 30°$, $\sigma_y = 0.45$ (values are given in $[kN, mm]$). It is assumed that the arch has simple supports at the ends. Consider a load imperfection in the form of an eccentricity. Find first yield for the arch using different imperfection amplitudes.

Problem 9.2 (Circular ring). An imperfect circular ring was studied in Problem 8.2. According to the manufacturer, the 0.1% proof stress is $270N/mm^2$. Evaluate for what pressure this value is reached for different imperfection amplitudes.

Problem 9.3 (Plane frame). Find first yield for the steel plane frame of Problem 8.4.

Problem 9.4 (Rectangular plate). Consider again Problem 8.17 (an aluminum rectangular plate under biaxial load). Use data about the material from Problem 9.2 and establish a curve of material "first yield" as a constraint.

Problem 9.5 (Ring on an elastic foundation). A ring has been filled with an elastic material with a modulus $K = 140N/mm^2$. The dimensions of the ring are $h = 5mm$, $A = 30.5mm^2$, and $E = 200GPa$. The load is radially distributed and is uniform. Find for what range of values of the radius will the ring have elastic buckling before first yield. Consider $\sigma_y = 0.002E$.

9.9 BIBLIOGRAPHY

[1] Almanzar, L., and Godoy, L. A., Failure of post-buckled thin-walled composite columns, in *Building to Last*, L. Kempner Jr. and C. B. Brown, Eds., ASCE, New York, 1997, 889–893.

[2] Bushnell, D., Plastic buckling, in *Pressure Vessel and Piping Division: Design Technology. A Decade of Progress*, S. Y. Zamrik and D. Dietrich, Eds., ASME, 1981, 47–117.

[3] Bushnell, D., Plastic buckling of various shells, *J. Pressure Vessel Technology*, ASME, 10454–10472, 1982.

[4] Bushnell, D., *Computerized Buckling Analysis of Shells*, Martinus Nijhoff, Dordrecht, 1985.

[5] Chen, W. F., and Liu, E. M., *Structural Stability*, Elsevier, New York, 1987.

[6] Chen, W. F., and Liu, E. M., *Stability Design of Steel Frames*, CRC Press, Boca Raton, FL, 1991.

[7] Croll, J. G. A., *Buckling of Metal Shells*, Lecture Notes, Laboratorio de Computacao Cientifica, Rio de Janeiro, 1983.

[8] Dowling, N. E., *Mechanical Behavior of Materials*, Prentice-Hall, Englewood Cliffs, NJ, 1993.

[9] Fukumoto, Y. (Ed.), *Structural Stability Design*, Pergamon Press, Oxford, 1997.

[10] Galambos, T. V. (Ed.), *Guide to Structural Design Criteria for Metal Structures*, Fifth Edition, Wiley, New York, 1998.

[11] Hill, R., *The Mathematical Theory of Plasticity*, Oxford, New York, 1983.

[12] Hutchinson, J. W., On the post-buckling behavior of imperfection-sensitive structures in the plastic range, *J. Appl. Mech.*, 39, 155–162, 1972.

[13] Hutchinson, J. W., Post-bifurcation behavior in the plastic range, *J. Mech. Phys. Solids*, 21, 163–190, 1973.

[14] Hutchinson, J. W., Imperfection-sensitivity in the plastic range, *J. Mech. Phys. Solids*, 21, 191–204, 1973.

[15] Hutchinson, J. W., Plastic buckling, in *Advances in Applied Mechanics*, vol. 14, C. S. Yih, Ed., Academic Press, New York, 1974, 67–144.

[16] Hyer, M. W., *Stress Analysis of Fiber-Reinforced Composite Materials*, WCB-McGraw-Hill, New York, 1998.

[17] Lubliner, J., *Plasticity Theory*, Macmillan, New York, 1990.

[18] Needleman, A. Post-bifurcation behavior and imperfection-sensitivity of elastic-plastic circular plates, *Internat. J. Mech. Sci.*, 17(1), 1–14, 1975.

[19] Prager, W., *An Introduction to Plasticity*, Addison-Wesley, Reading, MA, 1959.

[20] Sewell, M. J., A survey of plastic buckling, in *Stability* H. H. E. Leipholtz, Ed., University of Waterloo, 1972.

[21] Tsai, S. W., and Hahn, M. T., *Introduction to Composite Materials*, Technomic, Lancaster, PA, 1980.

CHAPTER

TEN

IMPERFECTION SENSITIVITY OF CRITICAL POINTS[28]

10.1 INTRODUCTION

The nonlinear postbuckling analysis of perfect structural systems was available since the early 1940s, when von Karman and Tsien [40] computed descending postbuckling paths for axially loaded cylindrical shells. It was not clear by then how the shell would actually behave having two close equilibrium paths in the vicinity of a bifurcation point. In 1950 Donnell and Wang published a classical paper about an imperfect cylinder under axial load [9]. For simplicity of the computations, the shape of the imperfection was the same as the deflected shape of the structure at each loading stage, but even for this simplified model the results showed a clear reduction in the maximum load that can bc reached.

The next major contribution to the understanding of the differences between the maximum loads of perfect and real shells was made by Koiter with the theory of initial postbuckling analysis [29]. This theory was applied to the axially loaded cylinder with axisymmetric geometric imperfections in 1963 [30], and for the first time imperfection sensitivity was directly computed.

But rather than computing the complete equilibrium path for the imperfect system, as was done in Chapter 8, a different approach is useful for problems in which the critical state is unstable. The limit point, the asymmetric bifurcation, and the unstable symmetric bifurcation all show a maximum load in the imperfect structure. Thus, the

[28]This chapter was written with Dr. Enrique G. Banchio, National University of Córdoba, Argentina.

information that is really needed in most cases is how the imperfection affects this maximum load that the system can reach. This is called imperfection sensitivity and is the subject of this chapter. The aim is to plot variations of the maximum load with amplitudes of specified imperfections without having to compute the complete equilibrium path for each imperfection considered.

Sections 10.2 to 10.4 deal with the limit point, for which regular perturbation expansions are possible to compute imperfections sensitivity. The rest of the chapter is concerned with the direct evaluation of the imperfection-sensitivity plot using singular perturbations. Section 10.5 deals with the methodology of analysis; and algorithms and examples are presented in section 10.6 for asymmetric bifurcations. Symmetric bifurcations are discussed in section 10.7. A simple experiment is discussed in section 10.8 to illustrate the different sensitivity of some structures to small and large imperfections. Other alternatives for imperfection-sensitivity analysis are mentioned in section 10.9.

10.2 PERTURBATION EQUATIONS FOR THE MAXIMUM LOAD IN LIMIT POINT SYSTEMS

10.2.1 Conditions in the Perturbation Analysis

What is most interesting for the present problem is the influence of imperfections on the limit point, rather than on the complete equilibrium path. Thus, only the curve joining the limit points of the imperfect path will be obtained, in a similar way to what we did in Figure 8.6. This curve is extremely important, since it allows one to identify the imperfection sensitivity of the system.

Unlike our previous perfect analysis, the influence of imperfections on the critical state will be tackled by perturbation of two conditions: equilibrium and critical state

$$V_i\left(Q_j, \Lambda, \xi\right) = 0, \tag{10.1}$$

$$V_{ij}\left(Q_j, \Lambda, \xi\right) x_j\left(\xi\right) = 0. \tag{10.2}$$

The simultaneous solution of the above conditions is called Λ^M, Q^M, and x_j^M, and identifies the maximum value reached along the equilibrium path with an imperfection. Our goal is to represent Λ^M and Q^M as a function of the imperfection parameter ξ.

In solving the above equations, use should be made of the conditions that the energy satisfies at a limit point, presented in Chapter 6, i.e.,

$$V_i' x_i \neq 0, \qquad C = V_{ijk} x_i x_j x_k \neq 0.$$

10.2.2 The Imperfection as Perturbation Parameter

We start by accepting that at the state with $\xi = 0$ all required derivatives with respect to ξ exist, so that we can proceed using a regular perturbation analysis. In section 10.5

we investigate problems in which a regular perturbation expansion is not possible in terms of ξ.

If the perturbation parameter s is set as the imperfection amplitude, $s = \xi$, then

$$\Lambda^M = \Lambda^c + \dot{\Lambda}\xi + \ddot{\Lambda}\xi^2 + \cdots, \tag{10.3}$$

$$Q_j^M = Q_j^c + \dot{q}_j\xi + \ddot{q}_j\xi^2 + \cdots, \tag{10.4}$$

$$x_j^M = x_j^c + \dot{x}_j\xi + \ddot{x}_j\xi^2. \tag{10.5}$$

The symbol $(\ )^M$ is used to denote that the values are associated with a maximum point (a limit point). The unknowns are the changes in the critical displacements $\dot{q}_k$, the changes in the critical load $\dot{\Lambda}$, and the changes in the eigenvector $\dot{x}_j$.

As usual, the coefficients of derivatives with respect to the perturbation parameter ξ can be obtained from perturbation equations of equilibrium and stability.

From equilibrium, the following perturbation equations are obtained:

$$V_{ij}\dot{q}_j + \dot{V}_i\dot{\xi} + V_i'\dot{\Lambda} = 0. \tag{10.6}$$

The condition of critical state leads to the first-order perturbation equation

$$(V_{ijk}\dot{q}_k + \dot{V}_{ij} + V_{ij}'\dot{\Lambda})x_j + V_{ij}\dot{x}_j = 0, \tag{10.7}$$

where a dot denotes the derivatives with respect to ξ. Notice that this is a partial derivative for V, but for q_i, x_i, and Λ they represent total derivatives.

For a problem with N degrees of freedom, there are $(2N + 1)$ unknowns and only $2N$ equations. Furthermore, in both equations we find terms $V_{ij}\dot{q}_j$ and $V_{ij}\dot{x}_j$ with singular matrices V_{ij} at the critical state. To increase the number of available equations it is necessary to employ contracted equations.

The following equation is contracted from first-order perturbation of equilibrium

$$\underbrace{V_{ij}x_i}\,\dot{q}_j + \dot{V}_i x_i + V_i' x_i \dot{\Lambda} = 0. \tag{10.8}$$

The scalar condition of contracted first-order perturbation of critical state is

$$V_{ijk}x_i x_j \dot{q}_k + \dot{V}_{ij}x_i x_j + \underbrace{V_{ij}'x_i}\, x_j \dot{\Lambda} + \underbrace{V_{ij}x_i}\, \dot{x}_j = 0. \tag{10.9}$$

Further conditions depend on the relations that the energy V satisfies at a specific critical state.

The only new derivatives that need to be computed with respect to the perfect analysis are $\dot{V}_i$ and $\dot{V}_{ij}$. Furthermore, if the imperfection is such that it only affects the linear terms in displacements in V, then $\dot{V}_{ij} = 0$ and the only new derivative to be computed is $\dot{V}_i$.

10.3 SENSITIVITY OF LIMIT POINTS TO IMPERFECTIONS. CASE 1: ($\dot{V}_i x_i|^c \neq 0$)

10.3.1 Formulation

For a limit point we may adopt $s = \xi$ without any difficulties.

(a) By assuming $x_1 = 1$, $\dot{x}_1 = 0$, contraction of the equilibrium equation (10.6) leads to

$$\underbrace{V_{ij} x_i}\, \dot{q}_j + \dot{V}_i x_i + V_i' x_i \dot{\Lambda}\ |^c = 0. \tag{10.10}$$

The first term vanishes at the critical state.

Let us first assume that

$$\dot{V}_i x_i\ |^c \neq 0; \tag{10.11}$$

that is, the imperfection has a component in the direction of critical stability. If the product is zero, one should consider case 2, discussed in section 10.4.

At a limit point, $V_i' x_j \neq 0$ is valid, so that $\dot{\Lambda}$ may be calculated as

$$\dot{\Lambda} = -\frac{\dot{V}_i x_i}{V_i' x_i}\Bigg|^c. \tag{10.12}$$

(b) The first-order perturbation equation of equilibrium, equation (10.6), is

$$V_{ij} \dot{q}_j\ |^c = -\dot{V}_i - V_i' \dot{\Lambda}\ |^c. \tag{10.13}$$

Equation (10.13) is a set of (N) equations; however, since V_{ij}^c is singular, we are left with only $(N-1)$ independent equations. The unknowns are the (N) components of the sensitivity of displacements $\dot{q}_j^M$. Notice that we do not know the value of $\dot{q}_1$.

The solution can be obtained in a convenient way by writing

$$\dot{q}_j = \dot{q}_1 x_j + w_j, \tag{10.14}$$

where w_j is an unknown displacement vector. The first component (assuming that $x_1 = 1$) in q_j^M is, from (10.14),

$$\dot{q}_1 = \dot{q}_1 x_1 + w_1. \tag{10.15}$$

Clearly, for this to be satisfied,

$$w_1 = 0,$$

and we now know one of the components of w_j. The $(N-1)$ remaining components of w_j can be obtained by substitution of (10.14) into (10.6):

$$V_{ij}\left(\dot{q}_1 x_j + w_j\right) |^c = -\dot{V}_i - V_i' \dot{\Lambda}\ |^c$$

or else

$$\underbrace{V_{ij}x_j}\ \dot{q}_1 + V_{ij}w_j\ |^c = -\left(\dot{V}_i + V_i'\dot{\Lambda}\right)\ |^c.$$

This leads to

$$V_{ij}w_j = -\dot{V}_i - V_i'\Lambda. \tag{10.16}$$

Equation (10.16) can be solved to obtain the $(N-1)$ components of w_j. Substitution of w_j in (10.14) would not yet yield the vector $\dot{q}_j$, since $\dot{q}_1$ is still unknown.

(c) To obtain $\dot{q}_1$, the equation of critical state (10.7) is contracted as

$$V_{ijk}x_i x_j \dot{q}_k\ |^c = -\left(\dot{V}_{ij} + V_{ij}'\dot{\Lambda}\right) x_i x_j\ |^c. \tag{10.17}$$

Substitution of (10.14) into (10.17) leads to

$$\underbrace{V_{ijk}x_i x_j x_k}_{C}\ \dot{q}_1\ |^c = -\left(\dot{V}_{ij} + V_{ij}'\dot{\Lambda}\right) x_i x_j - V_{ijk}x_i x_j w_k\ |^c. \tag{10.18}$$

This is a scalar equation with one unknown. The value of $\dot{q}_1$ may be obtained, since the rest of the variables in (10.18) are known at this stage

$$\dot{q}_1 = -\frac{\left(\dot{V}_{ij} + V_{ij}'\dot{\Lambda}\right) x_i x_j + V_{ijk}x_i x_j w_k}{C}\Bigg|^c. \tag{10.19}$$

The change in the displacement field at the critical state is given by (10.14).

(d) The change in the eigenvector $\dot{x}_j$ is calculated next from the first-order perturbation equation of stability, equation (10.7)

$$V_{ij}\dot{x}_j\ |^c = -(V_{ikl}\dot{q}_l + \dot{V}_{ik} + V_{ik}'\dot{\Lambda})x_k\ |^c, \tag{10.20}$$

where $\dot{x}_1 = 0$.

The next set of derivatives is obtained using a similar procedure and is left to the reader as Problem 10.1 at the end of this chapter.

10.3.2 Algorithm for Imperfect Limit Point Behavior

A summary and sequence of equations to be solved for sensitivity of limit points with $\dot{V}_i x_i \neq 0$ is as follows:

1. Compute $\dot{V}_i$ and $\dot{V}_{ij}$.
2. Evaluate load coefficient $\dot{\Lambda}$

$$\dot{\Lambda} = \frac{\dot{V}_i x_i}{V_i' x_i}\Bigg|^c.$$

3. Solve
$$V_{ij}w_j = -\left(\dot{V}_i + V_i'\Lambda\right)|^c \qquad \text{with} \qquad w_1 = 0.$$
4. Evaluate
$$\dot{q}_1 = -\frac{\left(\dot{V}_{ij} + V_{ij}'\dot{\Lambda}\right)x_ix_j + V_{ijk}x_ix_jw_k}{C}\Bigg|^c.$$
5. Evaluate
$$\dot{q}_j = \dot{q}_1 x_j + w_j.$$
6. Solve
$$V_{ij}\dot{x}_j\,|^c = -(V_{ikl}\dot{q}_l + \dot{V}_{ik} + V_{ik}'\dot{\Lambda})x_k\,|^c \qquad \text{with} \qquad \dot{x}_1 = 0.$$
7. The first-order sensitivity of a limit point to imperfections with amplitude ξ is given by
$$\Lambda^M(\xi) = \Lambda^c + \dot{\Lambda}\xi + \cdots,$$
$$Q_j^M(\xi) = Q_j^c + \dot{q}_j\xi + \cdots,$$
$$x_j^M = x_j^c + \dot{x}_j\xi + \cdots.$$

Example 10.1 (Changes in the central angle of an arch). *In this example we consider changes in the central angle of an arch but using the perturbation analysis to compute the sensitivity diagram. The central angle is written as*

$$\theta_0 = \alpha\,(1+\xi)\,, \tag{10.21}$$

so that the energy coefficients become

$$A_1 = -1,$$

$$A_{11} = \frac{\delta}{(1+\xi)^4}c^4 + \frac{8}{\pi^2}, \qquad A_{22} = \frac{16\delta}{(1+\xi)^4}c^4,$$
$$A_{111} = -\frac{6}{\pi}\frac{1}{(1+\xi)^2}c^2, \qquad A_{122} = -\frac{8}{\pi}\frac{1}{(1+\xi)^2}c^2,$$
$$A_{1111} = \frac{3}{2}\frac{1}{(1+\xi)^4}c^4, \qquad A_{2222} = \frac{6}{(1+\xi)^2}c^2, \qquad A_{1122} = 2\frac{1}{(1+\xi)^4}c^2,$$

where
$$\delta = \frac{I}{AR^2}, \qquad c = \frac{\pi}{2\alpha},$$
and the energy V has been divided by a factor $(EAR\alpha)$.

The eigenvector of this problem considered as a perfect system is

$$x = \begin{Bmatrix} 1 \\ 0 \end{Bmatrix}.$$

The following derivatives are required:

$$V_i' = \begin{Bmatrix} -1 \\ 0 \end{Bmatrix} \quad \textit{and} \quad \dot{V}_i = \begin{Bmatrix} \dot{V}_1 \\ \dot{V}_2 \end{Bmatrix}.$$

These derivatives need to be evaluated at the critical state, for which

$$Q_1 = Q_1^c = \frac{4}{\pi}\left(\frac{2\alpha}{\pi}\right)^2 \left[1 - \frac{1}{2\sqrt{3}}\sqrt{4 - \delta\pi^2\left(\frac{\pi}{2\alpha}\right)^4}\,\right],$$

$$\Lambda = \Lambda^c = A_{11}Q_1^c + \frac{1}{2}A_{111}\left(Q_1^c\right)^2 + \frac{1}{6}A_{1111}\left(Q_1^c\right)^3 \,|_{\xi=0}$$

as given in Chapter 6. The evaluated derivatives are

$$\dot{V}_1 \,|^c = 4\delta\,(1+\xi)^{-5}\left(\frac{\pi}{2\alpha}\right)^4 Q_1^c + \frac{6}{\pi}\left(\frac{\pi}{2\alpha}\right)^2 (1+\xi)^{-3}\left(Q_1^c\right)^2 - \left(\frac{\pi}{2\alpha}\right)^4 (1+\xi)^{-5}\left(Q_1^c\right)^3.$$

Then

$$\dot{V}_i x_i = \dot{V}_1 x_1 = \dot{V}_1,$$

$$V_i' x_i = V_1' x_1 = -1.$$

Sensitivity of the load results in

$$\dot{\Lambda} = -\frac{\dot{V}_i x_i}{V_i' x_i} = \dot{V}_1$$

and

$$\Lambda^M(\xi) = \Lambda^c + \dot{\Lambda}\,\xi + \cdots.$$

The above example has been computed for a specific arch structure with $R = 19.4$, $I = 2.49 \times 10^{-4}$, $A = 7.74 \times 10^{-3}$, *and* $\alpha = 0.5$ *rad and is plotted in Figure* 10.1 *for* $-0.1 \leq \xi \leq 0.1$. *The results agree very well with the exact values from a nonlinear analysis and show a linear sensitivity plot.*

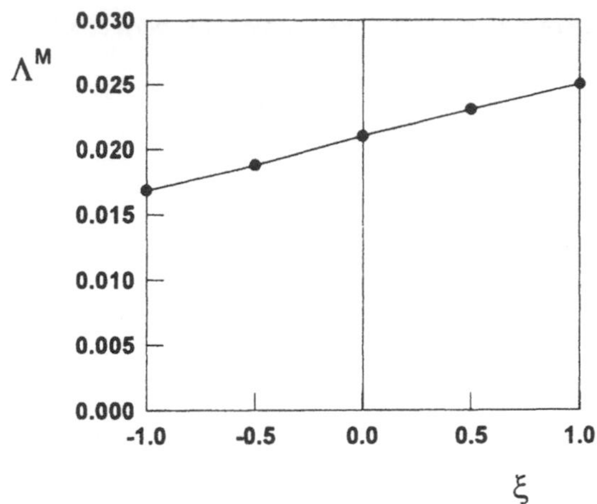

Figure 10.1 Imperfection sensitivity for a circular arch with changes in the central angle.

10.4 SENSITIVITY OF LIMIT POINTS TO IMPERFECTIONS. CASE 2: $(\dot{V}_i x_i|^c \neq 0)$

10.4.1 Formulation

We consider now the case in which the imperfection is orthogonal to the eigenvector considered, i.e.,

$$\dot{V}_i x_i \mid^c = 0.$$

Again, we adopt $s = \xi$ and proceed as follows:

(a) By assuming $x_1 = 1$, $\dot{x}_1 = 0$, contraction of the equilibrium equation (10.6) leads to

$$\underbrace{V_{ij} x_i}_{0} \, \dot{q}_j + \underbrace{\dot{V}_i x_i}_{0} + \underbrace{V_i' x_i}_{\neq 0} \, \dot{\Lambda} \mid^c = 0. \tag{10.22}$$

The first-order derivative of the load with respect to the imperfection $\dot{\Lambda}$ results in

$$\dot{\Lambda} = 0. \tag{10.23}$$

(b) The first-order perturbation equation of equilibrium equation (10.6) is

$$V_{ij} \dot{q}_j \mid^c = -\dot{V}_i \mid^c. \tag{10.24}$$

Because V_{ij}^c is singular, we are again left with only $(N-1)$ independent conditions in (10.13) and write the solution as we did before in case 1, i.e., (10.14)

$$\dot{q}_j = \dot{q}_1 x_j + w_j \tag{10.25}$$

with

$$w_1 = 0.$$

The components of w_j can be obtained by substitution in the first-order equilibrium condition

$$V_{ij}\left(\dot{q}_1 x_j + w_j\right)|^c = -\dot{V}_i|^c$$

or else

$$V_{ij} w_j |^c = -\dot{V}_i |^c. \tag{10.26}$$

Equation (10.26) can be solved to obtain the $(N-1)$ components of w_j.

(c) Notice that $\dot{q}_1$ is still unknown. To obtain $\dot{q}_1$, the equation of critical state (10.7) is contracted as

$$V_{ijk} x_i x_j \dot{q}_k |^c = -\dot{V}_{ij} x_i x_j |^c, \tag{10.27}$$

from which we get

$$\underbrace{V_{ijk} x_i x_j x_k}_{C}\, \dot{q}_1 |^c = -\dot{V}_{ij} x_i x_j - V_{ijk} x_i x_j w_k |^c \tag{10.28}$$

and

$$\dot{q}_1 = -\frac{\left(\dot{V}_{ij} x_i x_j + V_{ijk} x_i x_j w_k\right)}{C}\Bigg|^c. \tag{10.29}$$

(d) The change in the eigenvector $\dot{x}_j$ is calculated next from the first-order perturbation equation of stability equation (10.7)

$$V_{ij}\dot{x}_j |^c = -(V_{ikl}\dot{q}_l + \dot{V}_{ik}) x_k |^c, \tag{10.30}$$

where $\dot{x}_1 = 0$.

(e) We have to continue with the analysis, because we do not know the value of any derivative of the load. Consider the second-order equilibrium equation

$$2\dot{V}_{ij}\dot{q}_j + V_{ijk}\dot{q}_j\dot{q}_k + \ddot{V}_i + V_{ij}\ddot{q}_j + V_i'\ddot{\Lambda} = 0.$$

The new unknowns are $\ddot{\Lambda}$ and $\ddot{q}_j$.

(f) The contracted second-order equilibrium equation is

$$2\dot{V}_{ij} x_i \dot{q}_j + V_{ijk} x_i \dot{q}_j \dot{q}_k + \ddot{V}_i x_i + \underbrace{V_{ij} x_i}_{0}\, \ddot{q}_j + \underbrace{V_i' x_i}_{\neq 0}\, \ddot{\Lambda} = 0.$$

The second derivative of the load can be computed now in the form

$$\ddot{\Lambda} = -\frac{\left(2\dot{V}_{ij} x_i \dot{q}_j + V_{ijk} x_i \dot{q}_j \dot{q}_k + \ddot{V}_i x_i\right)}{V_i' x_i}.$$

We interrupt the analysis at this point, and can compute the sensitivity as

$$\Lambda^M = \Lambda^c + \ddot{\Lambda}\xi^2 + \cdots$$

$$Q_j^M = Q_j^c + \dot{q}_j\xi + \cdots$$

$$x_j^M = x_j^c + \dot{x}_j\xi + \cdots.$$

Higher-order terms could be computed if we follow the analysis, but they are usually not required.

10.4.2 Algorithm for Imperfect Limit Point Behavior

A summary and sequence of equations to be solved for sensitivity of limit points with $\dot{V}_i x_i = 0$ is as follows:

1. Compute $\dot{V}_i$, $\dot{V}_{ij}$, and $\ddot{V}_i$.
2. Define $\dot{\Lambda}$
$$\dot{\Lambda} = 0.$$
3. Solve
$$V_{ij} w_j = -\dot{V}_i \,|^c \qquad \text{with} \qquad w_1 = 0.$$
4. Evaluate
$$\dot{q}_1 = -\left.\frac{\dot{V}_{ij}x_i x_j + V_{ijk}x_i x_j w_k}{C}\right|^c.$$
5. Evaluate
$$\dot{q}_j = \dot{q}_1 x_j + w_j.$$
6. Solve
$$V_{ij}\dot{x}_j \,|^c = -(V_{ikl}\dot{q}_l + \dot{V}_{ik})x_k \,|^c \qquad \text{with} \qquad \dot{x}_1 = 0.$$
7.
$$\ddot{\Lambda} = -\frac{\left(2\dot{V}_{ij}x_i\dot{q}_j + V_{ijk}x_i\dot{q}_j\dot{q}_k + \ddot{V}_i x_i\right)}{V_i' x_i}.$$

The sensitivity of a limit point to imperfections with amplitude ξ is given by

$$\Lambda^M = \Lambda^c + \ddot{\Lambda}\xi^2 + \cdots,$$

$$Q_j^M = Q_j^c + \dot{q}_j\xi + \cdots,$$

$$x_j^M = x_j^c + \dot{x}_j\xi + \cdots.$$

Example 10.2 (Circular arch with load imperfection). *We study again our example of a circular arch with an eccentricity in the load at the apex.*

1. *The following derivatives result*

$$\dot{V}_i = \begin{Bmatrix} 0 \\ \Lambda B_2 \end{Bmatrix}, \qquad where \qquad B_2 = -\pi \frac{e}{R\theta_0},$$

$$\dot{V}_{ij} = 0, \qquad and \qquad \ddot{V}_i = 0.$$

2. *Assume* $\dot{\Lambda} = 0$.
3. *Solve* w_j

$$w_2 = -\frac{\dot{V}_2}{V_{22}}\bigg|^c,$$

where

$$V_{22}\,|^c = A_{22} + A_{122} Q_1^c + \frac{1}{2}\left[A_{2222}\left(Q_2^c\right)^2 + A_{1122}\left(Q_1^c\right)^2\right]$$

and

$$Q_1^c = \frac{4}{\pi}\left(\frac{2\alpha}{\pi}\right)^2\left[1 - \frac{1}{2\sqrt{3}}\sqrt{4 - \delta\pi^2\left(\frac{\pi}{2\alpha}\right)^4}\right],$$

$$\Lambda^c = A_{11} Q_1^c + \frac{1}{2} A_{111}\left(Q_1^c\right)^2 + \frac{1}{6} A_{1111}\left(Q_1^c\right)^3 \,|_{\xi=0}\,.$$

4. *The coordinate* $\dot{q}_1$ *reduces to*

$$\dot{q}_1 = -\frac{V_{112} w_2}{V_{111}}\bigg|^c,$$

where

$$V_{112}^c = A_{1122} \underbrace{Q_2^c}_{0} = 0,$$

$$V_{111}^c = A_{111} + \frac{1}{2} A_{1111} Q_1^c \neq 0.$$

Then we have

$$\dot{q}_1 = 0.$$

5. *Derivatives of displacements are*

$$\dot{q}_j = \begin{Bmatrix} 0 \\ w_2 \end{Bmatrix}.$$

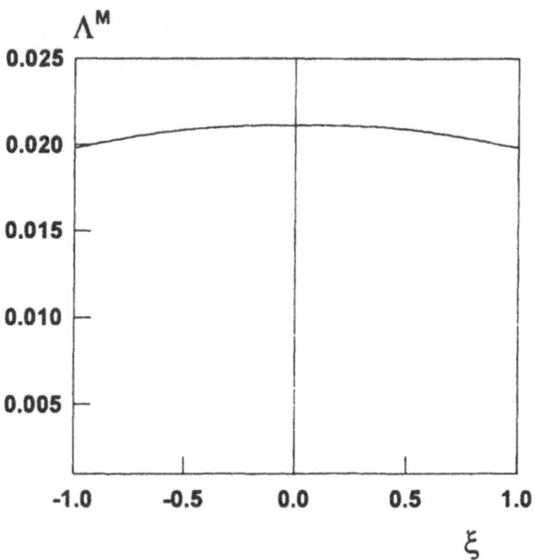

Figure 10.2 Imperfection sensitivity for a circular arch with load imperfection.

6. *To compute the second derivative of the eigenvalue we need*

$$V_{122}^c = A_{122} + A_{1122} Q_1^c.$$

Then

$$\ddot{\Lambda} = -\frac{V_{122}^c x_1 \dot{q}_2 \dot{q}_2}{V_1' x_1} = V_{122}^c (w_2)^2 .$$

The results for this problem represent a parabolic variation of Λ^M *with respect to the imperfection parameter* ξ

$$\Lambda^M = \Lambda^c + \ddot{\Lambda}\xi^2.$$

The sensitivity is quadratic and has been plotted in Figure 10.2. *For small values of e the sensitivity curve is almost flat, showing small influence of the eccentricity on the limit point.*

10.5 IMPERFECTION SENSITIVITY IN BIFURCATION BEHAVIOR

10.5.1 Asymptotic Approximation

Similar to what was done for a limit point, we now seek to obtain the imperfection-sensitivity curve directly, rather than through several nonlinear analyses. Our first attempt could be to choose a regular perturbation approximation as we did in section 10.2. However, this is not possible because the imperfection-sensitivity curve is singular at the critical state. The problem is that in bifurcation analysis the relation

$\Lambda^M = \Lambda^M(\xi)$ has an infinite slope at $\xi = 0$. This means that the required derivatives do not exist at the critical state, and expansion from there is not possible using regular perturbations.

An approach using singular perturbations was first presented in [16] for one-degree-of-freedom problems and extended to multiple degree-of-freedom systems in [5, 3]. This section is based on the above papers.

A consequence of the singularity at the critical state is that the expansions do not go in integral powers of ξ. Indeed the powers of ξ cannot be assumed a priori and must be determined as part of the solution. Similar problems arise in the analysis of dynamical systems when two or more equilibrium points coalesce. Moreover, in analyzing boundary layer phenomena, whether in plates, electrical conductors, or high Reynolds number flows, one must stretch one or more coordinates. Typically the stretching goes according to a nonintegral power of a naturally occurring, small parameter. Determining the proper stretching in these singular perturbation problems is a key element of the analysis, and that procedure is quite similar to the one employed here.

We assume that the expansions for the maximum load Λ^M and associated displacement Q_j^M and eigenvector x_j^M have the following form:

$$\begin{aligned} \Lambda^M &= \lambda_c + \dot{\lambda}\xi^\alpha + \cdots, \\ Q_j^M &= q_j^c + \dot{q}_j\xi^\beta + \cdots, \\ x_j^M &= x_j^c + \dot{x}_j\xi^\gamma + \cdots, \end{aligned} \tag{10.31}$$

where α, β, and γ are positive exponents (unknown at this stage), and $\dot{\lambda}$, $\dot{q}_j$, and $\dot{x}_j$ are not derivatives but unknown coefficients. We repeat that the coefficients in (10.31) are not related to derivatives and will be computed using perturbation of equilibrium and stability conditions. Evaluation of the exponents is done so as to obtain the least degenerate two-term approximation, as discussed in the following section.

10.5.2 Perturbation Equations

First, we substitute (10.31) into the equilibrium and stability conditions, i.e.,

$$\frac{\partial V}{\partial q_i} \equiv V_i\left(\dot{\lambda}, \dot{q}, \xi, \alpha, \beta\right) = 0, \tag{10.32}$$

$$\frac{\partial V}{\partial q_i \partial q_j} x_j \equiv V_{ij}\left(\dot{\lambda}, \dot{q}, \xi, \alpha, \beta\right) x_j\left(\dot{x}_j, \gamma\right) = 0. \tag{10.33}$$

Then we simplify the resulting equations by means of the conditions that the energy satisfies at a bifurcation point, i.e.,

$$V_i^c = 0, \qquad V_{ij}^c x_j^c = 0, \qquad V_i'^c x_i = 0.$$

Although it is not strictly necessary, we can also assume that

$$\dot{V}_i^c x_i \neq 0.$$

The equilibrium condition reduces to

$$\underbrace{V_i^c}_{0} + \dot{V}_i^c \xi^1 + \dot{\lambda} V_i'^c \xi^\alpha + V_{ij}^c \dot{q}_j \xi^\beta + \dot{\lambda} V_{ij}'^c \dot{q}_j \xi^{\alpha+\beta} + \frac{1}{2!} V_{ijk}^c \dot{q}_j \dot{q}_k \xi^{2\beta}$$

$$+ \frac{1}{2!} \dot{\lambda} V_{ijk}'^c \dot{q}_j \dot{q}_k \xi^{\alpha+2\beta} + \frac{1}{3!} V_{ijkl}^c \dot{q}_j \dot{q}_k \dot{q}_l \xi^{3\beta} + \frac{1}{3!} \dot{\lambda} V_{ijkl}'^c \dot{q}_j \dot{q}_k \dot{q}_l \xi^{\alpha+3\beta}$$

$$+ \dot{V}_{ij}^c \dot{q}_j \xi^{1+\beta} + \frac{1}{2!} \dot{V}_{ijk}^c \dot{q}_j \dot{q}_k \xi^{1+2\beta} + \frac{1}{3!} \dot{V}_{ijkl}^c \dot{q}_j \dot{q}_k \dot{q}_l \xi^{1+3\beta} + \dot{\lambda} \dot{V}_i'^c \xi^{1+\alpha}$$

$$+ \dot{\lambda} \dot{V}_{ij}'^c \dot{q}_j \xi^{1+\alpha+\beta} + \frac{1}{2!} \dot{\lambda} \dot{V}_{ijk}'^c \dot{q}_j \dot{q}_k \xi^{1+\alpha+2\beta} + \frac{1}{3!} \dot{\lambda} \dot{V}_{ijkl}'^c \dot{q}_j \dot{q}_k \dot{q}_l \xi^{1+\alpha+3\beta} = 0. \quad (10.34)$$

The condition of stability is given by

$$\underbrace{V_{ij}^c x_j}_{0} + \dot{V}_{ij}^c x_j \xi^1 + \dot{\lambda} V_{ij}'^c x_j \xi^\alpha + V_{ij}^c \dot{x}_j \xi^\gamma + \dot{\lambda} V_{ij}'^c \dot{x}_j \xi^{\alpha+\gamma} + V_{ijk}^c x_j \dot{q}_k \xi^\beta$$

$$+ \dot{\lambda} V_{ijk}'^c x_j \dot{q}_k \xi^{\alpha+\beta} + V_{ijk}^c \dot{x}_j \dot{q}_k \xi^{\beta+\gamma} + \dot{\lambda} V_{ijk}'^c \dot{x}_j \dot{q}_k \xi^{\alpha+\beta+\gamma} + \frac{1}{2!} V_{ijkl}^c x_j \dot{q}_k \dot{q}_l \xi^{2\beta}$$

$$+ \frac{1}{2!} \dot{\lambda} V_{ijkl}'^c x_j \dot{q}_k \dot{q}_l \xi^{\alpha+2\beta} + \frac{1}{2!} V_{ijkl}^c \dot{x}_j \dot{q}_k \dot{q}_l \xi^{2\beta+\gamma} + \frac{1}{2!} \dot{\lambda} V_{ijkl}'^c \dot{x}_j \dot{q}_k \dot{q}_l \xi^{\alpha+2\beta+\gamma}$$

$$+ O\left(\xi^{1+\varepsilon}\right) = 0. \quad (10.35)$$

To carry out the analysis we also need the contracted form of the above equations. The contracted perturbation equation is obtain by multiplication of (10.34) by the eigenvector of the perfect system x_i and results in

$$\dot{V}_i^c x_i \xi^1 + \dot{\lambda} \underbrace{V_i'^c x_i}_{0} \xi^\alpha + \underbrace{V_{ij}^c x_i}_{0} \dot{q}_j \xi^\beta + \dot{\lambda} V_{ij}'^c x_i \dot{q}_j \xi^{\alpha+\beta} + \frac{1}{2!} V_{ijk}^c x_i \dot{q}_j \dot{q}_k \xi^{2\beta}$$

$$+ \frac{1}{2!} \dot{\lambda} V_{ijk}'^c x_i \dot{q}_j \dot{q}_k \xi^{\alpha+2\beta} + \frac{1}{3!} V_{ijkl}^c x_i \dot{q}_j \dot{q}_k \dot{q}_l \xi^{3\beta} + \frac{1}{3!} \dot{\lambda} V_{ijkl}'^c x_i \dot{q}_j \dot{q}_k \dot{q}_l \xi^{\alpha+3\beta}$$

$$+ O\left(\xi^{1+\varepsilon}\right) = 0. \quad (10.36)$$

A similar contraction is performed on the stability equation (10.35), leading to

$$\dot{V}^c_{ij}x_i x_j \xi^1 + \dot{\lambda} V'^c_{ij} x_i x_j \xi^\alpha + \underbrace{V^c_{ij} x_i \dot{x}_j}_{0} \xi^\gamma + \dot{\lambda} V'^c_{ij} x_i \dot{x}_j \xi^{\alpha+\gamma} + V^c_{ijk} x_i x_j q_k \xi^\beta$$

$$+ \dot{\lambda} V'^c_{ijk} x_i x_j \dot{q}_k \xi^{\alpha+\beta} + V^c_{ijk} x_i \dot{x}_j \dot{q}_k \xi^{\beta+\gamma} + \dot{\lambda} V'^c_{ijk} x_i \dot{x}_j \dot{q}_k \xi^{\alpha+\beta+\gamma}$$

$$+ \frac{1}{2!} V^c_{ijkl} x_i x_j \dot{q}_k \dot{q}_l \xi^{2\beta} + \frac{1}{2!} \dot{\lambda} V'^c_{ijkl} x_i x_j \dot{q}_k \dot{q}_l \xi^{\alpha+2\beta} + \frac{1}{2!} V^c_{ijkl} x_i \dot{x}_j \dot{q}_k \dot{q}_l \xi^{2\beta+\gamma}$$

$$+ \frac{1}{2!} \dot{\lambda} V'^c_{ijkl} x_i \dot{x}_j \dot{q}_k \dot{q}_l \xi^{\alpha+2\beta+\gamma} + O\left(\xi^{1+\varepsilon}\right) = 0. \qquad (10.37)$$

10.5.3 Least Degenerate Solution of Exponents and Solution of Coefficients of the Expansion

These are cumbersome expressions; however, some reduction in the number of terms can be made immediately. The imperfection parameter ξ is assumed to be small, so that to obtain the exponents we need to look at the most significant terms in the perturbation equations. The most significant terms are those associated with lower exponents of the imperfection amplitude, and those associated with ξ with a larger exponent will make a negligible contribution. Thus, many terms can be ignored regardless of the choice of the exponents because they are smaller than those with ξ at a lower exponent.

For example, let us consider the perturbation equation of equilibrium, equation (10.34). One of the terms has the exponent 1, so that terms with $\xi^{1+\beta}$, $\xi^{1+2\beta}$, $\xi^{1+3\beta}$, $\xi^{1+\alpha}$, $\xi^{1+\alpha+\beta}$, $\xi^{1+\alpha+2\beta}$, $\xi^{1+\alpha+3\beta}$ should lead to smaller contributions than ξ^1.

In any case, we need to try values of α and β and consider the conditions that arise as a consequence of the choice made.

10.6 IMPERFECTION SENSITIVITY IN ASYMMETRIC BIFURCATION

10.6.1 Formulation

In asymmetric bifurcation we have

$$V^c_{ijk} x_i x_j x_k \neq 0.$$

Let us consider the following choice of exponents:

$$\alpha = \frac{1}{2}, \qquad \beta = \frac{1}{2}, \qquad \gamma = \frac{1}{2}. \qquad (10.38)$$

(a) Substitution of the above exponents in (10.34) leads to

$$\left[\dot{\lambda} V_i'^c + V_{ij}^c \dot{q}_j\right] \xi^{\frac{1}{2}} + \left[\dot{V}_i^c + \dot{\lambda} V_{ij}'^c \dot{q}_j + \frac{1}{2} V_{ijk}^c \dot{q}_j \dot{q}_k\right] \xi^1$$

$$+ O\left(\xi^{\frac{3}{2}}\right) = 0.$$

From the main term, we get

$$V_{ij}^c \dot{q}_j = -\dot{\lambda} V_i'^c. \tag{10.39}$$

The solution of (10.39) can be written as

$$\dot{q}_j = k_1\, x_j + \dot{\lambda}\, y_j. \tag{10.40}$$

Substitution into (10.39) leads to

$$V_{ij}^c \left(k_1\, x_j + \dot{\lambda}\, y_j\right) = -\dot{\lambda} V_i'^c$$

or else

$$V_{ij}^c\, y_j = -V_i'^c \qquad \text{with} \qquad y_1 = 0. \tag{10.41}$$

(b) Next we substitute the exponents in the contracted form of equilibrium equation (10.36) and get

$$\left[\dot{V}_i^c x_i + \dot{\lambda} V_{ij}'^c x_i \dot{q}_j + \frac{1}{2} V_{ijk}^c x_i \dot{q}_j \dot{q}_k\right] \xi + O\left(\xi^{\frac{3}{2}}\right) = 0. \tag{10.42}$$

The main term in the contracted equation of equilibrium provides the following scalar condition:

$$\dot{V}_i^c x_i + \dot{\lambda} V_{ij}'^c x_i \dot{q}_j + \frac{1}{2} V_{ijk}^c x_i \dot{q}_j \dot{q}_k = 0.$$

The value of $\dot{q}_j$ is substituted from (10.40) to obtain a nonlinear equation in terms of k_1 and λ

$$\dot{V}_i^c x_i + \dot{\lambda} V_{ij}'^c x_i \left(k_1\, x_j + \dot{\lambda}\, y_j\right)$$

$$+ \frac{1}{2} V_{ijk}^c x_i \left(k_1\, x_j + \dot{\lambda}\, y_j\right)\left(k_1\, x_k + \dot{\lambda}\, y_k\right) = 0. \tag{10.43}$$

(c) The perturbation equation of stability (10.35) is

$$\left[\dot{\lambda} V_{ij}'^c x_j + V_{ij}^c \dot{x}_j + V_{ijk}^c x_j \dot{q}_k\right] \xi^{\frac{1}{2}} + O\left(\xi\right) = 0. \tag{10.44}$$

The main term leads to the condition for $\dot{x}_j$

$$V_{ij}^c \dot{x}_j = -\left(V_{ijk}^c \dot{q}_k x_j + \dot{\lambda} V_{ij}^{\prime c} x_j\right). \tag{10.45}$$

(d) The contracted equation of stability becomes

$$\left[\dot{\lambda} V_{ij}^{\prime c} x_i x_j + V_{ijk}^c x_i x_j q_k\right] \xi^{\frac{1}{2}} + O(\xi) = 0.$$

The main term here is

$$\left(\dot{\lambda} V_{ij}^{\prime c} x_j + V_{ijk}^c x_j q_k\right) x_i = 0.$$

But the term in parentheses is the same term as the right-hand side of (10.45). Thus, the right-hand side of (10.45) is orthogonal to the eigenvector.

Substitution of q_k yields

$$\dot{\lambda} V_{ij}^{\prime c} x_i x_j + k_1 V_{ijk}^c x_i x_j x_k + \dot{\lambda} V_{ijk}^c x_j x_i y_k = 0. \tag{10.46}$$

We now have two conditions, (10.43) and (10.46), for the evaluation of k_1 and $\dot{\lambda}$.

In this section we employed the exponents of (10.38); however, other choices could have been made. A systematic procedure for the determination of the possible exponents was made in [3] and will not be included here for the sake of brevity. But it can be easily shown that any other choice made would lead to inconsistent results; i.e., the equations would not lead to the determination of the coefficients. In fact, the only choice for the present case leading to a consistent set of equations is (10.38). This procedure of selecting exponents based on the conditions that emerge is called the least degenerate equation.

Higher-order terms can be obtained using the same procedure. The results can be found in [4] and are of the form

$$\Lambda^M = \Lambda^c + \dot{\lambda}\, \xi^{\frac{1}{2}} + \ddot{\lambda}\, \xi + \cdots,$$

$$Q_j^M = q_j^c + \dot{q}_j\, \xi^{\frac{1}{2}} + \ddot{q}_j\, \xi + \cdots,$$

$$x_j^M = x_j + \dot{x}_j\, \xi^{\frac{1}{2}} + \ddot{x}_j\, \xi + \cdots.$$

In this problem, the sensitivity is usually called the **1/2 power law**, because of the exponent 1/2 in the first-order approximation in the sensitivity of the eigenvalue Λ^M.

10.6.2 Algorithm for Asymmetric Bifurcation

A summary of the equations of the previous section can be arranged in the form of an algorithm:

1. Solve
$$V_{ij}^c \, y_j = -V_i'^c \qquad \text{with} \qquad y_1 = 0.$$
2. Compute
$$C = V_{ijk} x_i x_j x_k \,|^c.$$
3. Solve two simultaneous equations
$$k_1 \dot{\lambda} \left(V_{ij}'^c x_i \, x_j + V_{ijk}^c x_i x_j y_k \right) + \frac{k_1^2}{2} C + (\dot{\lambda})^2 \left(V_{ij}'^c x_i y_j + \frac{1}{2} V_{ijk}^c x_i y_j y_k \right) + \dot{V}_i^c x_i = 0,$$
$$k_1 C + \dot{\lambda} \left(V_{ij}'^c x_i x_j + V_{ijk}^c x_j x_i y_k \right) = 0.$$
4. Evaluate
$$\dot{q}_j = k_1 \, x_j + \dot{\lambda} \, y_j.$$
5. Solve
$$V_{ij}^c \dot{x}_j = -\left(V_{ijk}^c \, x_j \, \dot{q}_k + \dot{\lambda} \, V_{ij}'^c \, x_j \right) \qquad \text{with} \qquad \dot{x}_1 = 0.$$
6. Results for first-order imperfection sensitivity are
$$\Lambda^M = \Lambda^c + \dot{\lambda} \xi^{\frac{1}{2}} + \cdots,$$
$$Q_j^M = q_j^c + \dot{q}_j \xi^{\frac{1}{2}} + \cdots,$$
$$x_j^M = x_j + \dot{x}_j \xi^{\frac{1}{2}} + \cdots.$$

Example 10.3 (Two-bar frame). *To demonstrate the concepts presented in the previous sections we study an example of the two-bar frame with an eccentricity in the application of the load. The problem was studied in section 7.8 under the condition that the bars are inextensible, similar to the assumptions made by Roorda* [31]. *The energy coefficients are given in section 7.8 using one finite element for each member of the frame. The problem has three degrees of freedom: the rotations of the three joints. A new term should be added to account for the imperfection, i.e., the product of the moment due to the eccentricity and the rotation of the central node. This yields*

$$V^{imp} = B_2 Q_2 = -\Lambda \xi L Q_2. \tag{10.47}$$

It was shown in Chapter 7 that this problem is an asymmetric bifurcation, with a trivial fundamental path, a critical load given by $\Lambda^c = 18.606$, *and an eigenvector*

$$x_j = \begin{Bmatrix} 1 \\ -0.58 \\ 0.29 \end{Bmatrix}.$$

The coefficient of third-order derivatives is

$$C = V_{ijk} x_i x_j x_k \,|^c = -1.4821.$$

We now proceed to use the algorithm for asymmetric bifurcation. The following derivative is obtained

$$V_j' \,|^c = 0,$$

so that

$$y_j = 0.$$

Let us call

$$\phi_1 = V_{ij}' x_i x_j \,|^c = V_{11}' (x_1)^2 + V_{22}' (x_2)^2 + V_{33}' (x_3)^2 = -0.216853,$$

$$\phi_2 = \dot{V}_j x_j \,|^c = \dot{V}_2 x_2 \,|^c = 10.79148.$$

Then we get the two simultaneous equations:

$$k_1 \dot{\lambda} \phi_1 + \frac{1}{2} k_1^2 C + \phi_2 = 0,$$

$$k_1 C + \dot{\lambda} \phi_1 = 0.$$

The solution yields

$$k_1 = 3.816, \qquad \dot{\lambda} = -26.08.$$

The imperfection sensitivity of the load is the most interesting aspect and results in

$$\Lambda^M = 18.606 - 26.08\, \xi^{\frac{1}{2}} + \cdots. \tag{10.48}$$

This has been plotted in Figure 10.3 *to illustrate the drop in the maximum load that can be expected in this frame problem.*

In the above case we only employed one finite element per bar to illustrate the analysis. The same problem has been computed using an increasing number of elements in each bar, and the results [4] *are shown in Table* 10.1, *for both first- and second-order sensitivity analysis.*

From the results, it is clear that sensitivity analysis of the present problem using just one finite element for each bar leads to significant errors in the first-order

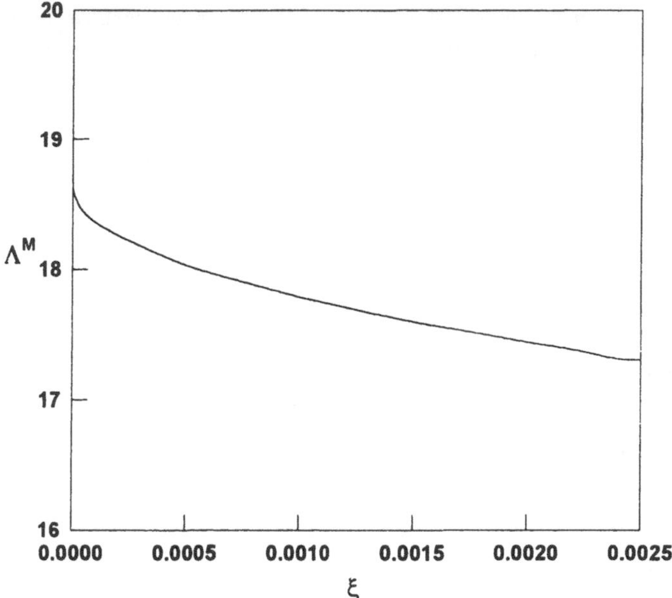

Figure 10.3 Imperfection sensitivity for a two-bar frame with eccentric load.

Table 10.1 Two-bar frame with load imperfection

Elements per bar	Λ^c	$\dot{\lambda}$	$\ddot{\lambda}$
1	18.606	26.083	−97.782
2	14.026	16.287	−14.778
3	13.919	16.067	−16.024
4	13.897	16.015	−15.878
5	13.890	15.999	−15.850
6	13.888	15.994	−15.841

sensitivity (63%) *and are unacceptable for second-order sensitivity. Two finite elements in each member, on the other hand, reduce the first-order error to* 1.8% *and the second-order error to* 6.7%.

10.7 IMPERFECTION SENSITIVITY IN SYMMETRIC BIFURCATION

10.7.1 Solution for Symmetric Bifurcation

The following conditions apply in symmetric bifurcation:

$$V_i' x_i = 0, \qquad C = 0.$$

We also assume that the imperfection is such that

$$\dot{V}_i x_i \neq 0.$$

The least degenerate solution in this case results in

$$\alpha = \frac{2}{3}, \qquad \beta = \frac{1}{3}, \qquad \gamma = \frac{1}{3}. \tag{10.49}$$

(a) Let us first consider the equilibrium equation. With the choice of (10.49), we get

$$\left[V^c_{ij}\dot{q}_j\right]\xi^{\frac{1}{3}} + \left[\dot{\lambda}V'^c_i + \frac{1}{2!}V^c_{ijk}\dot{q}_j\dot{q}_k\right]\xi^{\frac{2}{3}} + O(\xi) = 0.$$

From the main term we conclude that

$$V^c_{ij}\dot{q}_j = 0,$$

so that $\dot{q}_j$ can be written as

$$\dot{q}_j = k_1 x_j. \tag{10.50}$$

(b) Consider the contracted equation of equilibrium

$$0 + \left[\dot{V}^c_i x_i + \dot{\lambda}k_1 V'^c_{ij}x_i x_j + \frac{k_1^3}{3!}V^c_{ijkl}x_i x_j x_k x_l\right]\xi + O\left(\xi^{\frac{4}{3}}\right) = 0.$$

The main term in the above equation provides the condition

$$\dot{V}^c_i x_i + \dot{\lambda}k_1 V'^c_{ij}x_i x_j + \frac{k_1^3}{3!}V^c_{ijkl}x_i x_j x_k x_l = 0, \tag{10.51}$$

which is a nonlinear equation in terms of k_1 and $\dot{\lambda}$.

(c) The stability condition becomes

$$\left[V^c_{ij}\dot{x}_j + V^c_{ijk}x_j\dot{q}_k\right]\xi^{\frac{1}{3}} + \left[\dot{\lambda}V'^c_{ij}x_j + V^c_{ijk}\dot{x}_j\dot{q}_k\frac{1}{2!}V^c_{ijkl}x_j\dot{q}_k\dot{q}_l\right]\xi^{\frac{2}{3}} + O(\xi) = 0.$$

Because of (10.50), the main term becomes

$$V^c_{ij}\dot{x}_j = -k_1 V^c_{ijk}x_j x_k, \tag{10.52}$$

which has to be solved using the condition

$$\dot{x}_1 = 0.$$

There are n unknowns in (10.52): the $(n-1)$ values of $\dot{x}_j$ (for $j = 2, \ldots, n$) and k_1. Solution can be achieved if we write $\dot{x}_j$ in the form

$$\dot{x}_j = k_1 y_j. \tag{10.53}$$

Equation (10.52) can now be solved in terms of y_j,

$$V_{ij}^c y_j = -V_{ijk}^c x_j x_k. \tag{10.54}$$

(d) The contracted stability condition is

$$0 + \left[\dot{\lambda} V_{ij}'^c x_i x_j + k_1 V_{ijk}^c x_i \dot{x}_j x_k + \frac{1}{2} k_1^2 V_{ijkl}^c x_i x_j x_k x_l\right] \xi^{\frac{2}{3}} + O(\xi) = 0.$$

Next we substitute (10.53) into the above condition to get

$$\dot{\lambda} V_{ij}'^c x_i x_j + k_1^2 \left(\frac{1}{2} V_{ijkl}^c x_i x_j x_k x_l + V_{ijk}^c x_i y_j x_k\right) = 0. \tag{10.55}$$

The values of k_1 and $\dot{\lambda}$ can be computed from the simultaneous solution of (10.51) and (10.55).

Higher-order terms in the asymptotic expansion can be computed using the same procedure and are reported in [4]. It can be shown that the next set of exponents in symmetric bifurcation is given by

$$\Lambda^M = \Lambda^c + \dot{\lambda}\, \xi^{\frac{2}{3}} + \ddot{\lambda}\, \xi^{\frac{4}{3}} + \cdots,$$

$$Q_j^M = q_j^c + \dot{q}_j\, \xi^{\frac{1}{3}} + \ddot{q}_j\, \xi^{\frac{2}{3}} + \cdots,$$

$$x_j^M = x_j + \dot{x}_j\, \xi^{\frac{1}{3}} + \ddot{x}_j\, \xi^{\frac{2}{3}} + \cdots.$$

In this problem, the sensitivity is usually called the **2/3 power law**, because of the exponent 2/3 in the first-order approximation of the eigenvalue sensitivity, Λ^M. Notice that this problem is less sensitive to imperfections than the asymmetric bifurcation, for which the first exponent was 1/2.

10.7.2 Algorithm for Symmetric Bifurcation

Symmetric bifurcations with imperfections lead to the following sequence of computations:

1. Solve

$$V_{ij}^c y_j = -V_{ijk}^c x_j x_k \qquad \text{with} \qquad y_1 = 0.$$

2. Solve two simultaneous nonlinear equations in k_1 and $\dot{\lambda}$:

$$\dot{V}_i^c x_i + \dot{\lambda} k_1 V_{ij}^{\prime c} x_i x_j + \frac{k_1^3}{3!} V_{ijkl}^c x_i x_j x_k x_l = 0,$$

$$\dot{\lambda} V_{ij}^{\prime c} x_i x_j + k_1^2 \left(\frac{1}{2} V_{ijkl}^c x_i x_j x_k x_l + V_{ijk}^c x_i y_j x_k \right) = 0.$$

3. Evaluate

$$\dot{q}_j = k_1 x_j.$$

4. Evaluate

$$\dot{x}_j = k_1 y_j.$$

5. Results for first-order imperfection sensitivity are

$$\Lambda^M = \Lambda^c + \dot{\lambda}\, \xi^{\frac{2}{3}} + \cdots,$$

$$Q_j^M = q_j^c + \dot{q}_j\, \xi^{\frac{1}{3}} + \cdots,$$

$$x_j^M = x_j + \dot{x}_j\, \xi^{\frac{1}{3}} + \cdots.$$

Example 10.4 (Column on elastic foundation). *We now turn our attention to an unstable symmetric bifurcation to illustrate the procedure exposed above. The case is the column on an elastic foundation, in which the stiffness of the foundation is set to $k = 2$, a value already justified in section 7.9. The lowest eigenvalue is $\Lambda^c = 3$, and the associated eigenvector is*

$$x_j = \begin{Bmatrix} 1 \\ 0 \end{Bmatrix}.$$

For this case we have $V_j' = 0$, so that

$$y_j = 0.$$

The two coefficients result from the nonlinear equations

$$-1 - 2k_1\dot{\lambda} - \frac{5}{4}\pi^2 k_1^3 = 0,$$

$$-2\dot{\lambda} - \frac{15}{4}\pi^2 k_1^2 = 0,$$

and yield

$$k_1 = 0.343, \qquad \dot{\lambda} = -2.1834. \tag{10.56}$$

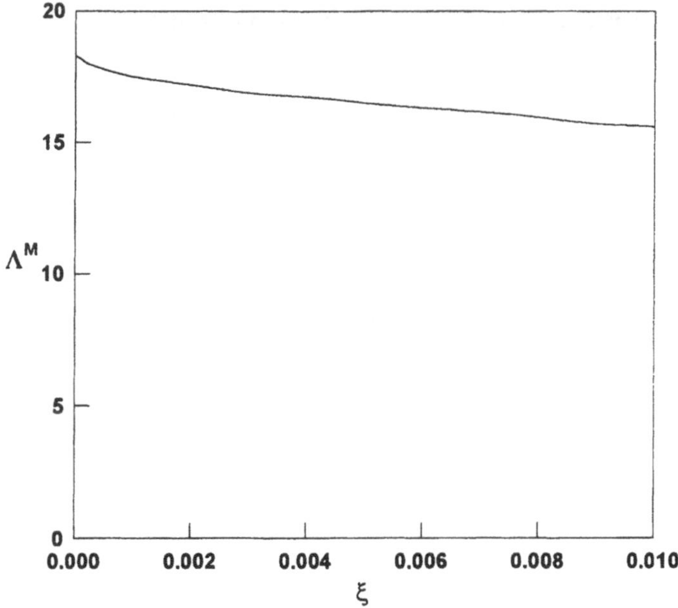

Figure 10.4 Imperfection sensitivity of a column on an elastic foundation, $k = 2$.

The change in the displacement vector is

$$q_j = \begin{Bmatrix} 0.3434 \\ 0 \end{Bmatrix},$$

and the eigenvector has no change due to the imperfection.

Results have been computed for first- and second-order sensitivity, leading to

$$\Lambda^M(\xi) = 3 - 2.18344\xi^{\frac{2}{3}} + 0.5085\xi^{\frac{4}{3}}, \tag{10.57}$$

$$Q_1^M = 0.3434\xi^{\frac{1}{3}} + 0.01\xi.$$

Figure 10.4 *shows plots of the imperfection sensitivity for this column on an elastic foundation.*

10.8 SENSITIVITY TO SMALL AND LARGE IMPERFECTIONS

"Real systems are probably sensitive to large imperfections in more complicated ways than 1/2 and 1/3 power-law rules. This may lead to systems, which although highly sensitive to small imperfections, may not be so sensitive to large imperfections, thereby leading to a range of imperfections for which a buckling load can be defined" [7, p. 53].

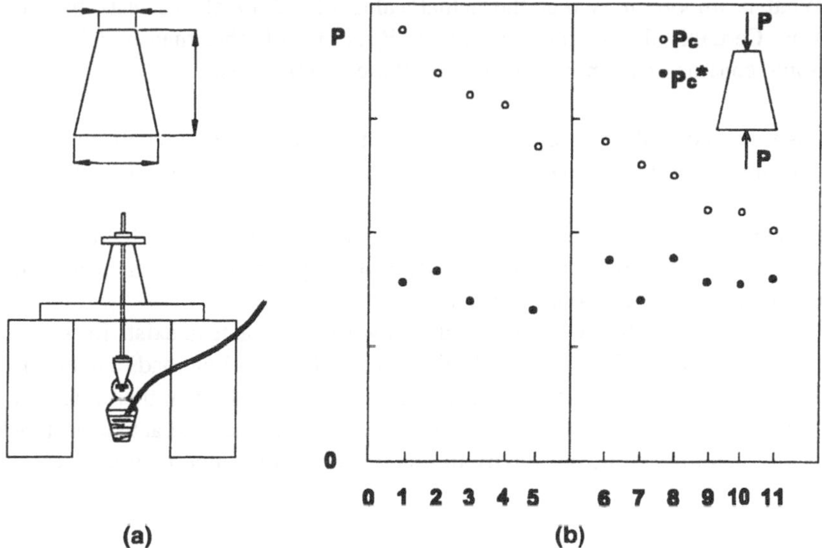

Figure 10.5 (a) Experimental setup for conical shells; (b) maximum loads for different coupons.

A simple, yet instructive, set of experiments to illustrate the sensitivity of shells to imperfections has been done as follows [15]: Tests were performed on thin-walled PVC conical frustra, as shown in Figure 10.5. Such shells were commercially available and were tested under axial load. Loads were applied to wood head platens at the top (smaller diameter end of the conical shell) by means of water poured on a bucket and transferred to the wood head by a bar. Thus, it was possible to increase the load in a continuous way.

Buckling was identified visually by formation of circumferential waves under constant load. This load is P_c. Under small increment of load, the shells showed collapse, with plasticity and cracks. Figure 10.5.b shows the results for different tests. There are significant differences in the results, with values ranging from $P = 187N$ in test 1 to $P = 104N$ in test 11. The differences are due to small imperfections among shells, both in the geometry and in the material. There are also sources of imperfections in the load system.

Next the shells were repaired using tape and tested again using the same procedure. In this new conditions, the shells had gross defects before the stability test. Again buckling was identified by circumferential waves, and the new buckling load was designated P_c^*.

The new results are of course lower than the value of P_c for the same shell. The reduction factor is

$$0.37 < \frac{P_c^* - P_c}{P_c} < 0.58.$$

For the first five shells it was of the order of 0.54, while for the last six shells it was around 0.3.

What is important is that the actual value of P_c^* in all 11 tests had a small variation. Certainly the scatter of results in P_c is much higher than in P_c^*.

Some conclusions can be drawn from these simple tests:

- Thin-walled conical shells are very sensitive to the presence of small imperfections. Such imperfections are unavoidable in practice and arise from material, geometry, and load.
- The sensitivity of buckling loads of conical shells to large imperfections is small. On the other hand, these are more manageable imperfections, in the sense that it is possible to have better control over them.
- "Stability theory is likely to be most useful when ... a plateau exists for sufficiently large values of imperfection. The plateau ensures that buckling loads can be specified with confidence over a range of imperfections. It is not sufficient therefore simply to determine whether postbuckling is stable or unstable, since the actual postbuckling behavior determines the sensitivity to randomly varying imperfections" [7].

10.9 FINAL REMARKS

The results of the nonlinear analysis are valid for even large imperfections and also away from the bifurcation state. The perturbation analysis for sensitivity, on the other hand, is valid in an asymptotic way: Imperfection amplitudes should be small (as they are taken as a perturbation parameter).

The approach using singular perturbations for imperfection sensitivity can be applied to other cases, such as the limit point, trifurcations, and coincident modes. For limit point analysis the same results obtained in Chapter 8 are reproduced by the present formulation. For trifurcations, in which three equilibrium paths are available at the critical state, and in coincident modes, in which more than one eigenvector is associated with the same eigenvalues, the analysis produces new results with the same approach.

For a number of years a different (indirect) solution was employed in the theory of stability (see, for example, [18, 37, 8, 19, 11]). In the indirect approach, the perturbation expansion is carried out with respect to another parameter, say, Q_1, and series are found in the form $\Lambda = \Lambda\,(Q_1)$, $Q_j = Q_j\,(Q_1)$, and $\xi = \xi\,(Q_1)$. Then the middle expansion is inverted to provide $Q_1 = Q_1\,(\xi)$, and the result is substituted into the two remaining equations to produce the desired expansions. The three primary expansions are regular and can be obtained in the usual straightforward way. The final expansions are in terms of fractional powers and ξ has an infinite slope at $\xi = 0$. In the approach described in this chapter there is no need to employ intermediate expansions, and the required relations are obtained directly.

In this chapter we considered deterministic imperfections in the sense that the shape of the imperfection is fixed and parametric or sensitivity studies are carried out with respect to the remaining parameters (usually the amplitude of the imperfection). Stochastic simulations have also been proposed in the literature [10], but this approach would seldom be chosen by an analyst.

A different procedure would be to determine the worst imperfection, associated with the largest reduction in the maximum load that the structure can take. This was pursued in [28] to determine the most undesirable imperfection.

Finally, there are studies of imperfection sensitivity in the plastic range, carried out within the framework of the asymptotic theory discussed in this book. The fundamental work in this field is due to Hutchinson, and the reader is referred to [24, 25, 27] for more details.

10.10 PROBLEMS

Review questions. (*a*) What is the meaning of the condition $\dot{V}_i x_i \neq 0$? (*b*) Is it possible to use regular perturbations for sensitivity of a limit point? (*c*) Explain the differences between imperfection sensitivity of a limit point when $\dot{V}_i x_i = 0$ and when $\dot{V}_i x_i \neq 0$. (*d*) Give reasons for the disagreement between theory and experiments in bifurcation buckling. (*e*) Discuss the statement, "In bifurcation buckling the critical load may be sensitive to imperfections, but the eigenvector is not affected by the imperfection." (*f*) What are the differences between singular and regular perturbation sensitivity analyses? (*g*) Is symmetric bifurcation more sensitive to imperfections than asymmetric bifurcation? (*h*) What is the "2/3 power law"?

Problem 10.1 (Theory). Calculate the second-order coefficients in the perturbation analysis for a limit point behavior with $\dot{V}_i x_i \,|^c \neq 0$. Follow section 10.3.1.

Problem 10.2 (Circular arch). Evaluate the first-order sensitivity for the arch with geometric deviation of Example 8.5 using the perturbation algorithm of section 10.3.2.

Problem 10.3 (Plane frame). This is the same as Problem 5.9 but with an imperfection in the form of a load applied to the center of one of the members, in the direction perpendicular to it.

Problem 10.4 (One degree-of-freedom mechanical model). A one degree-of-freedom problem is shown in Figure 10.6. The system is governed by the total potential energy

$$V = Q^2 - \frac{1}{3}Q^4 - \lambda\left(Q^2 - \frac{1}{12}Q^4\right) - \xi\left(2Q - \frac{1}{2}Q^3\right).$$

Evaluate the imperfection sensitivity using the theory presented in this chapter.

Problem 10.5 (One degree-of-freedom mechanical model). Figure 10.7 shows a one degree-of-freedom mechanical model, for which the energy is approximated as

$$V = Q^2 + \frac{1}{2}Q^3 - \frac{1}{48}Q^4 - \lambda\left(Q^2 - \frac{1}{12}Q^4\right) - \xi\left(2Q + \frac{1}{2}Q^2 - \frac{1}{12}Q^3 - \frac{1}{96}Q^4\right).$$

Plot the imperfection-sensitivity diagram using singular perturbations as explained in this chapter.

Problem 10.6 (Column on elastic foundation). Plot the imperfection-sensitivity curve of displacements and eigenvectors for the problems of Example 10.4.

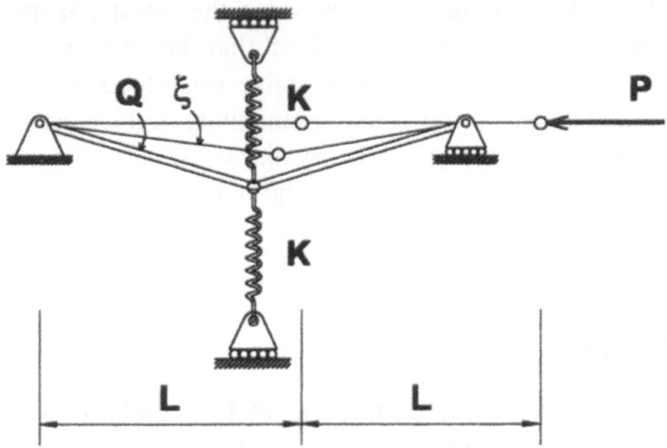

Figure 10.6 Problem 10.4.

Problem 10.7 (Two-degrees-of-freedom mechanical model) [6]. Let us consider again the two-degrees-of-freedom mechanical model presented in Problem 7.6, but include an imperfection $\varepsilon \neq 0$. The only changes in the total potential energy with respect to that given in Problem 7.6 are

$$(\Delta_1 + \Delta_2)/L = (Q_1^2 - Q_1 Q_2 + Q_2^2)$$

$$-\frac{1}{4}\left(Q_1^4 - 2Q_1^3 Q_2 + 3Q_1^2 Q_2^2 - 2Q_1 Q_2^3 + Q_2^4\right)$$

$$+\varepsilon\left[2\left(g_1 Q_1 + g_2 Q_2\right) - \left(g_2 Q_1 + g_1 Q_2\right)\right].$$

Obtain the imperfection sensitivity of the problem with respect to the imperfection parameter ε. Consider $g_1 = g_2 = 1$.

Problem 10.8 (Two-degrees-of-freedom mechanical model). Repeat Problem 10.7 using $g_1 = 1$, $g_2 = 0$. Compare this with the results of the previous problem.

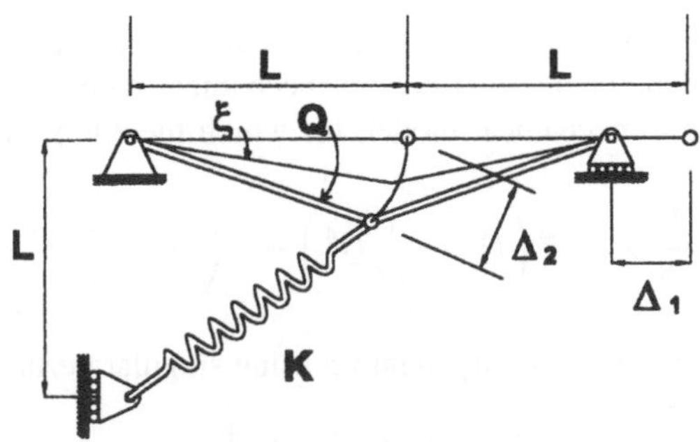

Figure 10.7 Problem 10.5.

10.11 BIBLIOGRAPHY

[1] Almroth, B. O., Post-buckling behavior of axially compressed circular cylinders, *AIAA J.*, 1, 630–633, 1963.

[2] Amazigo, J. C., Budiansky, B., and Carrier, G. F., Asymptotic analyses of the buckling of imperfect columns on non-linear elastic foundations, *Internat. J. Solids Structures*, 6, 1341, 1970.

[3] Banchio, E. G., and Godoy, L. A., Nueva formulacion de sensibilidad para estabilidad estructural, *Meccanica Computacional*, 17, 133–142, 1996.

[4] Banchio, E. G., and Godoy, L. A., New approach for imperfection sensitivity in buckling analysis, submitted for publication.

[5] Banchio, E. G., Godoy, L. A., and Mook, D. T., Un metodo de menor degeneracion para problemas de perturbaciones singulares, *Meccanica Computacional*, 15, 359–368, 1995.

[6] Brewer, A. T., and Godoy, L. A., Dynamic buckling of discrete structural systems under combined step and static loads, *Internat. J. Non-Linear Dynamics*, 9, 249–264, 1996.

[7] Chilver, A. H., The elastic stability of structures, in *Stability*, H. H. E. Leipholz, Ed., University of Waterloo, 1972.

[8] Croll, J. G. A., and Walker, A. C., *Elements of Structural Stability*, Macmillan, London, 1972.

[9] Donnell, L. H., and Wang, C. C., Effect of imperfections on buckling of thin cylinders and columns under axial compression, *J. Appl. Mech.*, 17, 73–83, 1950.

[10] Elisharkoff, I., Stochastic simulation of an initial imperfection data bank for isotropic shells with general imperfections, in *Buckling of Structures*, I. Elisharkoff et al., Eds., Elsevier, Amsterdam, 1988, 195–209.

[11] El Naschie, M. S., *Stress, Stability, and Chaos in Structural Engineering: An Energy Approach*, McGraw-Hill, London, 1990.

[12] Esslinger, M., and Geier, B., *Post-Buckling Behavior of Structures*, Springer-Verlag, Wien, 1975.

[13] Flores, F. G., and Godoy, L. A., Elastic post-buckling analysis via finite element and perturbation techniques. Part 1: Formulation, *Internat. J. Numer. Methods Engrg.*, 33, 1775–1794, 1992.

[14] Fok, W. C., Walker, A. C., and Rhodes, J., Buckling of locally imperfect stiffeners in plates, *J. Engrg. Mech. Div., ASCE*, 103, 895–911, 1977.

[15] Godoy, L. A., and Lagorio, J., *Inestabilidad del Equilibrio en Laminas Tronco-Conicas: Estudio Experimental*, Technical Report, Structures Department, National University of Cordoba, Argentina, December 1980.

[16] Godoy, L. A., and Mook, D. T., Higher order sensitivity to imperfections in bifurcation buckling analysis, *J. Solids Structures*, 33(4), 511–520, 1996.

[17] Ho, D., Higher order approximations in the calculation of elastic buckling loads of imperfect systems, *Internat. J. Non-Linear Mech.*, 6, 649, 1971.

[18] Hunt, G. W., *Perturbation Techniques in Elastic Stability*, Ph.D. thesis, University College, London, 1971.

[19] Huseyin, K., *Non-Linear Theory of Elastic Stability*, Noordhoff, Leyden, 1975.

[20] Hutchinson, J. W., Axial buckling of pressurized imperfect cylindrical shells, *AIAA J.*, 3, 1461, 1965.

[21] Hutchinson, J. W., Imperfection sensitivity of externally pressurized spherical shells, *J. Appl. Mech.*, March, 49–55, 1967.

[22] Hutchinson, J. W., and Amazigo, J. C., Imperfection-sensitivity of eccentrically stiffened cylindrical shells, *AIAA J.*, 5, 392, 1967.

[23] Hutchinson, J. W., Tennyson, R. C., and Muggeridge, D. B., Effect of a local axisymmetric imperfection on the buckling behavior of a circular cylindrical shell under axial compression, *AIAA J.*, 9, 48, 1971.

[24] Hutchinson, J. W., On the post-buckling behavior of imperfection-sensitive structures in the plastic range, *J. Appl. Mech.*, 39, 155–162, 1972.

[25] Hutchinson, J. W., Post-bifurcation behavior in the plastic range, *J. Mech. Phys. Solids*, 21, 163–190, 1973.

[26] Hutchinson, J. W., Imperfection-sensitivity in the plastic range, *J. Mech. Phys. Solids*, 21, 191–204, 1973.

[27] Hutchinson, J. W., Plastic buckling, in *Advances in Applied Mechanics*, vol. 14, C. S. Yih, Ed., Academic Press, New York, 1974, 67–144.
[28] Ikeda, K., and Murota, K., Critical initial imperfection of structures, *Internat. J. Solids Structures*, 26(8), 865–886, 1990.
[29] Koiter, W. T., *On the Stability of Elastic Equilibrium*, Doctoral Dissertation, Delft, 1945 (translated at NASA, Tech. Trans. F10, 833, 1967).
[30] Koiter, W. T., The effect of axisymmetric imperfections on the buckling of cylindrical shells under axial compression, *Proc. Konink. Nederl. Akad. Wetensch., Ser. B.*, 66, 265, 1963.
[31] Roorda, J., Stability of structures with small imperfections, *J. Engrg. Mech. Div., ASCE*, 91, 87, 1965.
[32] Roorda, J., The buckling behavior of imperfect structural systems, *J. Mech. Phys. Solids*, 13, 267, 1965.
[33] Roorda, J., Some thoughts on the Southwell plot, *J. Engrg. Mech. Div., ASCE*, 93, 37, 1967.
[34] Roorda, J., On the buckling of symmetric structural systems with first and second order imperfections, *Internat. J. Solids Structures*, 4, 1137, 1968.
[35] Seide, P., A reexamination of Koiter's theory of initial postbuckling behavior and imperfection sensitivity of structures, in *Thin-Shell Structures*, Y. C. Fung and E. E. Sechler, Eds., Prentice-Hall, Englewood Cliffs, NJ, 1974.
[36] Thompson, J. M. T., The branching analysis of perfect and imperfect discrete structural systems, *J. Mech. Phys. Solids*, 17, 1, 1969.
[37] Thompson, J. M. T.. and Hunt, G. W., *A General Theory of Elastic Stability* (Chap. 8), Wiley, London, 1973.
[38] Uchiyama, K., and Yamada, M., Buckling of clamped imperfect thin-shallow spherical shells under external pressure, I. The effects of geometrically symmetrical initial imperfections, 39, *Tech. Rep. Tohoku Univ.*, 101–132, 1974.
[39] Uchiyama, K., and Yamada, M., Buckling of clamped imperfect thin-shallow spherical shells under external pressure, II. The effects of geometrically asymmetrical initial imperfections, 40, *Tech. Rep. Tohoku Univ.*, 1–23, 1975.
[40] von Karman, T., and Tsien, H. S., The buckling of thin cylindrical shells under axial compression, *J. Aeronautical Sci.*, 8, 303–312, 1941.

CHAPTER

ELEVEN

DESIGN SENSITIVITY OF BIFURCATION STATES

11.1 INTRODUCTION

This chapter presents a general formulation for the sensitivity analysis of critical bifurcation states with respect to changes in a design parameter, denoted by τ. We assume that these changes do not destroy the bifurcation; this means that after a change in the design parameter takes place, the response is still a bifurcation. Cases in which a bifurcation is destroyed by changes in one of the parameters (imperfections) were considered in Chapter 10.

To further clarify the problem studied, let us consider the response shown in Figure 11.1. Equilibrium states are plotted in a load Λ-displacement Q_i diagram for bifurcation behavior. Solid lines represent stable equilibrium states, while dotted lines are associated with unstable states. The fundamental or primary path of equilibrium states arises from the origin and may in general be nonlinear; however, the presentation in this chapter is restricted to linear fundamental paths. A critical state is also plotted and assumed to occur at Λ^c and Q_i^c. A nonlinear path emerges from that state. In Figure 11.1, such secondary path is plotted as a stable path, in the sense that there are stable equilibrium states at values higher than the critical load. All the above may be computed for a specific value of a design parameter τ, for example, $\tau = 0$. But for $\tau \neq 0$ several changes may occur in the response:

- The primary path may change. This is considered, for example, in [7].
- The critical state may change. This is the subject of this chapter.
- The postcritical path may be modified in values and also in nature (changing from stable to unstable). This topic is explained in the next chapter.

There are many studies of sensitivity of eigenproblems, with particular application to the free vibration response (natural frequencies and modes of vibration) of

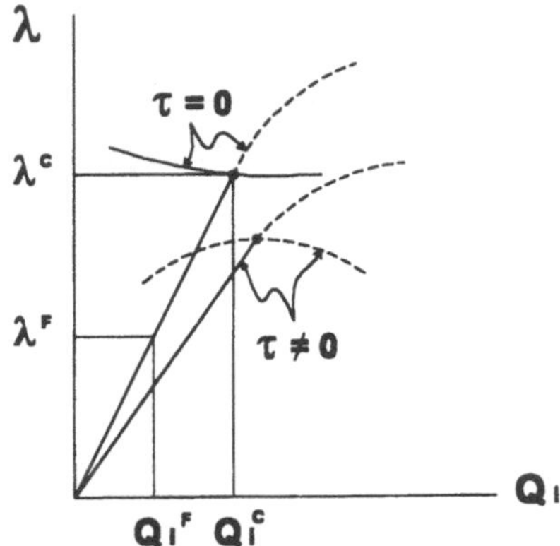

Figure 11.1 Examples of primary and secondary paths for a reference system and for a system with changes in a design parameter.

an unloaded structure, and an insight into the literature may be found in [4]. The simplest techniques currently employed in sensitivity analysis are based on the finite difference method. The application of perturbation techniques in sensitivity analysis in conjunction with generalized coordinates to model the structure is relatively new (see, for example, [1, 11]).

A distinctive feature of buckling problems is that the eigenproblem is subject to constraints: The solution should satisfy both equations of equilibrium (a linear or nonlinear condition) **and** critical state (an eigenproblem). This problem is more complex than the sensitivity of free vibration of an unloaded structure and is similar to the sensitivity of free vibrations of a preloaded structure.

There are several studies of sensitivity of eigenproblems in which equilibrium constraints are taken into account. The **direct method** was first proposed, and the **adjoint method** started with [3]. The adjoint method was extended to changes in the shape and boundary conditions of the system [15] and thermoelastic problems [2]. A summary of the contributions in this area may be found in [16, 17]. Most of the work done by these authors refers to first-order sensitivity analysis, in the sense that first-order derivatives of state variables with respect to the design parameters are involved. However, second-order sensitivity expressions are important in many cases depending on the particular applications. As stated in [14], the success of many computational algorithms relies directly upon an effective and accurate scheme to calculate higher-order sensitivity information.

This chapter explores the first- and second-order sensitivity of eigenvalue problems with respect to a chosen design parameter τ. It is assumed that changes in τ do not destroy the bifurcation point. Only one design parameter is considered, but the extension to multiple design parameters is possible with extra algebra involved.

The perturbation equations of the problem are presented in section 11.2, including first- and second-order perturbation of equilibrium, eigenproblem, normalization of the eigenvector, and contracted eigenproblem. The direct method of solution is explained in section 11.3, and the adjoint method is the subject of section 11.4. The material of these sections has been published by the author in [10, 6]. Simplifications to the formulation are studied in section 11.5. The direct and adjoint techniques are applied to a simple problem of a circular plate, an I-section thin-walled column, and a rectangular plate. Simplifications are also studied in this section for the circular plate and a finite strip analysis of folded plates; this follows the work of [9]. The results show that large errors are introduced by the simplifications considered.

11.2 DESIGN PARAMETERS

Design parameters may be modified in a structure; these are changes introduced by an engineer to improve a solution, and it is often necessary to evaluate the effects associated with this modification. We are interested not only in the quality of the changes (if improvements or changes for the worse occur) but also in the magnitude of the changes. These studies are known as design sensitivity, and a good account of this field is given in [13].

In some problems the design parameter is present in the formulation, perhaps in explicit form. This is the case of the modulus of elasticity, Poisson's ratio, etc. If we investigate sensitivity of the stability problem with changes in the material properties that affect the complete structural component, then we define the modulus E in the form

$$E = E_0 (1 + \tau) , \tag{11.1}$$

where τ is a design parameter and E_0 is a reference value of the modulus.

To investigate the influence of design changes we may have to define a new design parameter. This occurs whenever a specific feature of the design has a continuous variation in space, but we want to use only a few parameters to describe it, together with functions of the space coordinates.

For example, a certain modification in the shape of a structural component may affect an area of this component with a continuous variation. However, we employ a shape function for this modification and a scalar parameter (or several) so that the parametric definition of the design is now part of the energy functional.

Example 11.1 (Circular plate with thickness changes). *The energy coefficients of the plate have been obtained in Chapter 2 and are*

$$A_2 = \phi, \qquad A_{11} = k_1\, h^3, \qquad A_{22} = k_2\, h, \qquad A_{112} = m\, h,$$

and the rest of the coefficients are zero. The relation between the parameters and the plate flexural and extensional stiffness is given by

$$k_1 = \frac{\pi}{24}\frac{1.191 + \nu}{1 - \nu^2}\frac{E}{R^2}, \qquad k_2 = 2E\frac{1 + \nu}{1 - \nu^2},$$

$$m = \frac{\pi^2 + 4}{8}\frac{1 + \nu}{1 - \nu^2}\frac{E}{R}, \qquad \phi = -2\pi R,$$

where E and ν are the modulus of elasticity and of Poisson's ratio.

To investigate the influence of changes in the thickness h of the plate, we define the thickness in the parametric form

$$h = h_0(1 + \tau). \tag{11.2}$$

This leads to

$$V(Q_1, Q_2, \Lambda, \tau) = V(Q_1, Q_2, \Lambda) + \bar{V}(Q_1, Q_2, \tau), \tag{11.3}$$

where $V(Q_1, Q_2, \Lambda)$ *is the energy of the reference system with* $h = h_0$*; and*

$$\bar{V}(Q_1, Q_2, \tau) = \frac{1}{2}(B_{11}Q_1^2 + B_{22}Q_2^2) + \frac{1}{3!}(B_{112}Q_1^2 Q_2). \tag{11.4}$$

The new terms in the energy associated with the design parameter are

$$B_{11} = k_1\left(3\tau + 3\tau^2 + \tau^3\right), \qquad B_{22} = \tau\, k_2, \qquad B_{112} = \tau\, m.$$

There are linear, quadratic, and cubic terms in τ *in the total potential energy. Thus, a linear change in* τ *may have a nonlinear effect on the response.*

11.3 PERTURBATION EQUATIONS

We start by recalling the equations that are satisfied at a critical state: They are

$$V_i = 0 \quad \Longrightarrow \quad \text{equilibrium}, \tag{11.5}$$

$$V_{ij}x_j = 0 \quad \Longrightarrow \quad \text{critical state}, \tag{11.6}$$

where V is the total potential energy of the elastic system, written in terms of multiple generalized coordinates Q_i and a single load parameter Λ, so that $V = V[Q_i, \Lambda]$; and x_j is the eigenvector.

Let us look at the particular case of the specialized system, for which the energy results in the form

$$V = V[Q_i, \Lambda] = A_0 + A_i Q_i + \frac{1}{2}A_{ij}Q_iQ_j + \frac{1}{3!}A_{ijk}Q_iQ_jQ_k + \frac{1}{4!}A_{ijkl}Q_iQ_jQ_kQ_l. \tag{11.7}$$

The specialized system is defined in Chapter 2, as a system for which the load potential is linear in the load parameter Λ.

Because of (11.7), the linear equilibrium conditions of the fundamental path become

$$V_i = A_i + A_{ij}Q_j|^F = 0. \tag{11.8}$$

The solution of this equation is the fundamental path, and the fundamental displacements for a unit load are called $\bar{Q}_i$, so that

$$Q_i^F = \Lambda \bar{Q}_i.$$

The conditions of a critical state, (11.6), are

$$(A_{ij} + \Lambda_c A_{ijk} \bar{Q}_k) x_j^c = 0. \tag{11.9}$$

Finally, there are several ways to normalize the eigenvector x_j^c; in some cases the normalization

$$(A_{ijk} \bar{Q}_k) x_j^c x_j^c = 1 \tag{11.10}$$

is employed, especially in structural dynamics. In other cases it is more convenient to choose one of the coordinates of x_j and set it equal to unity, i.e.,

$$x_1^c = 1, \tag{11.11}$$

as has been done in previous chapters of this book.

Next, sensitivity of the critical state is required with respect to a design parameter τ. It is assumed that the coefficients A_i, A_{ij}, A_{ijk}, A_{ijkl} are functions of τ, and we want to find explicit expressions for

$$\bar{Q}_j = \bar{Q}_j(\tau), \qquad \Lambda_c = \Lambda_c(\tau), \qquad x_j^c = x_j^c(\tau), \tag{11.12}$$

that is, sensitivity of the fundamental path, of the critical load and the eigenvector with respect to changes in the design parameter τ.

It is possible to expand the variables of interest in terms of a perturbation expansion [18] and to choose τ as perturbation parameter, i.e.,

$$Q_j(\tau) = \bar{Q}_j + \dot{Q}_j \tau + \frac{1}{2} \ddot{Q}_j \tau^2 + \cdots, \tag{11.13}$$

$$\Lambda(\tau) = \Lambda_c + \dot{\Lambda}_c \tau + \frac{1}{2} \ddot{\Lambda}_c \tau^2 + \cdots,$$

$$x_j^c(\tau) = x_j^c + \dot{x}_j^c \tau + \frac{1}{2} \ddot{x}_j^c \tau^2 + \cdots.$$

Derivatives with respect to the design parameter will be denoted by a dot on the variable.

We assume here that the required derivatives exist (i.e., the variables in (11.13) are continuous and differentiable). This is typical of cases in which the eigenvalue problem is preserved even with changes in τ, and it evolves along the fundamental path.

Substitution of (11.13) into (11.8)–(11.10) leads to a set of equations in terms of τ^0, τ, τ^2, τ^3, etc. Collecting terms of equal powers in τ yields the perturbation equations of order 0, 1, 2, 3, etc., which are next evaluated at $\tau = 0$.

11.3.1 Equilibrium Conditions

The basic condition of equilibrium is given in (11.8). The first-order perturbation equations arise from

$$\frac{d}{d\tau}(V_i)|_{\tau=0} = 0$$

or else

$$A_{ij}\dot{Q}_j + \dot{A}_{ij}\bar{Q}_j + \dot{A}_{ij} = 0.$$

A convenient way to write this equation is

$$A_{ij}\dot{Q}_j = -u_i^1, \tag{11.14}$$

where

$$u_i^1 \equiv \dot{A}_{ij}\bar{Q}_j + \dot{A}_i \tag{11.15}$$

is the collection of terms not related to the unknowns, which can be computed from the reference state and the derivatives of coefficients.

The second-order perturbation equations

$$\frac{d^2}{d\tau^2}(V_i)|_{\tau=0} = 0$$

may be shown to result in

$$A_{ij}\ddot{Q}_j = -u_i^2, \tag{11.16}$$

where

$$u_i^2 \equiv (\ddot{A}_{ij}\bar{Q}_j + \ddot{A}_i) + 2(\dot{A}_{ij}\dot{Q}_j). \tag{11.17}$$

11.3.2 The Eigenproblem

The bifurcation loads and critical modes (or, similarly, the vibration frequencies and modes of vibration) arise from the linear eigenproblem (11.9). The first-order perturbation equation is

$$\frac{d}{d\tau}(V_{ij}x_j)|_{\tau=0} = 0$$

or else

$$(A_{ij} + \Lambda_c A_{ijk}\bar{Q}_k)\dot{x}_j + \dot{\Lambda}(A_{ijk}\bar{Q}_k x_j^c) + (\Lambda A_{ijk}x_j)\dot{Q}_k = -v_i^1 \tag{11.18}$$

with

$$v_i^1 = (\dot{A}_{ij} + \Lambda\dot{A}_{ijk}\bar{Q}_k)x_j^c. \tag{11.19}$$

The second-order perturbation equation

$$\frac{d^2}{d\tau^2}(V_{ij}x_j)|_{\tau=0} = 0$$

results in

$$(A_{ij} + \Lambda_c A_{ijk}\bar{Q}_k)\ddot{x}_j + \ddot{\Lambda}(A_{ijk}\bar{Q}_k x_j^c) + (\Lambda A_{ijk} x_j)\ddot{Q}_k = -v_i^2, \tag{11.20}$$

where

$$\begin{aligned} v_i^2 = & \left[\ddot{A}_{ij} + \Lambda_c(\ddot{A}_{ijk}\bar{Q}_k)\right] x_j^c + 2\dot{\Lambda}\left[(\dot{A}_{ijk}\bar{Q}_k + A_{ijk}\dot{Q}_k)\right] x_j^c \\ & +2\left[\dot{A}_{ij} + \Lambda_c(\hat{V}_{ijk}\bar{Q}_k + A_{ijk}\dot{Q}_k)\right]\dot{x}_j \\ & +2\Lambda\left[A_{ijk}\bar{Q}_k\right]\dot{x}_j + \Lambda(\dot{A}_{ijk}\dot{Q}_k)x_j^c. \end{aligned} \tag{11.21}$$

The terms in v_i^2 depend on the solution of the previous perturbation systems.

11.3.3 The Condition of Normalization of the Eigenvector

Let us consider two possibilities of normalization of the eigenvector:

(a) If we have chosen to normalize the eigenvector in the form of (11.10), i.e.,

$$(A_{ijk}\bar{Q}_k x_i^c x_j^c)|_{\tau=0} = 1,$$

then perturbation of this condition results in

$$\frac{d}{d\tau}(A_{ijk}\bar{Q}_k x_i^c x_j^c)|_{\tau=0} = 0$$

or else

$$2(A_{ijk}\bar{Q}_k x_j^c)\dot{x}_i + (A_{ijk} x_i^c x_j^c)\dot{Q}_k = -w_1, \tag{11.22}$$

where

$$w_1 \equiv \dot{A}_{ijk}\bar{Q}_k x_i^c x_j^c. \tag{11.23}$$

The second-order perturbation equation

$$\frac{d^2}{d\tau^2}(A_{ijk}\bar{Q}_k x_i^c x_j^c)|_{\tau=0} = 0$$

results in

$$2\left(A_{ijk}\bar{Q}_k x_j^c\right)\ddot{x}_i + \left(A_{ijk} x_i^c x_j^c\right)\ddot{Q}_k = -w_2 \tag{11.24}$$

with

$$\begin{aligned} w_2 \equiv & \left(\ddot{A}_{ijk}\bar{Q}_k\right) x_i^c x_j^c + 4\left(\dot{A}_{ijk}\bar{Q}_k + A_{ijk}\dot{Q}_k\right) x_i^c x_j^c \\ & +2\left(\dot{A}_{ijk}\dot{Q}_k\right) x_i^c \dot{x}_j + 2\left(A_{ijk}\bar{Q}_k\right)\dot{x}_i\dot{x}_j. \end{aligned} \tag{11.25}$$

This normalization has been used in [10, 6].

(b) A simpler form may be obtained if (11.11) is used for the normalization instead of (11.10). In this case, one has

$$x_1 = 1, \tag{11.26}$$

$$\dot{x}_1 = 0, \qquad \ddot{x}_1 = 0. \tag{11.27}$$

11.3.4 The Contracted Eigenproblem

Because the eigenproblem has a singularity of order one, then another scalar equation is required to solve the complete system of perturbation equations. A convenient choice is to use the contraction mechanism on the eigenproblem to produce a new condition. Contraction is achieved by premultiplication of the perturbation equations of the eigenproblem by the eigenvector and has been used in most chapters of this book.

The first-order contracted equation is

$$\dot{\Lambda}\left(A_{ijk}\bar{Q}_k x_i^c x_j^c\right) + \left(\Lambda A_{ijk} x_j^c\right) x_i^c \dot{Q}_k = -x_i^c v_i^1, \tag{11.28}$$

and the second-order contracted equation results in

$$\ddot{\Lambda}\left(A_{ijk}\bar{Q}_k x_i^c x_j^c\right) + \left(\Lambda_c A_{ijk} x_j^c\right) x_i^c \ddot{Q}_k = -x_i^c v_i^2. \tag{11.29}$$

If the dependence of (11.9) on $\bar{Q}_j$ is nonlinear, a more complete set of perturbation equations should be found, and the reader is referred to [10].

11.4 SOLUTION BY THE DIRECT METHOD

The complete set of the first-order perturbation equations is given by (11.14), (11.18), (11.22) or (11.27), and (11.28); and there are $(2N+1)$ unknowns, where N is the number of degrees of freedom of the problem. The total number of equations is $2(N+1)$, but one of the systems exhibits a singularity of order one, so that the total number of useful equations is $(2N+1)$. In the direct approach, the perturbation equations are solved sequentially and without the introduction of any further conditions.

The first stage is the solution of (11.8)–(11.10) to obtain the basic values of $\bar{Q}_j$, Λ_c, and x_j^c for $\tau = 0$. This is the classical eigenvalue problem of the reference system, with a given value of the design parameter, and should be done following the procedures of Chapter 5.

11.4.1 Solution of First-Order Perturbation Systems

Equation (11.14) is a function of only $\dot{Q}_j$. By definition, A_{ij} is nonsingular along the fundamental path; then $\dot{Q}_j$ can be solved from (11.14). The value of $\dot{\Lambda}$ is next obtained from (11.28)

$$\left(A_{ijk}x_i x_j Q_k\right)\dot{\Lambda} = -\left(\Lambda A_{ijk}x_j^c \dot{Q}_k + v_i^1\right)x_i^c$$

as

$$\dot{\Lambda} = -\frac{\left(A_{ijk}x_j \dot{Q}_k + v_i^1\right)x_i}{\left(A_{ijk}x_i x_j Q_k\right)}\Bigg|^c. \tag{11.30}$$

For the solution of $\dot{x}_i$ we have the uncontracted equation (11.18), that is,

$$\left(A_{ij} + \Lambda^c\, A_{ijk}\bar{Q}_k\right)\dot{x}_j = -g_i^1 \tag{11.31}$$

with

$$g_i^1 \equiv v_i^1 + \dot{\Lambda}\left(A_{ijk}\bar{Q}_k\right)x_j^c + \left(\Lambda A_{ijk}x_j^c\right)\dot{Q}_k. \tag{11.32}$$

The condition of normalization, (11.26), leads to

$$\dot{x}_1 = 0.$$

The solution $\dot{x}_i$ is written as the sum of a particular solution y_i^1, plus the solution of the homogeneous system x_i^c multiplied by a constant k_1 to be evaluated:

$$\dot{x}_i = k_1 x_i^c + y_i^1. \tag{11.33}$$

Substitution of (11.33) into (11.31) yields

$$\left(A_{ij} + \Lambda A_{ijk}\bar{Q}_k\right)y_i^1 = -g_i^1. \tag{11.34}$$

The matrix in this system is singular, and it may be necessary to choose a component of vector y_i^1 and eliminate an equation in (11.34). According to (11.26) one should have

$$y_1^1 = 1. \tag{11.35}$$

Once the solution of (11.34) has been obtained we only need to compute k_1 in (11.33). To do that, let us consider the first component in (11.33):

$$\dot{x}_1 = 0 = k_1 x_1 + y_1 = k_1 + 1;$$

therefore,

$$k_1 = -1. \tag{11.36}$$

Finally, the complete set of (11.33) provides sensitivity of the eigenvector

$$\dot{x}_i = -x_i^c + y_i^1. \tag{11.37}$$

If normalization of the eigenvector follows condition (11.10), instead of (11.11), then a more involved procedure needs to be followed to compute k_1; this is done in [10].

11.4.2 Solution of Second-Order Perturbation Systems

Let us proceed with the direct solution of the second-order perturbation coefficients. From (11.16) it is possible to compute $\ddot{Q}_j$.

The eigenvalue problem in (11.20) requires the use of the contraction mechanism to obtain a solution for $\ddot{\Lambda}$; this is obtained from (11.29) as

$$\ddot{\Lambda} = -\frac{\left(\Lambda A_{ijk} x_j^c \ddot{Q}_k + v_i^2\right) x_i^c}{A_{ijk} x_i^c x_j^c \bar{Q}_k}. \tag{11.38}$$

Equation (11.38) represents the second-order sensitivity of the eigenvalue.

The second-order sensitivity of the eigenvector is evaluated from the equation

$$\left(A_{ij} + \Lambda_c A_{ijk} \bar{Q}_k\right) \ddot{x}_j = -g_i^2, \tag{11.39}$$

where

$$g_i^2 = v_i^2 + \ddot{\Lambda}(A_{ijk} \bar{Q}_k) x_j^c + (\Lambda^c A_{ijk} x_j) \ddot{Q}_k \tag{11.40}$$

and

$$\ddot{x}_1 = 0. \tag{11.41}$$

The vector $\ddot{x}_i$ is written in a form similar to (11.33)

$$\ddot{x}_i = k_2 x_i^c + y_i^2. \tag{11.42}$$

Substitution of (11.42) into (11.39) leads to

$$\left(A_{ij} + \Lambda A_{ijk} \bar{Q}_k\right) y_i^2 = -g_i^2. \tag{11.43}$$

Solution of this singular system is possible by assuming

$$y_1^2 = 0. \tag{11.44}$$

Finally, k_2 is computed from the first part of (11.42) as

$$\ddot{x}_1 = 0 = k_2 x_1^c + y_1^2 = k_2 x_1^c;$$

therefore,

$$k_2 = 0. \tag{11.45}$$

Equation (11.42) can now be used as second-order sensitivity of the eigenvector in the form

$$\ddot{x}_i = y_i^2. \tag{11.46}$$

Again, a more involved computation of k_2 is required if the eigenvector is normalized with (11.10).

11.4.3 An Algorithm of Sensitivity Using the Direct Method

The above procedure can be summarized in the form of an algorithm as follows:

1. Compute the reference state for $\tau = 0$.
2. Solve $\dot{Q}_i$
$$A_{ij}\dot{Q}_j = -u_i^1, \qquad \text{where} \qquad u_i^1 = \dot{A}_{ij}\bar{Q}_j + \dot{A}_i.$$
There is no constraint on $\dot{Q}_j$, since A_{ij} is not a critical state.
3. Compute
$$v_i^1 = (\dot{A}_{ij} + \Lambda \dot{A}_{ijk}\bar{Q}_k)x_j^c.$$
4. Calculate
$$\dot{\Lambda} = -\frac{\left(A_{ijk}x_j\dot{Q}_k + v_i^1\right)x_i}{\left(A_{ijk}x_i x_j Q_k\right)}\Bigg|^c.$$
5. Calculate
$$g_i^1 \equiv v_i^1 + \dot{\Lambda}\left(A_{ijk}\bar{Q}_k\right)x_j^c + \left(\Lambda A_{ijk}x_j^c\right)\dot{Q}_k.$$
6. Solve y_i^1 using
$$\left(A_{ij} + \Lambda A_{ijk}\bar{Q}_k\right)y_i^1 = -g_i^1 \qquad \text{with} \qquad y_1^1 = 1.$$
7. Compute
$$\dot{x}_i = -x_i^c + y_i^1.$$
8. First-order sensitivity is
$$Q_i(\tau) = \bar{Q}_i + \dot{Q}_i(\tau) + \cdots,$$
$$\Lambda(\tau) = \Lambda^c + \dot{\Lambda}\,\tau + \cdots,$$
$$x_i(\tau) = x_i^c + \dot{x}_i\,\tau + \cdots.$$
9. For second-order sensitivity, compute
$$u_i^2 \equiv (\ddot{A}_{ij}\bar{Q}_j + \ddot{A}_i) + 2(\dot{A}_{ij}\dot{Q}_j).$$
10. Solve Q_j using
$$A_{ij}\ddot{Q}_j = -u_i^2.$$
11. Compute v_i^2
$$v_i^2 = \left[\ddot{A}_{ij} + \Lambda_c(\ddot{A}_{ijk}\bar{Q}_k)\right]x_j^c + 2\dot{\Lambda}\left[(\dot{A}_{ijk}\bar{Q}_k + A_{ijk}\dot{Q}_k)\right]x_j^c$$
$$+2\left[\dot{A}_{ij} + \Lambda_c(\hat{V}_{ijk}\bar{Q}_k + A_{ijk}\dot{Q}_k)\right]\dot{x}_j + 2\Lambda\left[A_{ijk}\bar{Q}_k\right]\dot{x}_j + \Lambda(\dot{A}_{ijk}\dot{Q}_k)x_j^c.$$

12. Calculate

$$\ddot{\Lambda} = -\frac{\left(\Lambda A_{ijk} x_j^c \ddot{Q}_k + v_i^2\right) x_i^c}{A_{ijk} x_i^c x_j^c \bar{Q}_k}.$$

13. Compute

$$g_i^2 = v_i^2 + \ddot{\Lambda}(A_{ijk}\bar{Q}_k)x_j^c + (\Lambda^c A_{ijk} x_j)\ddot{Q}_k.$$

14. Solve y_i^2 using

$$\left(A_{ij} + \Lambda A_{ijk}\bar{Q}_k\right) y_i^2 = -g_i^2 \qquad \text{with} \qquad y_1^2 = 0.$$

15. Define

$$\ddot{x}_i = y_i^2 \qquad \text{with} \qquad \ddot{x}_1 = 0.$$

16. Second-order sensitivity is given by

$$Q_i(\tau) = \bar{Q}_i + \dot{Q}_i\, \tau + \frac{1}{2}\ddot{Q}_i\, \tau^2 + \cdots,$$

$$\Lambda(\tau) = \Lambda^c + \dot{\Lambda}\, \tau + \frac{1}{2}\ddot{\Lambda}\, \tau^2 + \cdots,$$

$$x_i(\tau) = x_i^c + \dot{x}_i\, \tau + \frac{1}{2}\ddot{x}_i\, \tau^2 + \cdots.$$

11.5 SOLUTION BY AN ADJOINT METHOD

A first-order sensitivity analysis based on the adjoint method has been developed by Dems and Mróz [3]; this technique avoids the computation of the sensitivity of the equilibrium condition (11.8) as a step previous to the calculation of sensitivity of the critical state (i.e., avoids the computation of $\dot{Q}_j$ before the evaluation of $\dot{\Lambda}$).

Similar to what was done in the direct solution, the initial results required are Λ^c, $\bar{Q}_i$, and x_i^c for $\tau = 0$. Let us consider a vector of Lagrange multipliers μ_i and multiply the equilibrium condition by this vector:

$$\mu_i \left(A_{ij} Q_j + A_i\right) = 0. \tag{11.47}$$

The vector of Lagrange multipliers is also expanded as a function of τ in the form of a series of powers

$$\mu_i(\tau) = \mu_i + \dot{\mu}_i \tau + \frac{1}{2}\ddot{\mu}_i \tau^2 + \cdots. \tag{11.48}$$

Perturbation of (11.47) leads to a set of scalar perturbation equations; the equation of order zero is

$$\mu_i \cdot \left(\underbrace{A_{ij}\bar{Q}_j + A_i}_{0} \right) = 0, \tag{11.49}$$

which does not provide information about μ_i.

11.5.1 Solution of First-Order Perturbation Systems

The first-order equation obtained from (11.47) is

$$A_{ij}\dot{Q}_j\mu_i + u_i^1\mu_i = 0. \tag{11.50}$$

The contracted equation (11.28) is here augmented using Lagrange multipliers (11.50) to obtain

$$u_i^1\mu_i + x_i^c v_i^1 + \left(A_{ijk}x_i^c x_j^c \bar{Q}_k\right)\dot{\Lambda} + \left(A_{ij}\mu_i + \Lambda^c A_{ijk}x_i^c x_k^c\right)\dot{Q}_j = 0. \tag{11.51}$$

The unknowns in the last equation are the values of $\dot{Q}_j$, $\dot{\Lambda}$, and μ_i ($2N + 1$ unknowns). But the Lagrange multipliers may be chosen in a convenient way so as to eliminate the term associated with $\dot{Q}_j$. Thus, let us impose the following condition on μ_i:

$$\boxed{A_{ij}\mu_j = -\Lambda^c A_{ijk}x_j^c x_k^c} \tag{11.52}$$

Equation (11.52) is known as the **adjoint system**. The term on the right-hand side is computed from the solution of the system of order zero.

The new contracted equation (11.51) reduces to

$$\left(A_{ijk}x_i^c x_j^c \bar{Q}_k\right)\dot{\Lambda} = -\left(u_i^1\mu_i + x_i^c v_i^1\right). \tag{11.53}$$

Notice that (11.53) is equivalent to (11.30) in the direct method. We can calculate

$$\dot{\Lambda} = -\frac{\left(u_i^1\mu_i + x_i^c v_i^1\right)}{\left(A_{ijk}x_i^c x_j^c \bar{Q}_k\right)}. \tag{11.54}$$

Up to this point, in the adjoint method it has not been necessary to calculate the sensitivity of the precritical state, given by $\dot{Q}_j$. But if sensitivity of the eigenvector is needed, then (11.31)–(11.36) could be used, requiring the solution of (11.14) to compute (11.32).

11.5.2 Solution of Second-Order Perturbation Systems

The second-order perturbation equation may be obtained from (11.16) and (11.47) added to (11.29):

$$A_{ij}\ddot{Q}_j\mu_i + u_i^2\mu_i + \left(A_{ijk}x_i^c x_j^c \bar{Q}_k\right)\ddot{\Lambda} + \left(\Lambda_c A_{ijk}x_j^c\right)x_i^c\ddot{Q}_k + x_i^c v_i^2 = 0$$

or else

$$\left(A_{ij}\mu_i + \Lambda_c A_{ijk}x_i^c x_k^c\right)\ddot{Q}_j + \left(A_{ijk}x_i^c x_j^c \bar{Q}_k\right)\ddot{\Lambda} = -\left(u_i^2\mu_i + x_i^c v_i^2\right).$$

Clearly the term associated with $\ddot{Q}_j$ is the adjoint problem of (11.52). This is a surprising result, and it means that **the adjoint problem for the second-order sensitivity is the same as the adjoint problem of the first-order sensitivity.**

The value of $\ddot{\Lambda}$ results in

$$\ddot{\Lambda} = -\frac{\left(u_i^2\mu_i + x_i^c v_i^2\right)}{\left(A_{ijk}x_i^c x_j^c \bar{Q}_k\right)}. \tag{11.55}$$

Notice that the vector of Lagrange multipliers was expanded in (11.48) with the idea that more than one adjoint system could be necessary. The present results show that this is not the case, because the adjoint system of order 2, 3, and higher are identical to the first one.

11.5.3 Algorithm for Sensitivity Using the Adjoint Method

A summary of the calculations required for the adjoint method presented here can be given as an algorithm in the following form:

1. Compute the reference state for
$$\tau = 0.$$
2. Solve μ_j using
$$A_{ij}\mu_j = -\Lambda^c A_{ijk}x_j^c x_k^c.$$
3. Compute
$$u_i^1 = \dot{A}_{ij}\bar{Q}_j + \dot{A}_i.$$
4. Compute
$$v_i^1 = (\dot{A}_{ij} + \Lambda\dot{A}_{ijk}\bar{Q}_k)x_j^c.$$
5. Calculate
$$\dot{\Lambda} = -\frac{\left(u_i^1\mu_i + x_i^c v_i^1\right)}{\left(A_{ijk}x_i^c x_j^c \bar{Q}_k\right)}.$$

11.6 SIMPLIFIED SENSITIVITY ANALYSES

The resulting equations in section 11.3 are cumbersome; furthermore, they require the solution of two linear systems of equations for each order of perturbation (i.e., computation of $\dot{Q}_j$ and y_j^1 for first-order perturbation analysis). It is therefore relevant to investigate simplified forms by neglecting some terms that could be of minor significance in the results. Two such simplifications are considered in the following: neglecting sensitivity of the fundamental path and neglecting sensitivity of the load-geometry matrix. These simplifications are tested in the section of examples, and it is shown that the results may be significantly worsened with respect to the complete solution.

11.6.1 Approximation (a): The Sensitivity of the Fundamental Path Is Neglected

Let us assume that $\dot{Q}_i = \ddot{Q}_i = 0$; i.e., the fundamental path does not change with modifications of the design parameter. Then it is not necessary to solve (11.14) and (11.31).

The sensitivity of the eigenvalue, as reflected by (11.38) and (11.30), reduces to

$$\ddot{\Lambda} = - v_i^1 x_i^c, \tag{11.56}$$

$$\ddot{\Lambda} = - v_i^2 x_i^c. \tag{11.57}$$

For sensitivity of the eigenvector, the computations become simpler.

11.6.2 Approximation (b): The Influence of Sensitivity of the Load-Geometry Matrix Is Neglected

This approximation means that

$$\dot{A}_{ijk} \bar{Q}_k = 0, \tag{11.58}$$

and only the stiffness matrix of the problem is assumed to be sensitive to changes in the design parameter.

This assumption affects the perturbation equations in the first-order analysis; that is, (11.19) and (11.23) become simpler.

11.7 SENSITIVITY OF A CIRCULAR PLATE

A simple two-degrees-of-freedom plate buckling problem, solved in previous chapters, is discussed here as an example to demonstrate the application of the sensitivity equations derived previously. The plate has a circular geometry, with radius R and uniform thickness h, and is simply supported around the circumferential boundary.

The load is radial and is here represented by a load vector

$$V_i' = \begin{Bmatrix} 0 \\ \phi \end{Bmatrix}$$

and a load parameter λ responsible for buckling when λ is negative.

The displacements (w, u) are approximated in the simple form

$$w = Q_1 \cos(\pi x/2R), \tag{11.59}$$

$$u = (x/R)\, Q_2,$$

where x is the radial coordinate.

This is a trivial problem, in the sense that the sensitivity can be computed in exact form, but it illustrates the use of the procedures presented in this chapter.

The energy coefficients of the plate have been obtained in Chapter 2 and Example 11.1.

The fundamental (reference) solution. We consider the thickness as a design variable in the form

$$h = h_0(1+\tau) \tag{11.60}$$

and seek perturbation expansions in terms of the design parameter τ.

Considering $\tau = 0$ one has the solution of the reference problem

$$\bar{Q}_1 = 0, \qquad \bar{Q}_2 = a_2,$$

$$A_{ijk}\, x_i x_j Q_k = m\, h\, a_2, \qquad \Lambda^c = -\frac{k_1 h_0^2}{m\, a_2},$$

$$x_1^c = 1, \qquad x_2^c = 0.$$

First-order perturbation, direct analysis. We start by computing the derivatives

$$\dot{A}_i = 0$$

and

$$\dot{A}_{ij} = \begin{bmatrix} 3k_1 h_0^3 & 0 \\ 0 & k_2 h_0 \end{bmatrix}.$$

The u_i^1 vector now results in

$$u_i^1 = \begin{Bmatrix} 0 \\ A_{22}\bar{Q}_2 \end{Bmatrix} = \begin{Bmatrix} 0 \\ k_2\, h_0\, a_2 \end{Bmatrix}.$$

Sensitivity of the fundamental path may be obtained from

$$\begin{bmatrix} A_{11} & 0 \\ 0 & A_{22} \end{bmatrix} \begin{Bmatrix} \dot{Q}_1 \\ \dot{Q}_2 \end{Bmatrix} = -k_2 h_0 a_2 \begin{Bmatrix} 0 \\ 1 \end{Bmatrix}$$

from which we can compute

$$\dot{Q}_1 = 0, \qquad \dot{Q}_2 = -a_2.$$

Next

$$v_i^1 = \begin{Bmatrix} 2k_1 h_0^3 \\ 0 \end{Bmatrix}$$

and

$$\dot{\Lambda} = -\frac{(-\Lambda^c A_{112} x_1 \dot{Q}_2 + v_1^1 x_1)}{m\ h\ a_2} = 3\Lambda^c.$$

Thus, the first-order sensitivity of the eigenvalue is three times the critical load itself. For first-order sensitivity of the eigenvector we use the condition of normalization

$$\dot{x}_1 = 0$$

and compute

$$g_i^1 = 0,$$

so that

$$y_i^1 = x_i^c.$$

Sensitivity of the eigenvector is now completed in the form

$$\dot{x}_2 = -x_2^c + y_2^1 = (-1)0 + 0 = 0.$$

Finally

$$\dot{x}_i = \begin{Bmatrix} 0 \\ 0 \end{Bmatrix}.$$

This means that in the first-order sensitivity the eigenvector retains the same critical direction as the original reference problem. The fundamental path and the eigenvalue, on the other hand, are sensitive to changes in the design parameter.

Second-order perturbation, direct analysis. We need some second derivatives to continue the analysis

$$\ddot{A}_{11} = 6k_1 h_0^3, \qquad \ddot{A}_{22} = 0, \qquad \ddot{A}_{112} = 0.$$

It is now possible to compute

$$u_i^2 = -2k_2 h_0 a_2 \begin{Bmatrix} 0 \\ 1 \end{Bmatrix}$$

and

$$\begin{bmatrix} A_{11} & 0 \\ 0 & A_{22} \end{bmatrix} \begin{Bmatrix} \ddot{Q}_1 \\ \ddot{Q}_2 \end{Bmatrix} = 2k_2 h_0 a_2 \begin{Bmatrix} 0 \\ 1 \end{Bmatrix}$$

from which the second-order sensitivity of the fundamental state becomes

$$\begin{Bmatrix} \ddot{Q}_1 \\ \ddot{Q}_2 \end{Bmatrix} = \begin{Bmatrix} 0 \\ 2a_2 \end{Bmatrix}.$$

Next we calculate

$$v_1^2 = 4k_1 h_0^3, \qquad v_2^2 = 0,$$

and $\ddot{\Lambda}$ reduces to

$$\ddot{\Lambda} = -\frac{\left(v_1^2 + \Lambda^c \, A_{112} x_1 Q_2\right)}{m \; h \; a_2} = 6\Lambda^c.$$

This means that the second-order sensitivity of the eigenvalue is six times the critical load.

Finally, we follow equations for second-order sensitivity of the eigenvector, which in this case lead to

$$\ddot{x}_i = -x_i^c.$$

The results can be summarized in terms of the perturbation expansions of the response variables as

$$Q_i(\tau) = \left[1 - 1\tau + \tau^2 + \cdots\right] \bar{Q}_i,$$

$$\lambda\,(\tau) = \left[1 + 3\tau + 3\tau^2 + \cdots\right] \Lambda^c,$$

$$x_i(\tau) = \left[1 + (0)\tau - \frac{1}{2}\tau^2 + \cdots\right] x_i^c.$$

Adjoint method. In the adjoint method, the Lagrangian multipliers result in

$$\mu_i = -\Lambda \frac{m}{k_2} \begin{Bmatrix} 0 \\ 1 \end{Bmatrix}.$$

The values of $\dot{\Lambda}$ and $\ddot{\Lambda}$ are identical to those obtained in the direct analysis.

Table 11.1 Influence of simplifications on the sensitivity of circular plates with thickness changes

	Complete analysis	Simplification (a)	Simplification (b)	Simplification (a + b)
$\dot{Q}_1$	0	0	0	0
$\dot{Q}_2$	$-Q_2^F$	0	$-Q_2^F$	0
$\dot{\lambda}$	$3\Lambda^c$	$2\Lambda^c$	$4\Lambda^c$	$3\Lambda^c$
$\ddot{Q}_1$	0	0	0	0
$\ddot{Q}_2$	$2Q_2^F$	0	$2Q_2^F$	0
$\ddot{\lambda}$	$6\Lambda^c$	$2\Lambda^c$	$12\Lambda^c$	$6\Lambda^c$

Simplified analysis of the circular plate. To illustrate the influence of the simplifications discussed in section 11.5, let us consider again the circular plate, simply supported at the edges, under in-plane loading. The thickness is written in parametric form (11.60), and sensitivity to thickness changes is investigated.

The results of the complete and approximate analysis are shown in Table 11.1.

If sensitivity of the fundamental path is neglected, approximation (a), the sensitivity of the buckling load is underestimated:

$$\dot{\lambda}_{ap} = \frac{2}{3}\dot{\lambda}_{exact}, \qquad \ddot{\lambda}_{ap} = \frac{1}{3}\ddot{\lambda}_{exact}.$$

In the approximation (b) (the load-geometry matrix is assumed to be insensitive to changes in τ), there is an overestimation of the buckling load sensitivity:

$$\dot{\lambda}_{ap} = \frac{4}{3}\dot{\lambda}_{exact}, \qquad \ddot{\lambda}_{ap} = \frac{6}{3}\ddot{\lambda}_{exact}.$$

11.8 SENSITIVITY OF A THIN-WALLED COLUMN

In this section we present results obtained for sensitivity analysis of a thin-walled I-section composite column under axial load, in which several parameters are considered for design sensitivity. The presentation follows [8].

Instability is investigated for local buckling modes, which are associated with periodic distortions of the cross section.[29] Details of the computation of the energy using the Ritz method are given in Chapter 14. The displacement field for the local buckling mode is represented here using the rotation q_3 as a degree of freedom. The second variation of the total potential energy is evaluated along the fundamental path, and only components that are quadratic in q_3 are retained in the analysis. These contributions are necessary to define the critical state, and they are a function of the

[29]Global modes, on the other hand, occur in long columns involving lateral displacements with negligible distortions of the cross section.

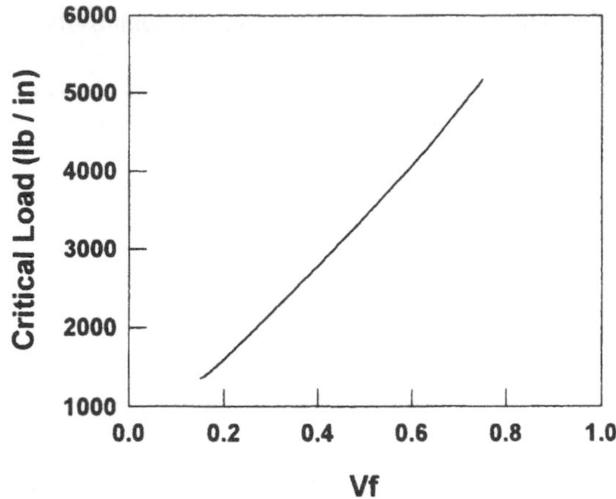

Figure 11.2 Sensitivity of a thin-walled I-column with changes in volume fraction.

number of waves considered in the local mode n. The lowest critical load is also a function of the geometric and constitutive parameters chosen for the design.

The influence of parameters of micromechanics can be studied on the buckling load, and this is shown in Figure 11.2 for a column with $L = 2540mm$, $b = h = 150mm$ as a function of the volume fraction V_f. The relation is almost linear.

The second study using the present perturbation analysis is shown in Figure 11.3 for $V_f = 0.5$ and changes in the angle of lamination θ. Here the critical load is

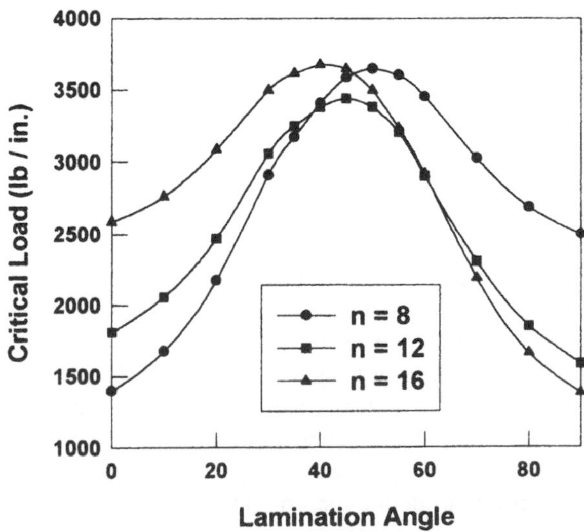

Figure 11.3 Sensitivity of a thin-walled I-column with changes in the lamination angle.

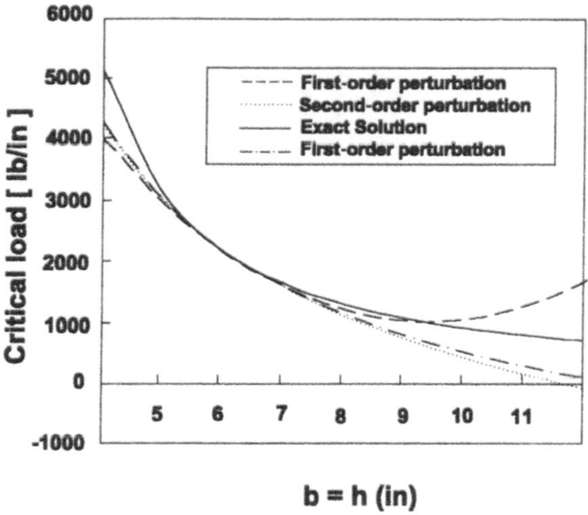

Figure 11.4 Sensitivity of an I-column with changes in the geometric dimensions.

maximum for lamination angles of 45° for the lowest mode identified by $n = 12$. For other modes, such as $n = 8$, the curve is not symmetric. Notice that the lowest critical load for lamination angles $\theta = 20°$ is associated with $n = 8$.

A final example refers to the actual dimensions of the I-column. The geometry has been kept in the form $b = h$, but the dimensions are modified from $100mm$ to $150mm$. Perturbation analysis of various orders have been employed in this case and are shown in Figure 11.4 and compared with the exact solution. The reference solution was chosen for $b = h = 150mm$. It is clear from this and other analysis that perturbation results cannot be extended far out of the reference solution, but their validity is restricted to its vicinity.

11.9 SENSITIVITY OF A RECTANGULAR PLATE

The sensitivity analysis explained in this chapter and the approximations of section 11.5 have been implemented in a finite strip code, which is fully described in [19]. Just a test case is presented here: A rectangular plate with step changes in thickness (Figure 11.5.a). The thickness is written in parametric form as in (11.60).

The coefficient k is related to the critical load by

$$P_c = k\, \sigma_0\, h_0, \qquad \text{where} \qquad \sigma_0 = \frac{E\pi^2}{12(1-\nu^2)} \left(\frac{h_0}{b}\right)^2.$$

Figure 11.5.b shows convergence of the solution for an increasing number of perturbation equations, NSP, employed in the simplified analysis. The coefficient k^* is the difference between the coefficients for constant and variable thickness, and

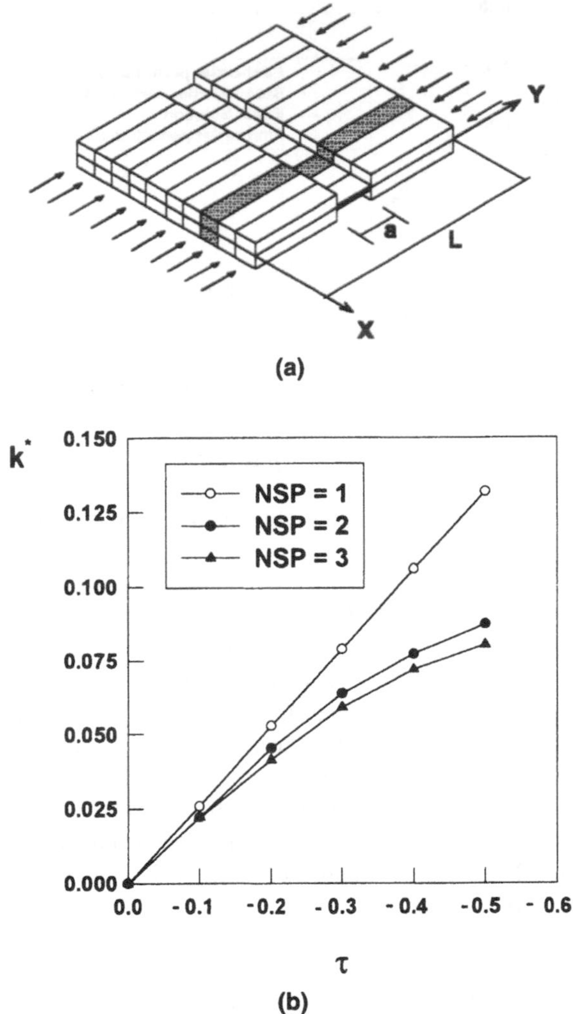

Figure 11.5 Rectangular plate with step change in thickness: (a) Geometry; (b) convergence of perturbation analysis. *Continued.*

b is the width. A solution with two perturbation systems leads to good results in the present context.

Further results are presented in Figure 11.5.c, in which the critical load is plotted versus the ratio a/L, the width of the plate with changes in the thickness, for a value of $\tau = -0.2$. The complete analysis is represented by hollowed circles, and the critical load is seen to decrease from $k = 43$ to $k = 37$ as the zone with a reduction in the thickness increases from 1.2% of L to 5% of L.

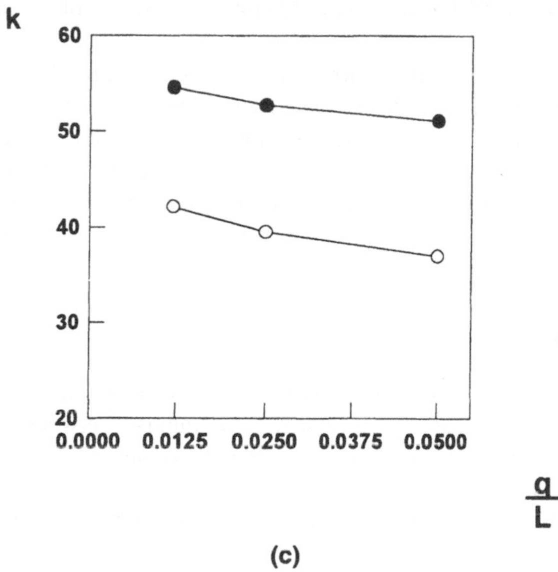

Figure 11.5 (Cont.) (c) Sensitivity for different ratios a/L.

The simplified results, associated with simplifications both in the fundamental path and in the load-geometry matrix in section 11.5, are shown with solid circles in Figure 11.5.c, and are compared with the complete analysis. It may be seen that there is a significant error (20% to 25%) in the approximate results for a range of plate dimensions. Thus, a complete solution is recommended on the basis of the present results.

11.10 FINAL REMARKS

This chapter includes a formulation for sensitivity analysis of buckling states, which is a necessary ingredient in stochastic buckling, in optimization, and in many other problems in structural mechanics. The formulation for the direct computation of sensitivity includes equilibrium constraints; it uses regular perturbation techniques and as such it requires derivatives of the parameters of the problem in terms of the design variables.

In the adjoint method for first-order sensitivity of eigenproblems, the form of the adjoint problem to be solved was well known in vibration analysis. In this chapter it is shown that no new adjoint problem is required for higher-order perturbations; thus the solution becomes even simpler.

Only one design parameter has been included in the derivations. The formulation can be extended to multiple design parameters changing simultaneously, and this would lead to more complex expressions than those presented here. This analysis can be implemented in the W formulation of Chapter 13 instead of the V formulation, as has been done here.

Two simplified versions of sensitivity analysis of buckling loads and modes have been investigated, and the results have been computed for two problems of plates with changes in the thickness. The results show that if sensitivity of the fundamental path is neglected, then sensitivity of the buckling load is underestimated. A second conclusion is that if sensitivity of the load-geometry matrix is neglected, then the results are overestimated, in both cases by differences of the order of 30%.

11.11 PROBLEMS

Review questions. (*a*) Explain the differences between design sensitivity analysis of buckling and of free vibration problems. (*b*) Distinguish between the direct and the adjoint approaches to design sensitivity. (*c*) Is it possible to neglect sensitivity of the fundamental path when computing the sensitivity of critical states? (*d*) How important is sensitivity of the load-geometry matrix on the sensitivity of a critical state?

Problem 11.1 (Column). The thickness of an axially loaded column in the central third of its length is allowed to increase by means of a cosine function. Write the energy functional so that the modification is considered as a design parameter.

Problem 11.2 (Circular arch). For a circular arch under a central point load, consider the first bifurcation state along a linear primary path. Use the direct method to compute sensitivity of the critical state. Assume R as design parameter in the form

$$R = R_0(1 + \tau).$$

Problem 11.3 (Circular arch). Repeat Problem 11.2 using ϕ as a design parameter. Use the adjoint method.

Problem 11.4 (Two-bar frame). The two-bar frame of section 5.9 in Chapter 5 is investigated and first-order sensitivity is required with respect to the cross-sectional area

$$A = A_0(1 + \tau).$$

Problem 11.5 (Rectangular plate). Find first-order sensitivity of a rectangular composite plate simply supported along all boundaries (see section 5.11 in Chapter 5), considering

$$D_{66} = D_{66}^0\ (1 + \tau).$$

Problem 11.6 (Stiffened plate). A simply supported square plate with stiffeners was studied in Problem 7.14. Use the thickness of the stiffener in the x_1 direction as a design variable. Obtain sensitivity of the critical load and of the critical direction. Use the adjoint method. The side of the plate has $4m$; the plate thickness is $h = 25mm$, $E = 205KN/mm^2$. The basic stiffeners in both directions have a spacing of $0.5m$, thickness $25mm$, and height of $200mm$.

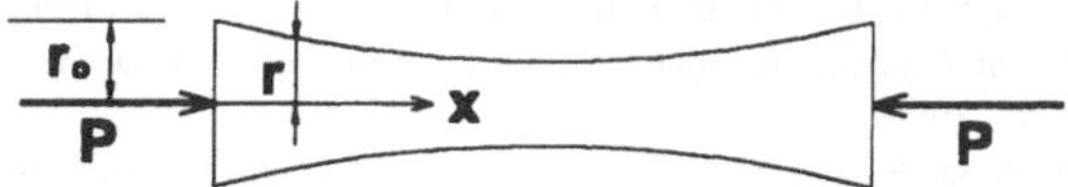

Figure 11.6 Nonprismatic bar of Problem 11.7.

Problem 11.7 (Nonprismatic column). The nonprismatic bar shown in Figure 11.6 has a circular cross section, with a radius given by

$$r = r_0 \left[1 + \tau \sin\left(\frac{\pi x}{L}\right) \right],$$

where r_0 is the radius at the ends of the bar. Use

$$w = Q_1 \sin\left(\frac{\pi x}{L}\right) + Q_2 \sin\left(\frac{2\pi x}{L}\right) + Q_3 \sin\left(\frac{3\pi x}{L}\right).$$

Find sensitivity of the critical state with respect to the design parameter τ.

Problem 11.8 (Thin-walled column). Carry out the details of the sensitivity analysis of the composite thin-walled column described in this chapter.

11.12 BIBLIOGRAPHY

[1] Benedettini, F., and Capecchi, D., A perturbation technique in sensitivity analysis of elastic structures, *Meccanica*, 23, 5–10, 1988.

[2] Dems, K., Sensitivity analysis in thermoelasticity problems, in *Computer Aided Optimal Design*, C. A. Mota Soares, Ed., Springer, Berlin, 1987, 563–573.

[3] Dems, K., and Mróz, Z., Variational approach by means of adjoint systems to structural optimization and sensitivity analysis, Part I: Variation of material parameters with fixed domain, *Internat. J. Solids Structures*, 19, 677–692, 1983.

[4] Eldred, M. S., Lerner, P. B., and Anderson, W. J., Higher order eigenpair perturbations, *AIAA J.*, 30(7), 1870–1876, 1992.

[5] Flores, F., and Godoy, L. A., Elastic post buckling analysis via finite element and perturbation techniques, Part I: Formulation, *Internat. J. Numer. Methods Engrg.*, 33, 1775–1794, 1992.

[6] Godoy, L. A., On stochastic buckling via sensitivity analysis, in *Proc. II Int. Conference on Stochastic Structural Dynamics*, H. Davoodi and A. Saffar, Eds., University of Puerto Rico Press, San Juan, January 1995.

[7] Godoy, L. A., *Thin-Walled Structures with Structural Imperfections*, Pergamon Press, Oxford, 1996.

[8] Godoy, L. A., and Almanzar, L., Improving design in composite thin-walled columns, in *Engineering Mechanics*, Vol. 1 (Proc. 11th. Conf.), Y. K. Lin and T. C. Su, Eds., American Society of Civil Engineers, New York, 1996, 1155–1158.

[9] Godoy, L. A., and Raichman, S. R., Design sensitivity of buckling states: Simplified analysis and errors, in *Applied Mechanics in the Americas*, L. Godoy et al., Eds., American Academy of Mechanics, Buenos Aires, 1995, 462–467.

[10] Godoy, L. A., Taroco, E. O., and Feijoo, R. A., Second order sensitivity analysis in vibration and buckling problems, *Internat. J. Numer. Methods Engrg.*, 37, 3999–4014, 1994.

[11] Godoy, L. A., Flores, F. G., Raichman, S. R., and Mirasso, E. A., *Perturbation Techniques in Nonlinear Finite Element Analysis* (in Spanish), Asociacion Argentina de Mecanica Computacional, Santa Fe, Argentina, 1990.

[12] Haftka, R. T., Cohen, G. A., and Mroz, Z., Derivatives of buckling loads and vibration frequencies with respect to stiffness and initial strain parameters, *J. Appl. Mech.*, 57, 18–24, 1990.

[13] Haug, E. J., Choi, K. K. and Komkov, V., *Design Sensitivity Analysis of Structural Systems*, Academic Press, London, 1985.

[14] Hou, G. J. W., and Sheen, J., Numerical methods for second order shape sensitivity analysis with application to heat conduction problems, *Internat. J. Numer. Methods Engrg.*, 36, 417–435, 1993.

[15] Mróz, Z., Sensitivity analysis and optimal design with account for varying shape and support conditions, in *Computer Aided Optimal Design*, C. A. Mota Soares, Ed., Springer, Berlin, 1987, 407–438.

[16] Mróz, Z., Sensitivity analysis for vibration and stability of structures, in *Optimization of Large Structural System*, G. Rozvany, Ed., NATO DFG ASI, Lecture Notes, vol. 2, 1992, 139–157.

[17] Mróz, Z., and Haftka, R. T., Design sensitivity analysis of nonlinear structures in regular and critical states, *Internat. J. Solids Structures*, 31(15), 2071–2098, 1994.

[18] Nayfeh, A. H., *Perturbation Methods*, John Wiley and Sons, New York, 1973.

[19] Raichman, S. R., and Godoy, L. A., A simplified perturbation technique in finite strips for the stability of non-prismatic plate assemblies (in Spanish), *Mecanica Computacional* (AMCA, Argentina), vol. 9, 1989, 317–332.

CHAPTER

TWELVE

DESIGN SENSITIVITY OF POSTCRITICAL STATES

12.1 INTRODUCTION

This chapter presents a theory to account for changes in the postbuckling path when design parameters are modified. This is a new field in the theory of elastic stability: the design sensitivity of postcritical states.

Sensitivity in buckling problems is a difficult nonlinear topic. But buckling loads, as computed from a bifurcation analysis, provide only limited information about the mechanics of the problem, especially in shell and shell-like structures. In such cases, there is a need to obtain not just critical but also postcritical states (that is, postcritical equilibrium paths emerging from the critical state). The sensitivity of postcritical states with respect to geometric or load imperfections has been the subject of research for some time and is a standard part of the general theory of elastic stability as presented in Chapter 10. Sensitivity of postbuckling states to changes in design parameters, on the other hand, has mainly been explored by numerical and analytical experimentation (see, for example, [8, 3, 4] and many others).

This chapter deals with a formulation for design sensitivity of postcritical equilibrium states. Examples of sensitivity using nonlinear analysis are presented in section 12.2. The framework of analysis in the rest of the chapter is the same employed throughout the book, so that use is made of perturbation techniques to find the critical and postcritical states for a reference configuration. Sensitivity of postcritical states is considered in sections 12.3 for symmetric and in section 12.4 for asymmetric bifurcations. Examples are presented in sections 12.5 and 12.6 to illustrate the procedure in a column with deformable cross section and a plate with changes in the thickness.

12.2 NONLINEAR POSTCRITICAL ANALYSIS

Let us consider the postbuckling behavior of a cylindrical narrow panel under axial load in order to identify some important features of postbuckling design sensitivity. Here an important design parameter Θ is the initial curvature of the panel. For $\Theta = 0$ the panel is flat and represents a rectangular plate under uniaxial load. Large values of curvatures are associated with Θ close to unity.

This problem was first solved by Koiter in 1956 [8], and a summary of the results for a linear fundamental path and different values of the curvature parameter Θ are reproduced in Figure 12.1. The plot is normalized with respect to the critical load P_{CL}, and the displacements δ are normalized in a similar way.

The postbuckling path changes dramatically with the design parameter Θ:

- For the flat panel ($\Theta = 0$) and for shallow panels ($\Theta < 0.6$), there is a stable postbuckling path.
- There is a significant change in the response for values of $0.65 \leq \Theta \leq 0.75$. The postcritical path becomes unstable, with decreasing values of the load as the displacements increase. In this range of Θ, the slope becomes negative.
- For $0.75 \leq \Theta \leq 1$ there is an unstable path with a decrease in both load and displacements.

Of course, design sensitivity does not provide the complete picture of this problem. In real structures we should also include imperfections, and this is expected to influence the maximum load as Θ increases. Thus, for flat panels we do not have sensitivity to imperfections. For deep narrow panels, on the other hand, the behavior is imperfection sensitivity, and the study should be combined with those of Chapter 10 for $\Theta > 0.65$.

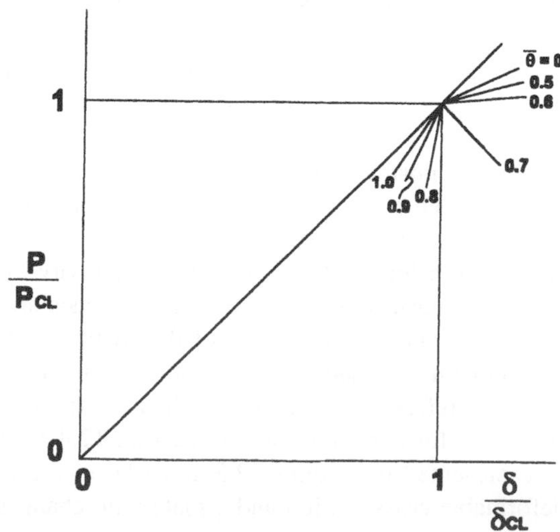

Figure 12.1 Postbuckling of a narrow panel under axial compression, obtained by Koiter in 1956 [8].

Rather than computing the postbuckling path for each design change, it is desirable to have a formulation that allows us to get sensitivity directly with respect to the design parameter chosen. This is the subject of the rest of this chapter.

12.3 SENSITIVITY OF SYMMETRIC BIFURCATION

12.3.1 Formulation

Let $\lambda^{(2)c}$ be the curvature of the postbuckling path evaluated from a stability analysis for a reference value of the design parameter τ, namely, $\tau = 0$. In the postcritical analysis, the sensitivity of the postbuckling state may be written in terms of the design parameter τ as

$$\lambda^{(2)}(\tau) = \lambda^{(2)c} + \dot{\lambda}^{(2)}\tau + \frac{1}{2}\ddot{\lambda}^{(2)}\tau^2 + \cdots, \tag{12.1}$$

$$q_j^{(2)}(\tau) = q_j^{(2)c} + \dot{q}_j^{(2)}\tau + \frac{1}{2}\ddot{q}_j^{(2)}\tau^2 + \cdots,$$

where, again, dots on top of a variable indicate derivation with respect to τ. Notice that $\dot{q}_i^{(1)} = \dot{x}_i$, and $\dot{x}_i$ was already computed.

To obtain the coefficients in the previous equation, let us write the curvature of the postbuckling path $\lambda^{(2)}$ in the form

$$3B\lambda^{(2)} + \tilde{V}_4|^c = 0. \tag{12.2}$$

First-order perturbation of this equation requires

$$3B\dot{\lambda}^{(2)}|^c = -3\dot{B}\lambda^{(2)} - \dot{\tilde{V}}_4|^c. \tag{12.3}$$

To solve $\dot{\lambda}^{(2)}$ it is necessary to obtain $\dot{B}$ and $\dot{\tilde{V}}_4$, and this will be done in the following for the specialized system.

Consider from Chapter 7

$$\tilde{V}_4 = A_{ijkl}x_ix_jx_kx_l + 3\left(A_{ijk} + A_{ijkl}Q_l^F\right)x_ix_jz_k,$$

$$B = \left(A_{ijk} + A_{ijkl}Q_l^F\right)x_ix_jy_k.$$

The first-order perturbation of $\tilde{V}_4$ with respect to the design parameter, called $\dot{\tilde{V}}_4$, is

$$\dot{\tilde{V}}_4 = \left(\dot{A}_{ijkl}x_l + 4A_{ijkl}\dot{x}_l\right)x_ix_jx_k + 3\eta_{ik}x_iz_k + 3\left[\left(A_{ijk} + A_{ijkl}Q_l^F\right)x_ix_j\right]\dot{z}_k, \tag{12.4}$$

where

$$\eta_{ik} = \left(\dot{A}_{ijk} + \dot{A}_{ijkl} Q_l^F + A_{ijkl} \dot{Q}_l\right) x_j + 2\left(A_{ijk} + A_{ijkl} Q_l^F\right) \dot{x}_j. \tag{12.5}$$

The first-order perturbation equation of B is

$$\dot{B} = \eta_{ik} y_k x_i + \left[\left(A_{ijk} + A_{ijkl} Q_l^F\right) x_i x_j\right] \dot{y}_k. \tag{12.6}$$

We now need the derivatives $\dot{z}_k$ and $\dot{y}_k$. They are obtained from derivation of the two conditions

$$V_{ij} y_j = -V_i',$$
$$V_{ij} z_j = -V_{ijk} x_j x_k.$$

The derivatives result in

$$\left(A_{ij} + \lambda^c A_{ijk} Q_k^F\right) \dot{y}_j = -\left(\gamma_{ij} y_j + \dot{A}_i\right), \tag{12.7}$$

where

$$\gamma_{ij} = \left(\dot{A}_{ij} + \dot{\lambda} A_{ijk} Q_k^F + \lambda \dot{A}_{ijk} Q_k^F + \lambda A_{ijk} \dot{Q}_k\right) \tag{12.8}$$

and

$$\left(A_{ij} + \lambda^c A_{ijk} Q_k^F\right) \dot{z}_j = -\left(\gamma_{ij} z_j + \eta_{ik} x_k\right). \tag{12.9}$$

At this stage, we need to set some values of $\dot{z}_1$ and $\dot{y}_1$ to be consistent with our earlier normalization of eigenvectors. Since we have already adopted $y_1 = 0$ and $z_1 = 0$, then it follows that $\dot{z}_1 = 0$ and $\dot{y}_1 = 0$.

It may also be required to compute the sensitivity of postcritical displacements. To do that, let us obtain the derivatives of

$$V_{ij} q_j^{(2)} = -V_{ijk}$$

leading to

$$\left(\dot{A}_{ij} + \dot{\lambda} A_{ijk} Q_k + \lambda \dot{A}_{ijk} Q_k + \lambda A_{ijk} \dot{Q}_k\right) q_j^{(2)} + \left(A_{ij} + \lambda A_{ijk} Q_k\right) \dot{q}_j^{(2)}$$
$$= -\left[\left(\dot{A}_{ijk} + \dot{A}_{ijkl} Q_l + A_{ijkl} \dot{Q}_l\right) x_j x_k + 2\left(A_{ijk} + A_{ijkl} Q_l\right) \dot{x}_i x_j + \dot{\lambda}^{(2)} A_i + \lambda^{(2)c} \dot{A}_i\right].$$

Now $\dot{q}_i^{(2)}$ can be obtained by solving the system

$$\left(A_{ij} + \lambda A_{ijk} Q_k\right) \dot{q}_i^{(2)} = -\left(\gamma_{ij} \dot{q}_j^{(2)} + \eta_{ik} x_k + \dot{\lambda}^{(2)} A_i + \lambda^{(2)c} \dot{A}_i\right). \tag{12.10}$$

A similar scheme of analysis can be followed to compute the second-order sensitivities of the postbuckling curvature $\ddot{\lambda}^{(2)c}$.

12.3.2 Algorithm for Symmetric Bifurcation

An algorithm for computation of $\dot{\lambda}^{(2)}$ for symmetric bifurcation is as follows:

1. Compute

$$\gamma_{ij} = \left(\dot{A}_{ij} + \dot{\lambda} A_{ijk} Q_k^F + \lambda \dot{A}_{ijk} Q_k^F + \lambda A_{ijk} \dot{Q}_k\right),$$

$$\eta_{ik} = \left(\dot{A}_{ijk} + \dot{A}_{ijkl} Q_l^F + A_{ijkl} \dot{Q}_l\right) x_j + 2\left(A_{ijk} + A_{ijkl} Q_l^F\right) \dot{x}_j.$$

2. Solve $\dot{y}_i$

$$\left(A_{ij} + \lambda^c A_{ijk} Q_k^F\right) \dot{y}_j = -\left(\gamma_{ij} y_j + \dot{A}_i\right) \qquad \text{with} \qquad \dot{y}_1 = 0.$$

3. Solve $\dot{z}_j$

$$\left(A_{ij} + \lambda^c A_{ijk} Q_k^F\right) \dot{z}_j = -\left(\gamma_{ij} z_j + \eta_{ik} x_k\right) \qquad \text{with} \qquad \dot{z}_1 = 0.$$

4. Compute

$$a_{ijk} = A_{ijk} + A_{ijkl} Q_l^F,$$

$$\dot{B} = \eta_{ik} x_i y_k + a_{ijk} x_i x_j \dot{y}_k.$$

5. Compute

$$\dot{\tilde{V}}_4 = \left(\dot{A}_{ijkl} x_l + 4 A_{ijkl} \dot{x}_l\right) x_i x_j x_k + 3 \eta_{ik} x_i z_k + 3 a_{ijk} x_i x_j \dot{z}_k.$$

6. Evaluate

$$\dot{\lambda}^{(2)} = -\frac{3\dot{B}\lambda^{(2)c} - \dot{\tilde{V}}_4}{3B}.$$

7. Evaluate first-order sensitivity

$$\lambda^{(2)}(\tau) = \lambda^{(2)c} + \dot{\lambda}^{(2)} \tau^2 + \cdots.$$

12.4 SENSITIVITY OF ASYMMETRIC BIFURCATION

12.4.1 Formulation

The coefficients of the perturbation expansion in this case can be obtained from

$$A\lambda^{(1)c} = -B + [B^2 - AC]^{\frac{1}{2}}.$$

First-order perturbation of this equation leads to

$$\dot{A}\lambda^{(1)c} + A\dot{\lambda}^{(1)} = -\dot{B} + \frac{1}{2}[B^2 - AC]^{-\frac{1}{2}}(2\dot{B}B - \dot{A}C - A\dot{C}). \tag{12.11}$$

The derivative $\dot{B}$ was previously obtained in (12.6). We need to calculate $\dot{A}$ and $\dot{C}$ from the derivatives of the coefficients

$$C = a_{ijk}x_i x_j x_k|^c, \tag{12.12}$$

$$A = a_{ijk}x_i y_j y_k|^c, \tag{12.13}$$

where $a_{ijk} = \left(A_{ijk} + A_{ijkl}Q_l^F\right)$. The first-order derivatives are

$$\dot{C} = \left(\dot{A}_{ijk} + \dot{A}_{ijkl}Q_l^F + A_{ijkl}\dot{Q}_l\right)x_i x_j x_k + 3a_{ijk}\dot{x}_i x_j x_k, \tag{12.14}$$

$$\dot{A} = \left(\dot{A}_{ijk} + \dot{A}_{ijkl}Q_l^F + A_{ijkl}\dot{Q}_l\right)x_i y_j y_k + a_{ijk}(\dot{x}_i y_j y_k + 2x_i\dot{y}_j y_k). \tag{12.15}$$

The computation of $\dot{y}_j$ is as in symmetric bifurcation (see (12.7)). Finally, from (12.11) one may obtain

$$\dot{\lambda}^{(1)} = \frac{-\dot{B} + \frac{1}{2}[B^2 - AC]^{-\frac{1}{2}}(2\dot{B}B - \dot{A}C - A\dot{C}) - \dot{A}\lambda^{(1)c}}{A}. \tag{12.16}$$

Notice that

- Only first-order sensitivity of the critical state is required to solve first-order sensitivity of the postcritical state.
- We have to solve two systems of linear equations for $\dot{y}$ and $\dot{z}$, and the rest of the algorithm is multiplication.

12.4.2 Algorithm for Asymmetric Bifurcation

An algorithm for the computation of $\dot{\lambda}^{(1)}$ in asymmetric bifurcation could proceed as follows:

1. Compute

$$\gamma_{ij} = \left(\dot{A}_{ij} + \dot{\lambda}A_{ijk}Q_k^F + \lambda\dot{A}_{ijk}Q_k^F + \lambda A_{ijk}\dot{Q}_k\right),$$

$$\eta_{ik} = \left(\dot{A}_{ijk} + \dot{A}_{ijkl}Q_l^F + A_{ijkl}\dot{Q}_l\right)x_j + 2\left(A_{ijk} + A_{ijkl}Q_l^F\right)\dot{x}_j.$$

2. Solve $\dot{y}_i$

$$\left(A_{ij} + \lambda^c A_{ijk} Q_k^F\right) \dot{y}_j = -\left(\gamma_{ij} y_j + \dot{A}_i\right) \qquad \text{with} \qquad \dot{y}_1 = 0.$$

3. Compute

$$\dot{A} = \eta_{ij} y_j y_k + a_{ijk}(-\dot{x}_i y_j y_k + 2x_i \dot{y}_j y_k),$$

$$\dot{B} = \eta_{ik} x_i y_k + a_{ijk} x_i x_j \dot{y}_k,$$

$$\dot{C} = \eta_{ik} x_i x_k + a_{ijk} \dot{x}_i x_j x_k.$$

4. Evaluate

$$\dot{\lambda}^{(1)} = \frac{-\dot{B} + \frac{1}{2}[B^2 - AC]^{-\frac{1}{2}}(2\dot{B}B - \dot{A}C - A\dot{C}) - \dot{A}\lambda^{(1)c}}{A},$$

$$\lambda^{(1)}(\tau) = \lambda^{(1)c} + \dot{\lambda}^{(1)}\tau + \cdots.$$

12.5 ANGLE SECTION COLUMN WITH DEFORMABLE CROSS SECTION

An academic but interesting example of sensitivity is the postbuckling response of an axially loaded, simply supported column with deformable cross section. The cross section considered is shown in Figure 12.2; this is a model of an angle section column in which the two plates are connected by a hinge and a moment spring, with stiffness K. In the undeflected configuration, the angle between the two plates is $\pi/2$. This example is considered in [2] and, more recently, in [7]. We assume that buckling in a torsional mode is prevented from occurring, so that only flexural buckling develops.

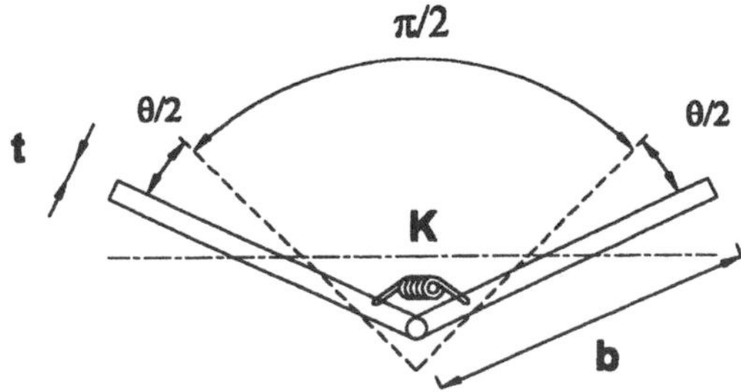

Figure 12.2 Geometry of an axially loaded column with deformable cross section. Reprinted from [5] with permission from Elsevier Science.

The axial (u) and transverse (w) displacements, and the rotation at the hinge (θ), are represented by

$$u(x) = Q_1\frac{x}{L}, \qquad w(x) = Q_2 \sin\left(\frac{\pi x}{L}\right), \tag{12.17}$$

$$\theta(x) = Q_3 \sin\left(\frac{\pi x}{L}\right),$$

where Q_1, Q_2, and Q_3 are the amplitudes of the assumed shape of u, w, and θ. We use nonlinear kinematic relations for column and linear constitutive equations, with a moment of inertia defined as

$$I(\theta) = I_0\left[1 - \left(\theta - \frac{\theta^3}{6}\right)\right], \tag{12.18}$$

where I_0 is the moment of inertia of the undeformed section. It is assumed that the moment of inertia about the weak axis decreases with deformations of the cross section θ.

The coefficients for the column under axial load, with a deformable cross section, are [7]

$$A_1 = -\Lambda L, \qquad A_1' = -L, \tag{12.19}$$

$$A_{11} = L, \qquad A_{22} = \frac{\pi^4}{2} d\,L, \qquad A_{33} = \frac{1}{2} f\,L,$$

$$A_{122} = \frac{\pi^2}{2}L, \qquad A_{223} = -\frac{4}{3}\pi^3\,d\,L, \qquad A_{2222} = \frac{9}{8}\pi^4 L,$$

where

$$\Lambda \equiv \frac{P}{EA_0}, \qquad d \equiv \frac{I_0}{A_0 L^2}, \qquad f \equiv \frac{K}{EA_0}.$$

The following results are obtained by use of the theory of elastic stability on the present model of the column and yield the fundamental path, critical state, and postcritical path.

The fundamental path results in

$$Q^F = \begin{Bmatrix} \Lambda \\ 0 \\ 0 \end{Bmatrix}.$$

The critical state is given by

$$\Lambda^c = -\pi^2 d,$$

$$Q^c = \begin{Bmatrix} -\pi^2 d \\ 0 \\ 0 \end{Bmatrix}, \qquad x^c = \begin{Bmatrix} 0 \\ 1 \\ 0 \end{Bmatrix}.$$

In writing V the load was initially assumed as tensile; for that reason, the critical load is negative. The perturbation parameter adopted to follow the postcritical path is the component Q_2. The two vectors required for the postbuckling path are

$$y = \begin{Bmatrix} 1 \\ 0 \\ 0 \end{Bmatrix}, \qquad z = \begin{Bmatrix} -\frac{\pi^2}{2} \\ 0 \\ \frac{8}{3}\pi^3 \frac{d}{f} \end{Bmatrix}.$$

The stability coefficient $\tilde{V}_4^c$ and B are

$$\tilde{V}_4^c = \frac{3}{8}\pi^4 L\xi, \qquad \text{where} \qquad \xi = 1 - \frac{159}{9}\pi^2 \frac{d^2}{f},$$

$$B = V_{122} = \frac{1}{2}\pi^2 L.$$

Finally, the second derivative of the load is

$$\lambda^{(2)c} = -\left(\frac{\pi}{2}\right)^2 \xi.$$

Notice that a positive value of ξ indicates a curvature with the same sign as the critical load and is thus a rising path.

Sensitivity analysis. Let us consider that the stiffness of the moment spring K is the design parameter, and write it in the parametric form

$$K = K_0(1 + \tau).$$

For convenience in the calculations, we shall assume a reference value K_0 such that $\xi = 0$. This means that

$$K_0 = \frac{256}{9}\pi^2 d^2 E A_0.$$

In the first stage, we make use of the sensitivity analysis of critical states outlined in Chapter 11. To obtain that, the derivatives of the energy coefficients with respect to τ are required; but since only A_{33} is a function of f (and thus of K and τ), the only nonzero derivative is

$$\dot{A}_{33} = \frac{1}{2}\frac{LK_0}{EA_0}.$$

It is simple to show that the fundamental path and the critical state are not sensitive to changes in τ. Thus

$$\dot{Q}_j = 0, \qquad \dot{x}_j = 0, \qquad \dot{\lambda} = 0.$$

Next, we want sensitivity of $\dot{\lambda}^{(2)}$ for a problem of symmetric bifurcation. Since $\dot{A}_i = 0 = \dot{A}_{ijk}$ and $\dot{\lambda} = 0 = \dot{Q}_k$, then $\gamma_{ij} = \dot{A}_{ij}$ and the only nonzero term becomes $\gamma_{33} = \dot{A}_{33}$. Furthermore, $\eta_{ik} = 0$, leading to $\dot{y}_j = 0$ and $\gamma_{33}\, y_3 = 0$.

To calculate the sensitivity of z we notice that

$$\gamma_{33}\, z_3 = \dot{A}_{33}\, z_3 = \frac{4}{3}\pi^3 L\, d.$$

The following system of equations should be solved, in which $\dot{z}_2 = 0$ (because the perturbation parameter adopted in this example is Q_2):

$$\begin{bmatrix} L & 0 & 0 \\ 0 & 0 & 0 \\ 0 & 0 & \frac{1}{2} f L \end{bmatrix} \begin{Bmatrix} \dot{z}_1 \\ 0 \\ \dot{z}_3 \end{Bmatrix} = -\frac{4}{3}\pi^3 L d \begin{Bmatrix} 0 \\ 0 \\ 1 \end{Bmatrix}.$$

The solution of this system is

$$\dot{z} = -\frac{8}{3}\pi^3 \frac{E A_0}{K} d \begin{Bmatrix} 0 \\ 0 \\ 1 \end{Bmatrix}.$$

The sensitivity of B reduces to

$$\dot{B} = \left(A_{ijk} + A_{ijkl} Q_l^F \right) x_i x_j \dot{y}_k = 0.$$

Finally, the sensitivity $\dot{\tilde{V}}_4$ may be calculated in the form

$$\begin{aligned} \dot{\tilde{V}}_4 &= 3\left(A_{ijk} + A_{ijkl}\bar{Q}_l\right) x_i x_j \dot{z}_k = 3\left(A_{223}\right) x_2 x_2 \dot{z}_3 \\ &= \left(-4\pi^3 d L\right)\left(-\frac{8}{3}\pi^3 \frac{E A_0}{K_0} d\right) = \frac{32}{3}\pi^6 d^2 \frac{E A_0 L}{K_0}. \end{aligned}$$

With the above results, it is possible to proceed with sensitivity of $\dot{\lambda}^{(2)c}$

$$\dot{\lambda}^{(2)} = -\frac{\dot{\tilde{V}}_4}{3B} = -\frac{64}{9}\pi^4 d^2 \frac{E A_0}{K_0}.$$

The final result for sensitivity of the curvature of the postbuckling path is

$$\lambda^{(2)}(\tau) = \lambda^{(2)c} - \left(\frac{64}{9}\pi^4 d^2 \frac{E A_0}{K_0}\right)\tau + \cdots.$$

The results are presented in Figure 12.3, and it is seen there that first-order sensitivity of the curvature of the postbuckling path changes sign with the value of τ.

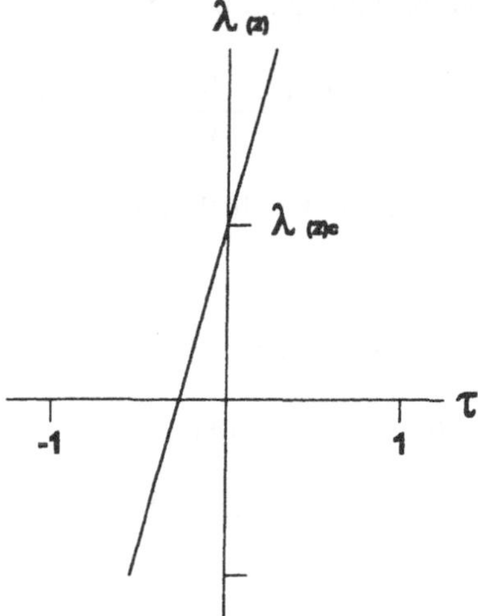

Figure 12.3 Sensitivity of the curvature of the postbuckling path for an angle section column. Reprinted with permission from [5] with permission from Elsevier Science.

The system is an unstable symmetric bifurcation with negative values of τ, while for positive values of τ the system is a stable symmetric bifurcation.

For this simple example, the values of sensitivity could have been obtained from the explicit expressions; however, in more complex problems involving, for example, finite element discretizations, there are no analytical solutions available and the present analysis would provide new information on sensitivity.

12.6 CIRCULAR PLATE UNDER IN-PLANE LOADING

The buckling of circular plates has been considered by several authors; the sensitivity of the critical state was solved in Chapter 11 and will be studied further here to extend the analysis to sensitivity of postcritical behavior. This case is more complex than the previous one, in the sense that all precritical, critical, and postcritical states are sensitive to changes in the design parameter (the thickness of the plate).

The fundamental path, the critical state, and the vectors required for the post-buckling path are given in Chapter 7.

The coefficients of the quadratic equation are

$$C = 0, \qquad B = 1.7337.$$

The stability coefficient is in this case

$$\tilde{V}_4 = A_{1111} + 3A_{112}z_2,$$

and the curvature of the postbuckling path becomes

$$\lambda^{(2)c} = \frac{A_{1111} - 3(A_{112})^2/A_{22}}{3A_{11}}.$$

Sensitivity analysis. Let us consider the thickness h as a design parameter, and write it in the form

$$h = h_0(1+\tau).$$

The derivatives of the energy coefficients become

$$\begin{aligned} &\dot{A}_2' = 0, \qquad \dot{A}_{11} = 3A_{11}, \qquad \dot{A}_{22} = A_{22}, \\ &\dot{A}_{112} = A_{112}, \qquad \dot{A}_{1111} = A_{1111}, \\ &\dot{f}_i = 0. \end{aligned} \tag{12.20}$$

Sensitivity of the fundamental path is given by

$$\dot{Q}_j = \begin{Bmatrix} 0 \\ -Q_2^F \end{Bmatrix},$$

and sensitivity of the critical state results in

$$\dot{\lambda} = 3\Lambda^c, \qquad \dot{x}_j = 0.$$

Next, we calculate sensitivity of the postcritical path. The matrices γ and η have coefficients equal to zero, except for

$$\gamma_{11}, \qquad \gamma_{22} = \dot{A}_{22}, \qquad \eta_{12} = \dot{A}_{112}x_1.$$

This leads to

$$\dot{y} = \begin{Bmatrix} 0 \\ -\gamma_{22}y_2/A_{22} \end{Bmatrix}, \qquad \dot{z} = \begin{Bmatrix} 0 \\ -(\gamma_{22}y_2 + \eta_{12})/A_{22} \end{Bmatrix}.$$

The sensitivity of B becomes zero. The derivative of the stability coefficient with respect to τ becomes

$$\dot{\tilde{V}}_4 = \dot{A}_{1111} + 3(\eta_{12}z_2 + A_{112}\dot{z}_2).$$

Finally, we can get

$$\dot{\lambda}^{(2)} = \lambda^{(2)}.$$

The result for sensitivity of the postbuckling curvature is

$$\lambda^{(2)}(\tau) = \lambda^{(2)c}(1+\tau)$$

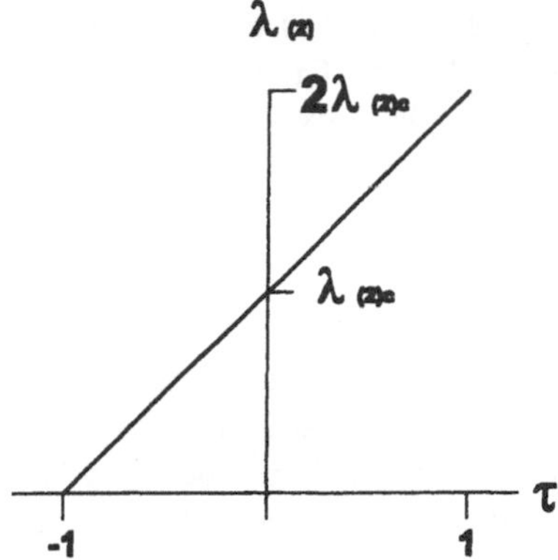

Figure 12.4 Sensitivity of the curvature of the postbuckling path for a circular plate with changes in the thickness. Reprinted from [5] with permission from Elsevier Science.

and is plotted in Figure 12.4. Contrary to what occurred in the first example, in the circular plate the postbuckling curvature cannot reverse the sign with changes in the design parameter.

12.7 FINAL REMARKS

A consistent derivation of first-order sensitivity of the postbuckling path in symmetric as well as asymmetric bifurcation has been presented in this chapter. The limitations of the analysis refer to a single design parameter and first-order sensitivity. Certainly, both limitations could be overcome, but this may be a good beginning to appreciate the difficulties and achievements in this field.

Once the information of design sensitivity of the critical state is obtained, following the present analysis it is possible to compute sensitivities of the curvature of the postbuckling path (in symmetric bifurcation) and sensitivity of the tangent to the postbuckling path (in asymmetric bifurcation). Algorithms for the computation of sensitivities were presented in both cases.

The results of the angle column show that although the critical state itself may be insensitive to the design parameter considered, the postbuckling path may be highly dependent on the parameter (stiffness coefficient) adopted. Sensitivity is also reflected in that the postbuckling behavior may change from stable to unstable depending on the values of the design parameter.

In the circular plate under in-plane loading, we notice that the fundamental path and the critical state are sensitive to changes in the design parameter chosen and that the postbuckling path is also sensitive to thickness changes. However, in this case there cannot be a change from stable to unstable behavior produced by changes in the design parameter.

For stable postbuckling behavior, it is expected that the limit state will be controlled by nonlinear material behavior. The results of this chapter can be extended to include material nonlinearity or plasticity as a constraint in the analysis, and this is reported in [6].

12.8 PROBLEMS

Review Questions. (*a*) Is it possible that the critical state of a structure is not sensitive to changes in a design parameter, and still the postcritical path is sensitive to the same parameter? (*b*) Explain how unstable postbuckling behavior develops in an angle section column in flexural modes. (*c*) Describe how the curvature of a curved panel under axial load affects the postbuckling response. (*d*) Explain the steps required to develop second-order design sensitivity in asymmetric bifurcation.

Problem 12.1 (Ring). Consider a ring with angle cross section, modeled by flat plates and a rotational spring, as in section 12.5. The ring is loaded by radial pressure. Obtain sensitivity with respect to changes in the stiffness of the central rotational spring.

Problem 12.2 (Angle section column). Consider an arch with angle cross section, modeled by flat plates, as in section 12.5. The arch is loaded by a central point load. Obtain sensitivity with respect to changes in the stiffness of the central rotational spring.

Problem 12.3 (Two-bar frame). For the two-bar frame with point load studied in Chapter 7, evaluate sensitivity of the postbuckling state with respect to the thickness of the bars.

Problem 12.4 (Nonprismatic column). A nonprismatic column is made with three circular parts with different radius but with the same central axis. Each part has a length $L/3$, the external parts have a radius r_0, and the central part has a radius r. We shall

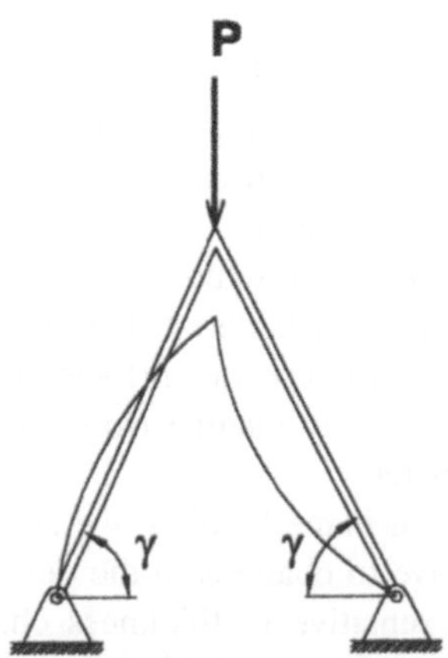

Figure 12.5 Plane frame for Problem 12.5.

employ the radius of the central part as a design parameter, i.e., $r = r_0(1-\tau)$. Find sensitivity of the critical and postcritical solution. Use simply supported boundary conditions. Use the following approximation for the out-of-plane displacement:

$$w = Q_1 \sin\left(\frac{\pi x}{L}\right) + Q_2 \sin\left(\frac{2\pi x}{L}\right).$$

Problem 12.5 (Plane frame). A plane steel frame is made with two bars as shown in Figure 12.5. The angle γ will be considered as a design variable, with a reference state given by $\gamma = 54.7°$. Assume extensional members.

Problem 12.6 (Theory). Develop second-order design sensitivity in asymmetric bifurcation.

Problem 12.7 (Theory). Develop second-order design sensitivity in symmetric bifurcation.

12.9 BIBLIOGRAPHY

[1] Almanzar, L., and Godoy, L. A., Design sensitivity of buckled thin-walled composite structures, *Appl. Mech. Rev.*, ASME, 50(11), 3–10, 1997.

[2] Eterovic, A., Godoy, L. A., and Prato, C. A., Initial post-critical behavior of thin-walled angle section columns, *J. Engrg. Mech.*, ASCE, 116(11), 2573–2577, 1990.

[3] Fitch, J. R., and Budianski, B., Buckling and post-buckling behavior of spherical caps under axisymmetric load, *AIAA J.*, 8(4), 686–693, 1970.

[4] Flores, F., and Godoy, L. A., Post-buckling of elastic cone-cylinder and sphere-cylinder complex shells. *Internat. J. Pressure Vessels and Piping*, 45, 237–258, 1991.

[5] Godoy, L. A., Sensitivity of post-critical states to changes in design parameters, *Internat. J. Solids Structures*, 33(15), 2177–2192, 1996.

[6] Godoy, L. A., and Taroco, E. O., Design sensitivity of post-buckling states including material constraints, *Comput. Methods Appl. Mech. Engrg.*, 1999.

[7] Lopez-Anido, R., and Godoy, L. A., Post-buckling of one-dimensional elements with thin-walled angle section using simplified models, *Mech. Structures Mach.*, 24(4), 473–495, 1996.

[8] Koiter, W. T., *Buckling and Post-Buckling of a Cylindrical Panel under Axial Compression*, National Aeronautical Research Institute, Amsterdam, Report S-476, May 1956.

CHAPTER

THIRTEEN

BIFURCATION STATES: THE *W*-FORMULATION

13.1 BIFURCATION ANALYSIS IN TERMS OF THE FUNDAMENTAL PATH

In previous chapters we have worked with the energy V constructed using generalized coordinates. Such coordinates were measured with respect to an initial configuration of the system, for example, the unloaded state. The energy was thus denoted as $V[Q_i, \Lambda]$, and each path was described in terms of the same initial configuration. Figure 13.1.a shows the equilibrium path for a one-degree-of-freedom system with limit point behavior, while a bifurcation is illustrated in Figure 13.1.b.

Thompson and Hunt [13] introduced a different bifurcation analysis, which we call here the W-formulation. This assumes that the fundamental path has been evaluated for loads higher than the lowest critical load Λ^c. Then a set of incremental coordinates, measured from the fundamental path, is defined as denoted in Figure 13.2. They are called sliding coordinates, in the sense that they seem to slide along the fundamental path as the value of the load parameter Λ is increased.

Thus, we can write

$$\Lambda(s) = \Lambda^F(s) + \lambda,$$

$$Q_i(s) = Q_i^F(s) + q_i,$$

in which Q_i^F, Λ^F define the known fundamental path; Q_i, Λ are the displacements and loads measured from the initial configuration; and q_i are changes in generalized

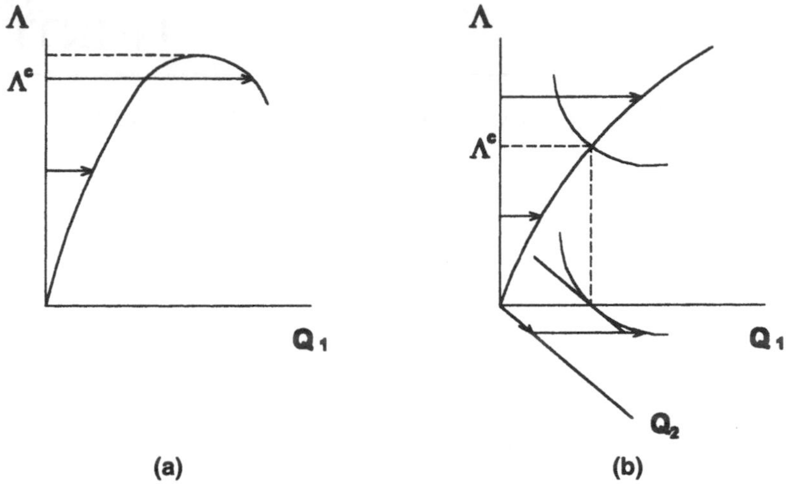

Figure 13.1 (a) Limit points; (b) bifurcation points.

coordinates measured from the fundamental path. Notice that if we choose a perturbation parameter s to follow the fundamental path, then Q_i^F depends on s, but q_i is independent of s. In a general case we may choose

$$s \equiv \Lambda$$

and

$$Q_i(\Lambda) = Q_i^F(\Lambda) + q_i.$$

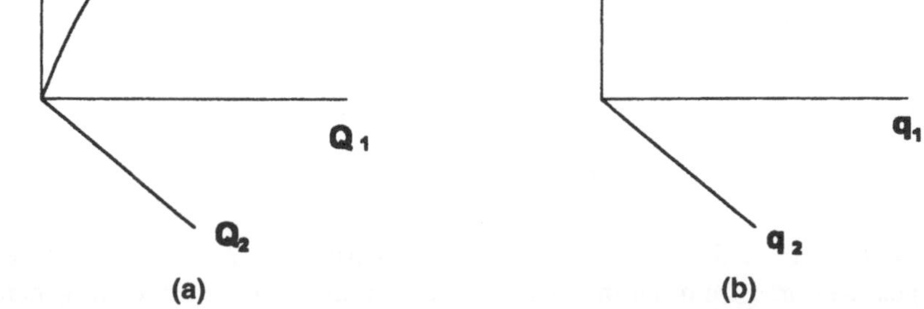

Figure 13.2 The V to W transformation. (a) Bifurcation in the original generalized coordinates; (b) bifurcation plotted in terms of incremental generalized coordinates.

The fundamental path is assumed to increase in Λ^F for increasing values of Q_i^F, so that there is no problem in choosing the load as perturbation parameter $s \equiv \Lambda$. Notice that a limit point is excluded from the analysis presented in this chapter.

The fundamental path could be a nonlinear path, and the presentation in [13] is made with this assumption. We restrict ourselves initially to the case of a linear fundamental path, for which the displacement can be written as

$$Q_i^F(\Lambda) = \Lambda \bar{Q}_i,$$

where $\bar{Q}_i(\Lambda)$ is the displacement vector in the fundamental path for a unit value of Λ. The displacements of a secondary path are

$$Q_i(\Lambda) = \Lambda \bar{Q}_i + q_i. \tag{13.1}$$

The assumption of linearity may seem somehow restrictive, but the W-formulation is particularly convenient in this case. If Q_i^F is nonlinear in Λ, perhaps a better choice would be to perform the analysis in terms of V as in the preceding chapters.

The equilibrium paths can also be plotted in terms of q_i, as in Figure 13.2.b. The fundamental path lies along the Λ-axis, and only secondary paths may now be seen.

13.2 THE *W* FUNCTIONAL

This method of analysis was clearly described, for example, by Sewell: "The basic assumption is that one entire equilibrium path through the $\lambda - q_i$ origin has been discovered already (or at least a finite portion of it containing the origin)." He identified "the excess potential energy at a general point of configuration space, over and above that on the fundamental path at any given value of λ" [11].

Let us start from the energy V written in the form

$$V[Q_i, \Lambda] = A_i Q_i + \frac{1}{2!} A_{ij} Q_i Q_j + \frac{1}{3!} A_{ijk} Q_i Q_j Q_k + \frac{1}{4!} A_{ijkl} Q_i Q_j Q_k Q_l. \tag{13.2}$$

Terms that are constant in Q_i are not considered, since they vanish when we differentiate V.

We assume that only A_i is a function of Λ, so that it takes the form

$$A_i = \hat{U}_i - \Lambda A_i' \tag{13.3}$$

in which A_i' is the load vector, and $\hat{U}_i$ represents the strain energy terms linear in Q_i and independent of the load. Notice that $\hat{U}_i = 0$ in most problems in structural mechanics. This is a **specialized system**, in the sense that the load parameter acts on a linear displacement Q_i.

Next, we substitute $Q_i(\Lambda)$ from (13.1) into V, leading to

$$V[\Lambda\bar{Q}_i + q_i, \Lambda] = (\hat{U}_i - \Lambda A_i')(\Lambda\bar{Q}_i + q_i)$$
$$+ \frac{1}{2}A_{ij}(\Lambda^2\bar{Q}_i\bar{Q}_j + 2\Lambda\bar{Q}_i q_j + q_i q_j)$$
$$+ \frac{1}{3!}A_{ijk}(\Lambda^3\bar{Q}_i\bar{Q}_j\bar{Q}_k + 3\Lambda^2\bar{Q}_i\bar{Q}_j q_k + 3\Lambda\bar{Q}_i q_j q_k + q_i q_j q_k)$$
$$+ \frac{1}{4!}A_{ijkl}(\Lambda^4\bar{Q}_i\bar{Q}_j\bar{Q}_k\bar{Q}_l + 4\Lambda^3\bar{Q}_i\bar{Q}_j\bar{Q}_k q_l$$
$$+ 6\Lambda^2\bar{Q}_i\bar{Q}_j q_k q_l + 4\Lambda\bar{Q}_i q_j q_k q_l + q_i q_j q_k q_l). \tag{13.4}$$

In (13.4) there are two terms that cancel each other from equilibrium of linear fundamental path, i.e.,

$$-\Lambda A_i' q_i + \frac{1}{2}A_{ij}2\Lambda\bar{Q}_i q_j = \Lambda\underbrace{[A_{ij}\bar{Q}_j - A_i']}_{0}q_i. \tag{13.5}$$

Because of (13.5), there is no energy contribution in W containing both Λ and q_i linearly.

In its new form, V depends on q_i and Λ, since the $\bar{Q}_i$ are known. The generalized coordinates employed to evaluate the energy are now different, and we denote this by changing the name of the functional to W, i.e.,

$$W[q_i, \Lambda] \equiv V[\Lambda\bar{Q}_i + q_i, \Lambda]. \tag{13.6}$$

Furthermore, we can neglect terms that are independent of q_i (because they will vanish in the search for equilibrium) and terms with nonlinear contribution from $\bar{Q}_i$ to be consistent with our assumption about linearity of the fundamental path. Thus, W reduces to

$$W[q_i, \Lambda] = \frac{1}{2}A_{ij}q_i q_j + \Lambda\frac{1}{2}A_{ijk}q_i q_j\bar{Q}_k + \frac{1}{3!}A_{ijk}q_i q_j q_k \tag{13.7}$$
$$+\frac{1}{3!}\Lambda A_{ijkl}\bar{Q}_i q_j q_k q_l + \frac{1}{4!}A_{ijkl}q_i q_j q_k q_l.$$

Equation (13.7) contains the total potential energy in terms of incremental displacements measured from the fundamental path. In the W-formulation one can identify linear, cuadratic, cubic, and quartic contributions in q_i in a similar fashion to what was done for the energy V. Thus, we write

$$W[q_i, \Lambda] = \hat{W}_i q_i + \frac{1}{2!}\hat{W}_{ij}q_i q_j + \frac{1}{3!}\hat{W}_{ijk}q_i q_j q_k + \frac{1}{4!}\hat{W}_{ijkl}q_i q_j q_k q_l \tag{13.8}$$

with

$$\hat{W}_i = \hat{U}_i, \tag{13.9}$$

$$\hat{W}_{ij} = A_{ij} + \Lambda A_{ijk}\bar{Q}_k,$$

$$\hat{W}_{ijk} = A_{ijk} + \Lambda A_{ijkl}\bar{Q}_l,$$

$$\hat{W}_{ijkl} = A_{ijkl}.$$

In the V-formulation, the load Λ acts on a linear displacement Q_i. In the W-formulation, a Λ acting on a linear q_i is not present, and the first term with Λ is associated with quadratic displacements $q_i q_j$. This is a general conclusion: The $V \longrightarrow W$ transformation is a standard technique to change a functional in which Λ acts on a linear displacement to another functional in which Λ acts on quadratic and cubic displacements [13]. Notice that the new load terms affect displacements that are incremental, so that they are not contained in the plane (or hyperplane) of the fundamental path.

From (13.8) it results that

$$\hat{W}_i = \left.\frac{\partial W}{\partial q_i}\right|_{q_i=0},$$

$$\hat{W}_{ij} = \left.\frac{\partial^2 W}{\partial q_i \partial q_j}\right|_{q_i=0},$$

etc.

A third approach consists in considering that the q_i are increments or variations with respect to an equilibrium state; then the total variation of W is given by

$$\Delta W = W[q_i, \Lambda] - W[0, \Lambda] = \delta W + \frac{1}{2}\delta^2 W + \frac{1}{3!}\delta^3 W + \frac{1}{4!}\delta^4 W. \tag{13.10}$$

Comparison of (13.10) with (13.8) indicates that the first- and higher-order variations of W are given by

$$\delta W = \hat{W}_i q_i,$$

$$\delta^2 W = \hat{W}_{ij} q_i q_j,$$

$$\delta^3 W = \hat{W}_{ijk} q_i q_j q_k,$$

$$\delta^4 W = \hat{W}_{ijkl} q_i q_j q_k q_l.$$

Example 13.1 (Circular plate). *As a simple example of the application of the W-formulation we consider here the buckling of a circular plate under radial load. This problem was investigated in previous chapters using the V-formulation. Following the Ritz discretization of the plate adopted in many examples of this book, the energy coefficients of V are listed in Example 2.4.*

Next, we use the $V \rightarrow W$ transformation and obtain the coefficients $\hat{W}$,

$$\hat{W}_i = 0,$$

$$\hat{W}_{11} = A_{11} + \Lambda\, A_{112}\bar{Q}_2 = \frac{\pi^2}{2}\frac{D}{R^2}(1.191 + \nu) + \Lambda\frac{(\pi^2+4)}{8},$$

$$\hat{W}_{22} = A_{22}, \qquad \hat{W}_{112} = A_{112}, \qquad \hat{W}_{1111} = A_{1111}.$$

Notice that the load parameter Λ only affects the quadratic term $\hat{W}_{11}$. The energy $\hat{W}$ becomes

$$W[q_i, \Lambda] = \frac{1}{2!}(\hat{W}_{11}q_1^2 + \hat{W}_{22}q_2^2) + \frac{1}{3!}(3\hat{W}_{112}q_1^2 q_2) + \frac{1}{4!}\left(\hat{W}_{1111}q_1^4\right).$$

13.3 SOME PROPERTIES OF *W*

13.3.1 Derivatives of *W*

Now that $W[q_i, \lambda]$ has been constructed in terms of the coefficients of V, we may explore the derivatives of W. In this section we are especially interested in derivatives evaluated at the fundamental path.

- The first derivatives of W yield

$$W_i[q_i, \Lambda] = \hat{W}_i + \hat{W}_{ij}q_j + \frac{1}{2}\hat{W}_{ijk}q_j q_k + \frac{1}{3!}\hat{W}_{ijkl}q_j q_k q_l. \tag{13.11}$$

If we evaluate W_i at the fundamental path, for which $q_i = 0$, we get a condition

$$\boxed{W_i[0, \Lambda] = \hat{W}_i = \hat{U}_i = 0} \tag{13.12}$$

- Next, consider

$$W_i'[q_i, \Lambda] = \hat{W}_i' + \hat{W}_{ij}'q_j + \frac{1}{2!}\hat{W}_{ijk}'q_j q_k + \frac{1}{3!}\hat{W}_{ijkl}'q_j q_k q_l,$$

$W_i' = 0$, because $\hat{W}_i$ does not contain Λ. Then W_i' evaluated along the fundamental path becomes zero

$$W_i'[0, \Lambda] = 0. \tag{13.13}$$

- Similarly, it can be shown that

$$W_i''[0, \Lambda] = 0,$$
$$W_i'''[0, \Lambda] = 0. \tag{13.14}$$

- The second derivatives of W result in

$$W_{ij}[q_i, \Lambda] = \hat{W}_{ij} + \hat{W}_{ijk}q_k + \frac{1}{2}\hat{W}_{ijkl}q_kq_l.$$

If we evaluate the second derivative at $q_i = 0$, we get

$$W_{ij}[0, \Lambda] = \hat{W}_{ij},$$
$$W_{ij}'[0, \Lambda] = A_{ijk}\bar{Q}_k.$$

Substituting $\hat{W}_{ij}$, we get

$$W_{ij}[0, \Lambda] = A_{ij} + \Lambda A_{ijk}\bar{Q}_k. \tag{13.15}$$

13.3.2 Critical State

From Chapter 5, we know that the condition of critical state in terms of the energy V is

$$V_{ij}x_j|^c = 0. \tag{13.16}$$

To get V_{ij}, let us now consider the expansion (13.2) and create the derivatives; we obtain

$$V_{ij} = A_{ij} + A_{ijkl}Q_k + \frac{1}{2}A_{ijkl}Q_kQ_l$$

or else (using (13.1)), for $q_i = 0$,

$$V_{ij} = A_{ij} + \Lambda\ A_{ijk}\bar{Q}_k,$$

which is identical to (13.15). Thus, because of (13.16), we may write the condition for critical state in terms of W:

$$W_{ij}[0, \Lambda]x_j|^c = 0 \qquad \Rightarrow \qquad \text{critical state.} \tag{13.17}$$

Equation (13.17) can also be written in terms of the coefficients A_{ij} and A_{ijk} as

$$[A_{ij} + \Lambda\ A_{ijk}\bar{Q}_k]x_j = 0 \qquad \Rightarrow \qquad \text{critical state.} \tag{13.18}$$

Equation (13.18) is an eigenvalue problem associated with the condition of critical state. The scalar Λ is the eigenvalue and x_j is the eigenvector. The matrix $\left[A_{ij}\right]$ is known as the linear stiffness matrix of the structure, and the matrix $\left[A_{ijk}\bar{Q}_k\right]$ is known as the load-geometry matrix, or geometric matrix. It is also known as initial stress matrix.

Example 13.2 (Circular plate). *For the plate of Example* 13.1, *let us obtain the critical state using the energy* W. *The condition of a critical state is now given by*

$$\hat{W}_{ij}x_j\,|^c= 0$$

or else

$$\begin{bmatrix} \hat{W}_{11} & 0 \\ 0 & \hat{W}_{22} \end{bmatrix} \begin{Bmatrix} x_1 \\ x_2 \end{Bmatrix} = \begin{Bmatrix} 0 \\ 0 \end{Bmatrix}.$$

This leads to the eigenvalue

$$\Lambda^c = -\frac{(1.191+\nu)}{3.4674}\left(\frac{\pi}{R}\right)^2 D.$$

If we substitute Λ^c *into the condition of critical state, we get the eigenvector*

$$x_i = \begin{Bmatrix} 1 \\ 0 \end{Bmatrix}.$$

13.3.3 Summary of the Properties of W

The new properties of the energy W can be summarized as follows:

(a) The $V \longrightarrow W$ transformation modifies the terms associated with the load parameter Λ, but the internal energy has the same form as before in terms of Q_i or q_i.

(b) In V, the load parameter acts on a linear displacement whenever we have a specialized system; in W, Λ acts on quadratic and cubic displacements. The transformation thus destroys the linearity of displacements in Λ even in the simplest systems.

(c) Equilibrium along the fundamental path is given by $W_i[0, \Lambda] = 0$.

(d) Since the load terms in W are at least quadratic in q_i, the first derivative contains at least a linear contribution Λq_i. Thus, there are no terms associated with the load that are independent of displacements in W_i. From this, it follows that $W_i'[0, \Lambda] = W_i''[0, \Lambda] = \cdots = 0$.

(e) In the V-formulation, at a critical state we have $|V_{ij}[Q_i^c, \Lambda^c]| = 0$. Now in the W-formulation we have $|W_{ij}[0, \Lambda^c]| = 0$.

(f) The W-formulation is useful for studying what occurs once the critical state has been identified as a bifurcation. It is not capable of dealing with limit points for which $V_i'x_i|^c \neq 0$.

(g) The W-formulation can be extended to account for situations in which the fundamental path is nonlinear.

13.4 POSTCRITICAL PATH IN TERMS OF W

The postcritical path will be investigated by means of a perturbation analysis that starts from the critical state. To carry out the perturbation analysis, the variables of load and displacements, λ and q_i, are expanded in terms of a suitable perturbation parameter s:

$$\lambda(s) = \Lambda^c + \lambda^{(1)} s + \frac{1}{2}\lambda^{(2)} s^2 + \cdots, \tag{13.19}$$

$$q_i = Q_i^c + q_i^{(1)} s + \frac{1}{2} q_i^{(2)} s^2 + \cdots.$$

A convenient choice for s is to take one of the components of vector q_j as a perturbation parameter. A condition required so that s can describe adequately the path is that each point of the path should be associated with a single value of s. It is not difficult to find a component of q_j that can serve as a perturbation parameter.

To expedite the notation, let us write $s \equiv q_1$ as the perturbation parameter, and check that the secondary path has a component $q_1 \neq 0$. The postcritical path may be derived again; or else the conditions already obtained for the V-formulation can be simplified because of the properties of W. Both cases are treated as problems left to the reader, and in this section we give a brief summary of the results.

The first-order, uncontracted equation is

$$W_{ij} q_j^{(1)} = 0. \tag{13.20}$$

The contracted first-order equation becomes an identity. The second-order uncontracted equation becomes

$$W_{ijk} q_j^{(1)} q_k^{(1)} + W_{ij} q_j^{(2)} + 2W'_{ij} q_j^{(1)} \lambda^{(1)} \,|^c = 0.$$

The second-order contracted equation is

$$W_{ij} x_i q_j^{(2)} + 2W'_{ij} x_i q_j^{(1)} \lambda^{(1)} \,|^c = 0.$$

From the second-order contracted equation we get only one solution for $\lambda^{(1)c}$:

$$\lambda^{(1)c} = \left.\frac{\tilde{W}_3}{2\tilde{W}'_2}\right|^c, \tag{13.21}$$

where

$$\tilde{W}_3^c \equiv W_{ijk} q_i^{(1)} q_j^{(1)} q_k^{(1)} |^c, \tag{13.22}$$

$$\tilde{W}_2'^c = W'_{ij} q_i^{(1)} q_j^{(1)} |^c. \tag{13.23}$$

13.4.1 Symmetric Bifurcation

If the coefficient C is zero

$$C \equiv \tilde{W}_3|^c = 0,$$

then we are in the presence of symmetric bifurcation. The first-order coefficients are

$$\lambda^{(1)c} = 0, \quad q_j^{(1)c} = x_j. \tag{13.24}$$

From the second-order, uncontracted equation, we get

$$W_{ij} q_j^{(2)}|^c = -W_{ijk} q_j^{(1)} q_k^{(1)}|^c, \tag{13.25}$$

and we can obtain $q_j^{(2)c}$ with the condition

$$q_1^{(2)c} = 0. \tag{13.26}$$

The second derivative of λ results in

$$\lambda^{(2)c} = -\frac{\tilde{W}_4}{3\tilde{W}_2'}\bigg|^c \tag{13.27}$$

with

$$\tilde{W}_4 = W_{ijkl} q_i^{(1)} q_j^{(1)} q_k^{(1)} q_l^{(1)} + 3W_{ijk} q_i^{(1)} q_j^{(1)} q_k^{(2)}|^c. \tag{13.28}$$

13.4.2 Asymmetric Bifurcation

If $\tilde{W}_3 = C \neq 0$, then $\lambda^{(1)c} \neq 0$ and is given by (13.21). The first-order uncontracted equation gives

$$W_{ij} q_j^{(1)}|^c = 0$$

with $q_1^{(1)c} = 1$.

The second-order uncontracted equation is

$$W_{ij} q_j^{(2)}|^c = -W_{ijk} q_j^{(1)} q_k^{(1)} - 2W_{ij}' q_j^{(1)} \lambda^{(1)}|^c, \tag{13.29}$$

and it can be solved with the constraint (13.26).

The third-order contracted equation allows one to obtain

$$\lambda^{(2)c} = -\frac{\tilde{W}_4 + 3\tilde{W}_3'}{3\tilde{W}_2'}, \tag{13.30}$$

where

$$\tilde{W}_3' \equiv \lambda^{(1)c}[W_{ijk}' q_i^{(1)} q_j^{(1)} q_k^{(1)} + W_{ij}' q_i^{(1)} q_j^{(2)} + W_{ij}'' q_i^{(1)} q_j^{(1)} \Lambda^{(1)}]. \tag{13.31}$$

It is possible to obtain higher-order derivatives of displacements and loads in the perturbation expansion, but we will not pursue this any further here.

Example 13.3 (Circular plate). *The first-order displacement field in the postcritical path is given by*

$$q_i^{(1)} = x_i = \begin{Bmatrix} 1 \\ 0 \end{Bmatrix}.$$

The coefficients necessary to compute the first-order load coefficient are

$$\tilde{W}_3^c = 0,$$

$$\left(\tilde{W}_2'\right)^c = 1.734.$$

This leads to a symmetric bifurcation with

$$\lambda^{(1)c} = 0.$$

The stability coefficient is

$$\tilde{W}_4^c = \frac{C}{R^2}(6.0498 - 4.51\nu) > 0.$$

The stability coefficient is positive for values of $\nu < 0.5$*, and this means that the bifurcation is stable symmetric. The curvature coefficient* $\lambda^{(2)c}$ *becomes*

$$\lambda^{(2)c} = -\frac{\tilde{W}_4^c}{3\tilde{W}_2'^c} = -\frac{C}{R^2}(1.163 - 0.867\nu).$$

Notice that the sign of the curvature should be interpreted in conjunction with the sign of the critical load. In this example both are negative, so that the load increases with q_1^2 *after buckling. The load-displacement equation is*

$$\Lambda = -\frac{(1.191 + \nu)}{3.4674}\left(\frac{\pi}{R}\right)^2 D - \frac{C}{R^2}(1.163 - 0.867\nu)\, q_1^2.$$

13.5 QUADRATIC FUNDAMENTAL PATH

Let us write the W-functional in the form of (13.8) for a class of problems in which the fundamental path is quadratic, i.e.,

$$Q_i(\Lambda) = Q_i^F(\Lambda) + q_i = \Lambda \bar{Q}_i + \frac{1}{2}\Lambda^2 \bar{\bar{Q}}_i + q_i.$$

Notice that nonlinear contributions arising from $\bar{Q}_i$, $\bar{\bar{Q}}_i$ cannot be neglected now. Are (13.13) and (13.14) valid for these new assumptions? What is the form of (13.18)? Is the postcritical analysis of section 13.4 valid?

13.5.1 Critical State

We start by writing V in the form

$$\begin{aligned} V[Q_i^F(\Lambda)+q_i,\Lambda] &= A_i(Q_i^F+q_i) \\ &+ \frac{1}{2}A_{ij}(Q_i^F Q_j^F + 2Q_i^F q_j + q_i q_j) \\ &+ \frac{1}{3!}A_{ijk}(Q_i^F Q_j^F Q_k^F + 3Q_i^F Q_j^F q_k + 3Q_i^F q_j q_k + q_i q_j q_k) \\ &+ \frac{1}{4!}A_{ijkl}(Q_i^F Q_j^F Q_k^F Q_l^F + 4Q_i^F Q_j^F Q_k^F q_l + 6Q_i^F Q_j^F q_k q_l \\ &+4Q_i^F q_j q_k q_l + q_i q_j q_k q_l). \end{aligned}$$

Next, substitute $Q_i^F = \Lambda\ \dot{Q}_i + \frac{1}{2}\Lambda^2 \ddot{Q}_i$, but we will retain up to terms that are quadratic in Q_i^F:

$$\begin{aligned} W &= A_i\left(\Lambda \dot{Q}_i + \frac{1}{2}\Lambda_i^2 \ddot{Q} + q_i\right) \\ &+\frac{1}{2}A_{ij}\left[\left(\Lambda^2 \dot{Q}_i \dot{Q}_j + \frac{\Lambda^4}{4}\dot{Q}_j \dot{Q}_i + \Lambda^3 \dot{Q}_i \dot{Q}_j\right) + q_i q_j + 2\Lambda\ \dot{Q}_i q_j + \Lambda_i^2 \ddot{Q} q_j\right] \\ &+\cdots. \end{aligned}$$

For consistency one must neglect products of third and fourth order in the fundamental path that were neglected in the approximation of the path itself. Thus

$$W_{ij} = A_{ij} + \Lambda A_{ijk}\bar{Q}_k + \frac{1}{2}\Lambda^2(A_{ijk}\ \bar{\bar{Q}}_k + A_{ijkl}\bar{Q}_k \bar{Q}_l).$$

The condition of critical equilibrium state is now a quadratic eigenvalue problem in Λ:

$$\left[A_{ij} + \Lambda(A_{ijk}\ \dot{Q}_k) + \frac{1}{2}\Lambda^2(A_{ijk}\ddot{Q}k + +A_{ijkl}\ \dot{Q}_k\ \dot{Q}_l)\right]x_j = 0.$$

Techniques for solving nonlinear eigenvalue problems are available in the literature. We discussed some of them in Chapter 5, and the reader is referred to section 5.4.1 for more details.

13.6 THE *W*-FORMULATION STARTING FROM A CONTINUOUS APPROACH

The fundamental path may also be introduced into the energy in the continuous formulation, before the problem is written in terms of generalized coordinates. There are advantages in such formulations because an approximation can be introduced in the fundamental path. To consider such an approach, let us study an energy functional for some thin-walled structure, such as

$$V = \frac{1}{2}\int (N_{ij}\varepsilon_{ij} + M_{ij}\chi_{ij})\,dx_1\,dx_2 - P_i u_i \tag{13.32}$$

with $i, j = 1, 2$.

We write the displacements, strains, and stress resultants as

$$\begin{aligned} u_i &= u_i^F \Lambda + u_i, \\ \varepsilon_{ij}(\Lambda) &= \Lambda\varepsilon_{ij}^F + \varepsilon'_{ij} + \varepsilon''_{ij}, \\ N_{ij}(\Lambda) &= \Lambda N_{ij}^F + N'_{ij} + N''_{ij}, \\ \chi_{ij}(\Lambda) &= \Lambda\chi_{ij}^F + \chi_{ij}, \\ M_{ij}(\Lambda) &= \Lambda M_{ij}^F + M_{ij}. \end{aligned}$$

The fundamental path is often assumed to be a membrane state; then

$$\chi_{ij}^F = M_{ij}^F = 0.$$

However, the postbuckling path includes bending, and the bending state is given by incremental curvatures and moments only

$$\begin{aligned} \chi_{ij}(\Lambda) &= \chi_{ij}, \\ M_{ij}(\Lambda) &= M_{ij}, \end{aligned}$$

$$V = V(\Lambda u_i^F + u_i, \Lambda) = W(u_i, \Lambda). \tag{13.33}$$

Substituting in V leads to the functional $W = W(u_i)$:

$$W - W^F = \frac{1}{2}\Lambda \int (N'_{ij}\varepsilon^F_{ij} + N^F_{ij}\varepsilon'_{ij} + M_{ij}\chi^F_{ij} + M^F_{ij}\chi_{ij})\, dx_1\, dx_2 - P_i u_i$$

$$+ \frac{1}{2}\int (\Lambda N''_{ij}\varepsilon^F_{ij} + N'_{ij}\varepsilon'_{ij} + \Lambda N^F_{ij}\varepsilon''_{ij} + M'_{ij}\chi'_{ij})\, dx_1\, dx_2$$

$$+ \frac{1}{2}\int (N''_{ij}\varepsilon'_{ij} + N'_{ij}\varepsilon''_{ij})\, dx_1\, dx_2$$

$$+ \frac{1}{2}\int (N''_{ij}\varepsilon''_{ij})\, dx_1\, dx_2.$$

A common approach is to describe the fundamental and the incremental states with different degrees of approximation (or even different thin-walled theories). For example, the fundamental state may be calculated using a membrane solution that satisfies simplified boundary conditions; and the incremental state may be obtained from a bending solution that satisfies the exact boundary conditions of the problem. The classical critical loads of cylindrical and spherical shells have been solved in such a way.

Notice terms such as

$$N'_{ij}\varepsilon^F_{ij} = N^F_{ij}\varepsilon'_{ij},$$

$$N''_{ij}\varepsilon^F_{ij} = N^F_{ij}\varepsilon''_{ij}.$$

However, in some cases it may be convenient to deal with each separate contribution individually.

The first terms in $\left(W - W^F\right)$ are the linear contributions in the incremental displacements. The second term is quadratic in u_i. Some terms are a function of Λ, while other terms do not depend on Λ:

$$W_2 = \frac{1}{2}\int [(N'_{ij}\varepsilon'_{ij} + M'_{ij}\chi'_{ij}) + \Lambda(N^F_{ij}\varepsilon''_{ij} + N''_{ij}\varepsilon^F_{ij})]\, dx_1\, dx_2.$$

The first contribution leads to the linear stiffness matrix of the problem, while the second one yields the load-geometry term.

The third and fourth terms in $\left(W - W^F\right)$ are independent of Λ and may be written as

$$W_3 = \int (N''_{ij}\varepsilon'_{ij} + N'_{ij}\varepsilon''_{ij})\, dx_1\, dx_2,$$

$$W_4 = \int (N''_{ij}\varepsilon''_{ij})\, dx_1\, dx_2.$$

Finally, if we substitute the fundamental path by an adequate solution and approximate the incremental displacements in terms of generalized coordinates, in the form

$$u_i = \phi_{ij} q_j \quad \text{for} \quad i = 1, 2, 3, \qquad j = 1, \ldots, N,$$

one gets a discrete version of $W[q_j, \Lambda]$.

The choice of one or another approach to obtain a discrete form of W depends on how easy it is to have a simple analytical solution of the fundamental path. When a constant fundamental state is a good approximation, it is convenient to use the alternative procedure presented in this section.

13.7 THE CYLINDRICAL SHELL

Cylindrical shells are used in many applications in engineering in which they are prone to buckling: they are components of silos and tanks in civil engineering [9], of airplanes and space vehicles in aeronautics and astronautics, of submarine vehicles in naval engineering [8], and many others. A comprehensive treatment of the buckling and postbuckling of cylindrical shells is given in the book by Yamaki [16].

The behavior of a cylindrical shell is very complex, and is only second to that of a spherical shell. We shall see that the tools that we have presented up to this point in the book are not sufficient to tackle the problem. However, we carry out the analysis for several reasons: first, to illustrate the application of the W-formulation, and, second, because this was the first approach followed years ago and led to important discussions that made further advances possible.

Thin-walled shells behave in a complex way: Even in the linear range of strain-displacement relation, there is coupling between bending and membrane actions. This coupling arises from the curvature in the original geometry, a feature that is not present in columns and plates. Several shell theories may be found in the literature for linear and nonlinear analysis of shells, and significant contributions may be found in the works of Koiter [5] and Sanders [10], among others. For the examples in this chapter even a simple kinematic theory is capable of modeling the essential features of the unstable response.

The geometry and displacements of a cylindrical shell are shown in Figure 13.3, with x_1 the axial direction along the meridian, $x_2 = R\theta$ is the circumferential coordinate, and x_3 is the normal to the midsurface, positive inward. The positive displacements u_1, u_2, and u_3 are also indicated in Figure 13.3.

The curvature of the undeformed shell is described by a curvature tensor k_{ij}, the components of which are

$$k_{ij} = \begin{bmatrix} 0 & 0 \\ 0 & -1/R \end{bmatrix}. \tag{13.34}$$

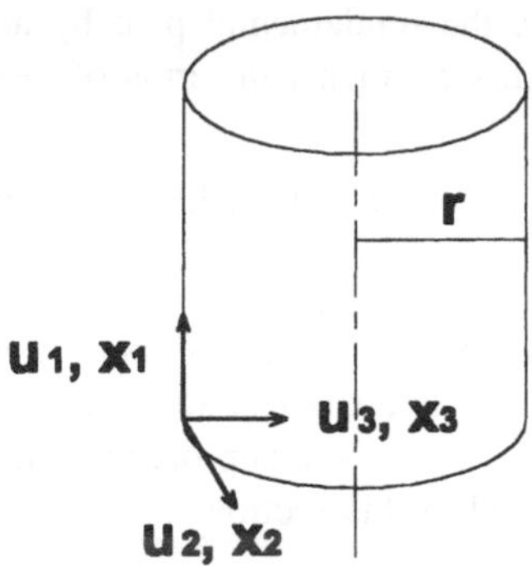

Figure 13.3 Geometry of a cylindrical shell.

The displacements u^* of a point outside the midsurface may be related to displacements of the midsurface u by

$$u_i^*(x_1, x_2, x_3) = u_i(x_1, x_2) + \beta_i x_3, \tag{13.35}$$

$$u_3^*(x_1, x_2, x_3) = u_3(x_1, x_2).$$

The kinematic relations following Donnell [2] are

$$\beta_1 = -\frac{\partial u_3}{\partial x_1}, \qquad \beta_2 = \Gamma\frac{u_2}{R} - \frac{\partial u_3}{\partial x_2}, \tag{13.36}$$

$$\varepsilon_{11} = \frac{\partial u_1}{\partial x_1} + \frac{1}{2}(\beta_1)^2,$$

$$\varepsilon_{22} = \left(\frac{\partial u_2}{\partial x_2} + \frac{u_3}{R}\right) + \frac{1}{2}(\beta_2)^2,$$

$$\varepsilon_{12} = \frac{1}{2}\left(\frac{\partial u_1}{\partial x_2} + \frac{\partial u_2}{\partial x_1}\right) + \frac{1}{2}(\beta_1\beta_2),$$

$$\chi_{11} = -\frac{\partial \beta_1}{\partial x_1} \qquad \chi_{22} = -\frac{\partial \beta_2}{\partial x_2}, \tag{13.37}$$

$$\chi_{12} = \frac{1}{2}\left(\frac{\partial \beta_1}{\partial x_2} + \frac{\partial \beta_2}{\partial x_1}\right).$$

The value of Γ is either 1 or 0, depending on the shell theory employed. Using Donnell's approximation we have $\Gamma = 0$.

The constitutive equations for elastic behavior are assumed in the linear form

$$N_{11} = C(\varepsilon_{11} + \nu\varepsilon_{22}), \qquad N_{22} = C(\nu\varepsilon_{11} + \varepsilon_{22}), \tag{13.38}$$
$$N_{12} = C(1-\nu)\varepsilon_{12},$$

$$M_{11} = D(\chi_{11} + \nu\chi_{22}), \qquad M_{22} = D(\nu\chi_{11} + \chi_{22}), \tag{13.39}$$
$$M_{12} = D(1-\nu)\chi_{12},$$

$$C = \frac{Et}{1-\nu^2}, \qquad D = \frac{Et^3}{12\left(1-\nu^2\right)}. \tag{13.40}$$

The internal energy of the shell may be computed as the sum of the membrane and bending contributions

$$U = U_m + U_b, \tag{13.41}$$

where

$$U_m = \frac{1}{2}\int_{x1=0}^{l}\int_{\theta=0}^{2\pi} \varepsilon_{ij} N_{ij} dx_1 R d\theta, \tag{13.42}$$

$$U_m = \frac{1}{2}\int_{x1=0}^{l}\int_{\theta=0}^{2\pi} \chi_{ij} M_{ij} dx_1 R d\theta. \tag{13.43}$$

The fundamental path. We shall consider several load cases, each having a different fundamental path. A simply supported shell under axial load (Figure 13.4.a) shows the displacement field in the fundamental path with bending at the ends to satisfy the simply supported boundary conditions, and the central part of the shell only expands. Consideration of the bending effects in the fundamental path produces a series of complications if we seek to obtain analytical solutions. To simplify the problem it is assumed that the fundamental path is governed by a membrane state of stresses, as shown in Figure 13.4.b. This hypothesis is very important and seems to be a rather crude one; however, for a range of geometries the results are not in great error with respect to a more refined fundamental path as the basis of the stability analysis.

Similarly, a membrane fundamental path will be used for lateral pressure, instead of the correct solution involving bending at the edges. This is shown in Figure 13.5.

Under axial load the simplified membrane state induces stress resultants given by

$$N_{11}^F = -\sigma_{11} t, \qquad N_{22}^F = N_{12}^F = 0. \tag{13.44}$$

For the laterally loaded cylinder under pressure p, one has

$$N_{11}^F = 0, \qquad N_{22}^F = p/(2\pi R), \qquad N_{12}^F = 0. \tag{13.45}$$

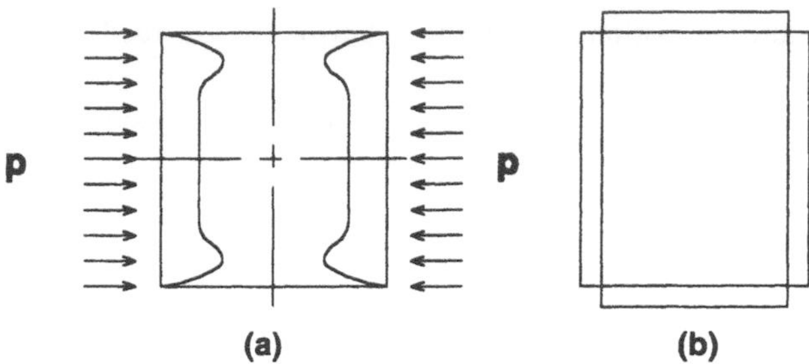

Figure 13.4 Boundary conditions for axially loaded cylinder. (a) Real deflections; (b) assumed deflections.

These are constant states of stresses that are thus independent of x_i and can be taken out of the integrations. The curvatures are zero in both cases, leading to zero bending energy U_b.

The *W*-formulation. Analytical solutions for shell problems are possible if one employs different levels of approximation for the fundamental and for the secondary equilibrium paths. This means that the fundamental path is approximated using a membrane solution, as indicated in the last section, while the secondary path is modeled using a bending solution. Such situation is possible if one employs the W-formulation, starting from a continuous approach, before discretization of the system is done (see section 13.7). This will be followed in this section in order to reproduce classical solutions for the buckling of cylinders.

Strain energy. We have seen that it is possible to separate the analysis of the fundamental path from the critical state by using the W-formulation of the energy and then proceed with the discretization.

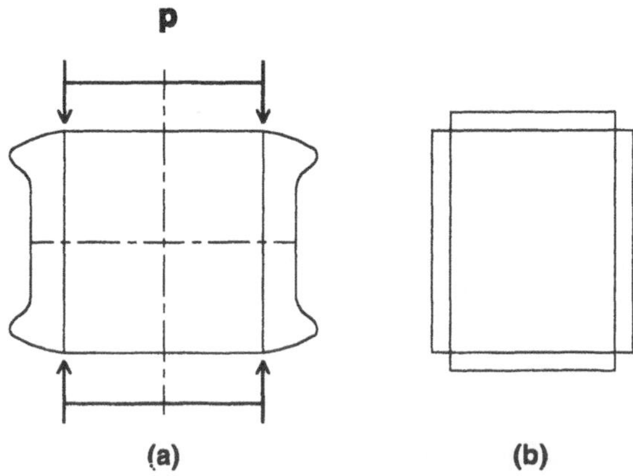

Figure 13.5 Boundary conditions for cylinder under pressure. (a) Real deflections; (b) assumed deflections.

For the postcritical path we assume the following Ritz solution:

$$u_1 = q_1^{mn} \cos(\mu x_1) \cos(\eta x_2), \tag{13.46}$$

$$u_2 = q_2^{mn} \sin(\mu x_1) \sin(\eta x_2),$$

$$u_3 = q_3^{mn} \sin(\mu x_1) \cos(\eta x_2),$$

where

$$\mu = m\frac{\pi}{l}, \qquad \eta = n\frac{1}{R}. \tag{13.47}$$

The integer m is the number of half waves in the longitudinal direction x_1, while n is the number of full waves in the circumferential direction. Such a displacement field satisfies simply supported boundaries at $x_1 = 0$ and $x_1 = l$. The generalized coordinates q_j^{mn} for $j = 1, 2, 3$, are the amplitudes of the functions chosen. Two values of (m, n) define a specific deformation of the cylinder. The displacements u_j are incremental with respect to the fundamental path, and they are obtained as the summation in m and n of the modes chosen to carry out the discretization. Of course, the more terms we include, the better the resulting solution is, but what terms one should consider in the analysis is a matter of some further thought.

For the axially loaded shell the coefficients of the W-functional are obtained by substitution of the approximate functions in the continuous version of W. This is done quite effectively using a symbolic manipulator (we used Maple V) and leads to

$$\hat{W}_{11} = C\frac{\pi}{2}lR\left(\mu^2 + \frac{1-\nu}{2}\eta^2\right), \tag{13.48}$$

$$\hat{W}_{22} = C\frac{\pi}{2}lR\left(\eta^2 + \frac{1-\nu}{2}\mu^2\right),$$

$$\hat{W}_{12} = C\frac{\pi}{2}lR\mu\eta,$$

$$\hat{W}_{13} = -C\frac{\nu}{r}\frac{\pi}{2}lR\mu,$$

$$\hat{W}_{23} = -C\frac{1}{r}\frac{\pi}{2}lR\eta,$$

$$\hat{W}_{33} = \left\{\left[D(\mu^2+\eta^2)^2 + \frac{C}{R^2}\right] + \Lambda N_{11}^F \mu^2\right\}\frac{\pi}{2}lR.$$

Higher-order terms include a quartic contribution

$$\hat{W}_{3333} = \frac{3}{16}C\{9(\mu^4+\eta^4) + 2\mu^2\eta^2\}\frac{\pi}{2}lR. \tag{13.49}$$

Notice that the $\hat{W}_{ij}$ coefficients are not in terms of a summation in m and n, but the strain energy for each mode results independent of the displacements in the other modes. This result is due to the choice of trigonometric interpolation functions and because the geometric twisting of the surface (k_{12}) is zero.

For a cylinder under lateral pressure, only one of the coefficients should be modified

$$\hat{W}_{33} = \left\{ \left[D(\mu^2 + \eta^2)^2 + \frac{1}{r^2} C \right] + \Lambda N_{22}^F \eta^2 \right\} \frac{\pi}{2} l R. \tag{13.50}$$

13.7.1 Critical State under Lateral Pressure

The eigenproblem

$$W_{ij}[0, \Lambda] x_j \ |^c = 0 \tag{13.51}$$

results in this case in the form

$$\hat{W}_{ij}[0, \Lambda] x_j \ |^c = 0. \tag{13.52}$$

For a specified pattern of deformation (m, n) the critical load results in

$$\det[\hat{W}_{ij}] = 0 \tag{13.53}$$

leading to

$$p^c = 2\pi R t \frac{D(\mu^2 + \eta^2)^2 + (1 - \nu^2) \frac{C}{R^2} \frac{\mu^4}{(\mu^2 + \eta^2)^2}}{\eta^2}. \tag{13.54}$$

If we input different values of m and n in this expression, it is possible to obtain the values of critical lateral pressure, or the critical stress in the circumferential direction.

Example 13.4. *As an example, let us consider an aluminum cylindrical shell with* $l/R = 1$ *and* $R/t = 400$. *The results of critical load as a function of the number of circumferential waves n is presented in Figure* 13.6. *Each curve is in fact a representation of discrete results for different values of the axial wave number m.*

Each curve for a different value of m has a distinct minimum (although it may not be an integer). Furthermore, the lowest critical load occurs for $m = 1$, *and for increasing values of m the minimum in the critical load increases. The lowest value of critical stress is given by* $m = 1$ *and* $n = 11$, *and other modes are associated with higher values of critical stresses.*

Some simplifications are often used for cylinders with length-to-radius ratio $0.2 \leq l/R \leq 5$, *for which* $\eta^2 \gg \mu^2$ *and* $(\mu^2 + \eta^2)^2$ *can be approximated by* η^4. *In this case the critical pressure is simplified to*

$$p^c = 2\pi R t \frac{D\eta^4 + C \frac{(1-\nu^2)}{R^2} \left(\frac{\mu}{\eta}\right)^2}{\eta^2}. \tag{13.55}$$

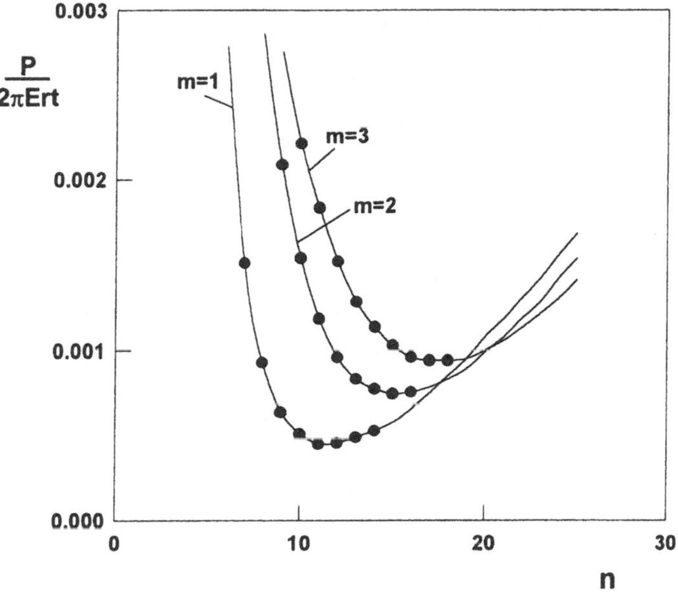

Figure 13.6 Bifurcation pressures for a cylinder under uniform lateral pressure. $l/R = 1$, $R/t = 400$.

The minimum of this expression occurs for $m = 1$, and the number of circumferential waves is given by

$$n_{\min} |^c = R\left[48(1-\nu^2)\left(\frac{\pi^2}{l^2 t}\right)^2\right]^{1/8}. \tag{13.56}$$

Introducing the circumferential wave number that leads to the minimum critical pressure into (13.55) *yields*

$$p^c = 2\pi Rt\frac{0.855E}{(1-\nu^2)^{0.75}(R/t)^{1.5}(l/R)}. \tag{13.57}$$

This approximate value of the critical pressure is known as Yamaki's equation.

13.7.2 Critical State under Axial Load

Under axial load N_{11}, the critical load is given by

$$\Lambda N_{11}^c = \frac{D(\mu^2+\eta^2)^2 + (1-\nu^2)\frac{C}{R^2}\frac{\mu^4}{(\mu^2+\eta^2)^2}}{R^2\mu^2}. \tag{13.58}$$

The eigenmode can be shown to be

$$x_1 = \frac{\mu}{R}\frac{\eta^2 - \nu\mu^2}{\left(\eta^2 - \mu^2\right)^2}, \tag{13.59}$$

$$x_2 = \frac{\eta}{r}\frac{\eta^2 + (2+\nu)\,\mu^2}{\left(\eta^2 - \mu^2\right)^2},$$

$$x_3 = 1.$$

A plot of the critical load versus m and n is shown in Figure 13.7 for the same data as in Figure 13.6. Unlike our previous study of lateral pressure, the minimum critical load for $m = 1$ and for $m = 2$ is the same value, and also for $m = 3$, $m = 4$, etc.

We say that there is a coincidence of critical modes for the same critical load, and this is a characteristic feature of axially loaded cylindrical shells. The minimum critical load occurs for modes that satisfy the condition

$$\mu^2 + \eta^2 = \mu\left[12\frac{(1-\nu^2)}{(rt)^2}\right]^{1/4}. \tag{13.60}$$

Plotting this equation, one obtains a circle in (η, μ) shown in Figure 13.8. This is known as **Koiter's circle**.

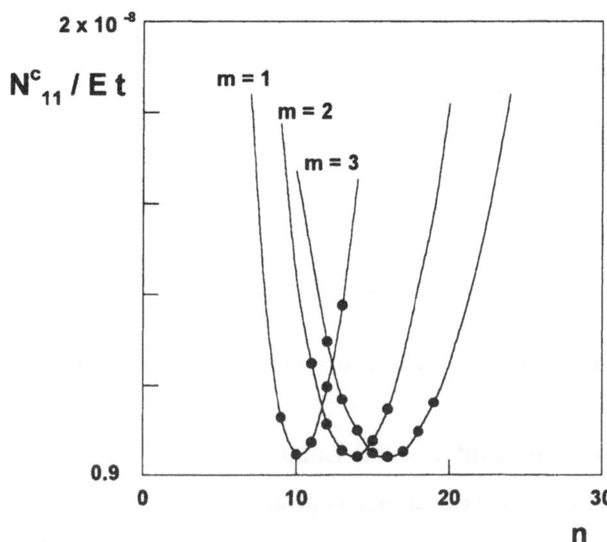

Figure 13.7 Bifurcation loads for a cylinder under uniform axial load. $l/R = 1$, $R/t = 400$.

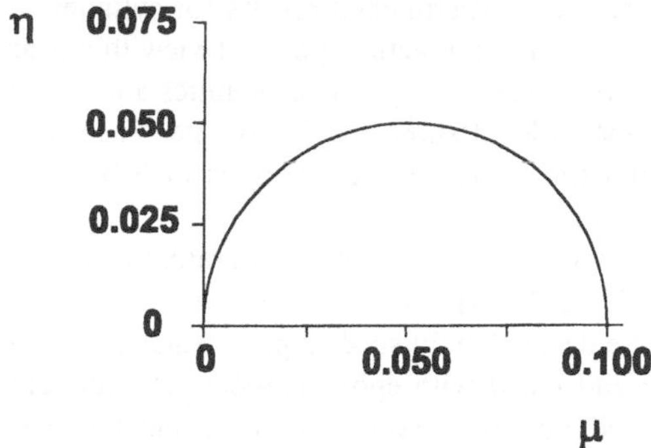

Figure 13.8 Koiter's circle for a cylinder under uniform axial load. $l/r = 1$, $r/t = 400$.

An important conclusion is that this is not a problem that can be classified as having "distinct bifurcation points." The theory that we have developed in Chapter 7 does not apply for this case, because there is a number of coincident eigenvalues associated with different eigenvectors.

Let us forget this point of coincidence for a moment and pursue the analysis for isolated modes as was studied in this chapter.

The lowest value of the critical load can be calculated by substitution of (13.60) into (13.58) and yields

$$N_{11}^{c} = \frac{Et^2}{Rc}, \qquad \text{where} \qquad c = \left[3\left(1 - \nu^2\right)\right]^{1/2}. \tag{13.61}$$

This is the "classical" critical load for a cylinder under uniform axial compression. However, we must mention at this stage that cylindrical shells buckle at loads lower than what is predicted by (13.61). The discrepancies have been an important point of discussion in the theory for a number of years, in which experimental researchers improved their testing facilities and theoretical researchers refined their theories. Nowadays it is accepted that modes couple and interact and that imperfections play a significant role in the determination of the buckling load in real cylinders.

13.8 EXPERIMENTS WITH CYLINDERS

Buckling loads and modes can be predicted using analytical tools, as was done in previous sections. However, there are assumptions that one introduces into the formulation, so that it is always necessary to check analytical or numerical results produced for the first time. We have chosen here to report experimental results for cylindrical shells because they differ from the analytical values obtained using the "classical" approach. The differences between classical and experimental results led to fruitful discussions that helped further develop the theory of elastic stability.

We review here selected experimental results for cylinders under combined axial load and lateral pressure. From a practical point of view this is an important load case in many engineering structures, notably in aeronautics and astronautics, in submarine vehicles, in semisubmersible offshore platforms, etc. From the theoretical point of view, this problem illustrates two different classes of behavior, depending on which load dominates the problem.

The number of experimental works in the literature is not very large and includes the works of [15, 6, 7, 12, 14, 4] among others.

Weigarten and Seide [15] published experimental results for 10 models made with Mylar polymer and glued with epoxy cement. The modulus of the material are $E = 4.8kN/mm^2$ and Poisson's ratio $\nu = 0.3$, and the geometries tested range between $1 \leq l/R \leq 3$ and $400 \leq R/t \leq 800$. These are very thin shells and are representative of aeronautic applications. The critical stresses are plotted in Figure 13.9 in the form of an interaction diagram for one particular case.

The hoop stress is representative of buckling under lateral pressure, while the axial stress resultant is associated with buckling under axial load. The values chosen for normalization for the pressure are experimental values.

All models buckled elastically. There is a large reduction of experimental values with respect to classical values in the zone dominated by axial load. This means that under axial load, the shell buckles at a load that is approximately 50% of the classical

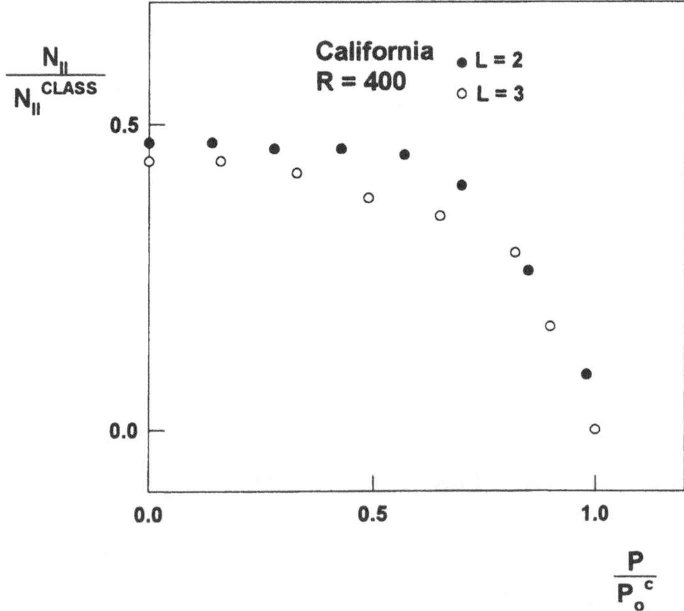

Figure 13.9 Experimental results of Weingarten and Seide for shells with $R/t = 400$, $l/R = 2$, and $l/R = 3$.

theoretical predictions. This is a difficult problem for the theoretical analyst, since the calculations led to higher-than-real buckling results and thus were on the unsafe side.

A second set of results was reported by Mungan [7] using ten Plexiglas models fabricated from thicker tubes. Material properties are $E = 2.9$ to $3.1 kN/mm^2$ and $\nu = 0.3$ to 0.36. The geometries studied had $73.5 \leq R/t \leq 200$ and $1.85 \leq l/R \leq 2.7$, and the models were clamped at both ends.

Typical elastic buckling loads are shown in Figure 13.10. For axially dominated loads, buckling occurred at 41% to 35% of the classical values, while values higher than the classical predictions were reported in the experiments for pressure-dominated cases.

Many other experiments could be presented here, but the main trend is clear. First, there is a significant reduction in buckling load under axial load. This reduction is not constant between experiments and shells, but as the ratio R/t and l/R increase, the experimental results become lower with respect to theoretical results. Second, the pressure-loaded cylinder does not have as strong a reduction as the axially loaded case. The boundary conditions in the experiments are close to clamped edges, thus increasing the pressure necessary for buckling.

The lack of agreement between theoretical and experimental results for cylindrical shells depends on the load case. Furthermore, this feature of lack of agreement occurs in some thin-walled structures, but not in all. Had we chosen to illustrate experiments

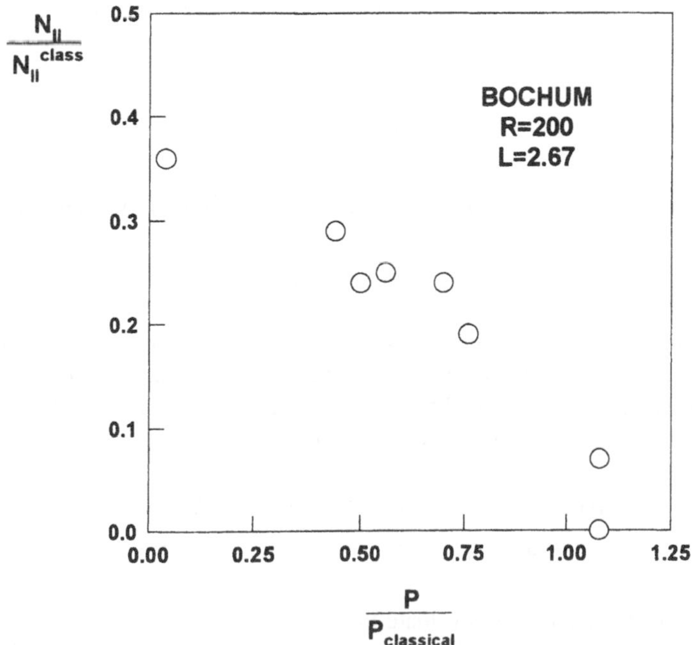

Figure 13.10 Experimental results of Mungan for shells with $R/t = 200, l/R = 2.67$.

with columns, plates, rings, and many other structural forms, we would have found good agreement.

The theory-versus-experiments problem dominated buckling discussions for many years. It is now clear that there are two factors that explain the problem: First, the influence of small imperfections produces a reduction in buckling load and destroys the nature of bifurcations. Second, when two or more modes (eigenvectors) have the same (or close) eigenvalues, then there is a possibility of interaction between modes that leads to a complex coupled behavior.

13.9 PROBLEMS

Review questions. (*a*) What are the differences between the V-formulation and the W-formulation? (*b*) Discuss advantages and disadvantages of the W-formulation. (*c*) Compare the load potentials in V and W energy forms. (*d*) What is the consequence of a trivial fundamental path on the W-formulation? (*e*) Is it possible to employ assumptions about the behavior of a system that are different for the primary and secondary paths? Discuss how.

Problem 13.1 (Theory). Derive the equations of symmetric bifurcation in the W-formulation using the V-formulation of Chapter 7.

Problem 13.2 (Theory). Derive the equations of symmetric bifurcation in the W-formulation starting from the equilibrium and stability conditions

$$W_i[0, \Lambda] = 0,$$

$$W_{ij}[0, \Lambda]x_j = 0.$$

Problem 13.3 (Theory). Discuss under what conditions the following term vanishes in the W-formulation:

$$\frac{1}{3!}A_{ijkl}\Lambda\bar{Q}_l q_i q_j q_k.$$

Problem 13.4 (Column with I cross section). An I-shaped composite column has been discretized using two degrees of freedom: Q_1 for the axial displacement and Q_2 for the transverse displacement. For a volume fraction of 0.15, and fiber orientation [45, −45], one gets the following energy coefficients:

$$A_1 = -P, \qquad A_{11} = 30,380, \qquad A_{22} = 296,$$

$$A_{112} = 1,497, \qquad A_{2222} = 1,173.$$

Use the W-formulation to evaluate the postbuckling path.

Problem 13.5 (Shallow shell). Develop the equations for critical and postcritical states for a shallow shell, as shown in Figure 13.11. Assume simply supported edges,

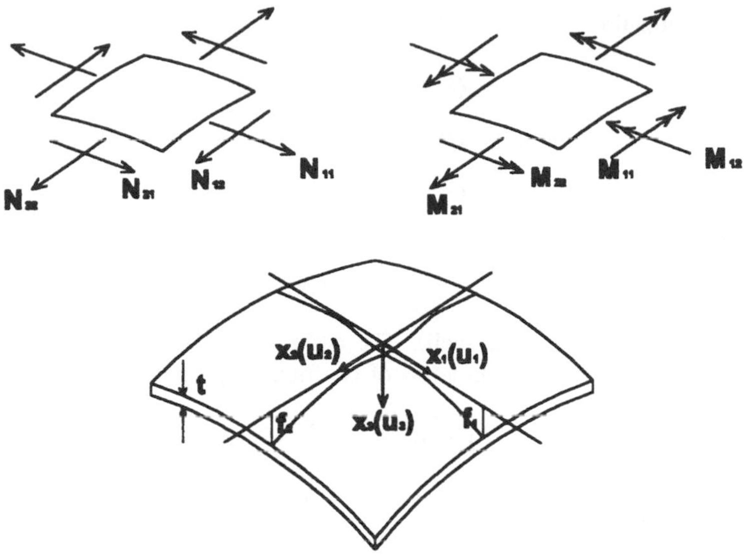

Figure 13.11 Shallow shell.

a curvature parameter of the midsurface $k_{12} = 0$, and a constant membrane state of stresses in the fundamental path.

Use the following kinematics:

$$u_i^* = u_i + \beta_i x_3, \qquad u_3^* = u_3,$$

$$\beta_i = -\frac{\partial u_3}{\partial x_i} - \Gamma k_{ii} u_i,$$

$$\varepsilon_{ij} = \frac{1}{2}\left(\frac{\partial u_i}{\partial x_j} + \frac{\partial u_j}{\partial x_i}\right) - k_{ij}u_3 + \frac{1}{2}\beta_i\beta_j,$$

$$\chi_{ij} = \frac{1}{2}\left(\frac{\partial \beta_i}{\partial x_j} + \frac{\partial \beta_j}{\partial x_i}\right).$$

Constitutive relations are

$$N_{ij} = \frac{Et}{1+\nu}\left(\varepsilon_{ij} + \frac{\nu}{1-\nu}\delta_{ij}\varepsilon_{mm}\right),$$

$$M_{ij} = \frac{Et^3}{12\,(1+\nu)}\left(\chi_{ij} + \frac{\nu}{1-\nu}\delta_{ij}\chi_{mm}\right).$$

The discretization could be in the form

$$u_1 = Q_1^{mn} \cos(\mu x_1) \sin(\eta x_2),$$
$$u_2 = Q_2^{mn} \sin(\mu x_1) \cos(\eta x_2),$$
$$u_3 = Q_3^{mn} \sin(\mu x_1) \sin(\eta x_2),$$

with

$$\mu = \frac{m\pi}{2a}, \qquad \eta = \frac{n\pi}{2b}.$$

Problem 13.6 (Cylindrical shell). The equations for biaxial compression of a cylindrical shell can be written in terms of coefficients γ_1 and γ_2, using

$$N_{11}^F = \gamma_1 \sigma_{11} t, \qquad N_{22}^F = \gamma_2 \sigma_{22} t.$$

For axially dominated load, $\gamma_1 = 1$ and $\gamma_2 = N_{22}^F / N_{11}^F$, and for pressure dominated load, $\gamma_2 = 1$ and $\gamma_1 = N_{11}^F / N_{22}^F$. Calculate the critical loads and draw an interaction diagram.

Problem 13.7 (Stiffened cylindrical shells). It is possible to study stiffened cylindrical shells using a "smeared-out" approach, in which the stiffeners are incorporated into the constitutive equations of the shell. Using the shell formulation for cylinders of this chapter, the new constitutive relations are

$$N_{11} = \left(C + E\frac{A_s}{d_s}\right)\varepsilon_{11} + C\varepsilon_{22} - \left(e_s E\frac{A_s}{d_s}\right)\chi_{11},$$

$$N_{22} = \left(C + E\frac{A_r}{d_r}\right)\varepsilon_{22} + C\varepsilon_{11} - e_r E\frac{A_r}{d_r}\chi_{22},$$

$$N_{12} = C(1-\nu)\varepsilon_{12},$$

$$M_{11} = \left(e_s E\frac{A_s}{d_s}\right)\varepsilon_{11} + \left(D + E\frac{I_s}{d_s}\right)\chi_{11} + \nu D\chi_{22},$$

$$M_{22} = \left(-e_r E\frac{A_r}{d_r}\right)\varepsilon_{22} + \left(D + E\frac{I_r}{d_r}\right)\chi_{22} + \nu D\chi_{11},$$

$$M_{12} = \left[D(1-\nu) + \frac{1}{2}G\left(\frac{J_s}{d_s} + \frac{J_r}{d_r}\right)\right]\chi_{12},$$

where subindex s refers to stringer stiffeners, and r refers to ring stiffeners; and the new variables are indicated in Figure 13.12, with

$$e_s = \frac{1}{2}(d_s + h), \qquad A_s = d_s h_s,$$

$$I_s = \frac{1}{12}h_s d_s^3 + e_s^2 A_s, \qquad J_s = \frac{1}{3}d_s h_s^3.$$

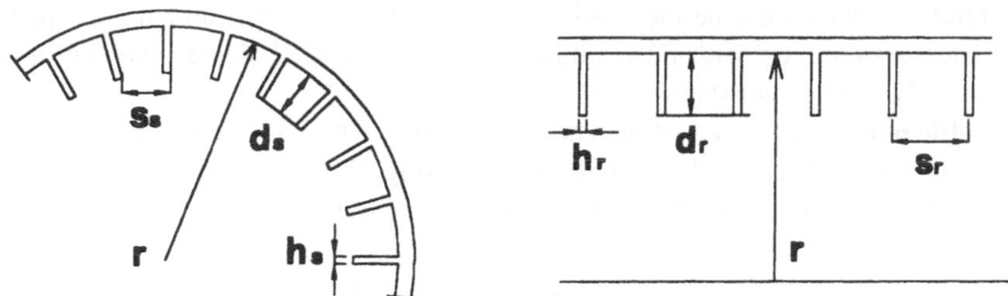

Figure 13.12 Stiffened cylindrical shell.

Use the W-formulation to obtain the critical state of the shell. Remember that this is not a design value, since the shell is sensitive to the influence of small imperfection and the critical load drops because of that.

Problem 13.8 (Ring stiffened cylinder). A steel shell used for the legs of a semisubmersible offshore platform has dimensions $L = 10m$, $r = 5m$, $h = 25mm$, $E = 205KN/mm^2$. Ring stiffeners on the inside of the shell are used with $e_r = 1.5m$, $h_r = h$, and $d_r = 8h$. (a) Evaluate the critical external pressure for different mode shapes. (b) Evaluate the buckling load of an isotropic shell with the same overall dimensions and with the same amount of steel. Compare with the solution of the ring stiffened shell.

Problem 13.9 (Stringer stiffened cylinder). The shell of Problem 13.8 is investigated but with stringer rather than ring stiffeners. Consider 20 stiffeners, $h_s = h$ and $d_s = 8h$. (a) Evaluate the critical axial load for different mode shapes. (b) Evaluate the buckling load of an isotropic shell with the same overall dimensions and with the same amount of steel. Compare with the solution of the stringer stiffened shell.

Problem 13.10 (Two-bar frame). Use the W-formulation to solve the postbuckling path of the two-bar problem shown in Figure 13.13. The bars can have bending

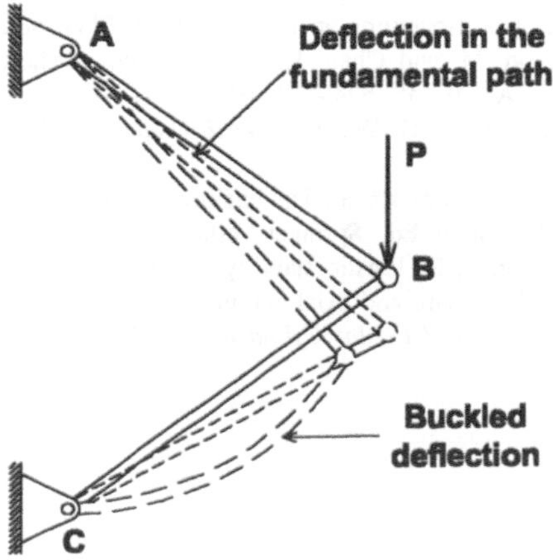

Figure 13.13 Structure of Problem 13.10.

and stretching and should be modeled as in Appendix B. In the fundamental path the bars undergo stretching with no bending, while the buckled shape involves bending of the bar BC. EI is constant.

Problem 13.11 (Torsion of a cylinder). A long circular cylinder has a torsional load at the edges. Because the cylinder is very long, the boundary conditions do not influence the buckling load. Assume a buckling mode in the form

$$w = Q^{mn} \sin\left(\frac{m\pi rx}{L} - n\theta\right),$$

where m is the number of axial half waves, and n is the number of circumferential full waves. Identify the lowest buckling load.

13.10 BIBLIOGRAPHY

[1] Croll, J. G. A., and Walker, A. C., *Elements of Structural Stability*, Macmillan, London, 1972.
[2] Donnell, L. H., *Beams, Plates, and Shells*, McGraw-Hill, New York, 1976.
[3] El Naschie, M., *Stress, Stability and Chaos: An Energy Approach*, McGraw-Hill, New York, 1990.
[4] Galletly, G. D., and Pemsing, K., On design procedures for the buckling of cylinders under combined axial compression and external pressure, *ASME 4th US Congress on Pressure Vessels and Piping Technology*, Portland, 1983. ASME paper 83 PVP-5.
[5] Koiter, W. T., On the non-linear theory of thin elastic shells, *Proc. Konik. Nederl. Akad. Wetensch.*, B69, 1–54, 1966.
[6] Lee, L. H. N., Inelastic buckling of cylindrical shells under axial compression and external pressure, in *Proc. 4th US National Congress on Applied Mechanics*, ASME, 1962, 989–998.
[7] Mungan, I., Buckling stress states of cylindrical shells, *J. Structural Division*, ASCE, 100(11), 2289–2306, 1974.
[8] Nash, W. A., *Hydrostatically Loaded Structures: The Structural Mechanics, Analysis and Design of Powered Submersibles*, Pergamon Press, Oxford, 1995.
[9] Rajagopalan, K., *Finite Element Buckling Analysis of Stiffened Cylindrical Shells*, Balkema, Rotterdam, 1993.
[10] Sanders, J. L., Non-linear theories for thin-elastic shells, *Quart. Appl. Math.*, 21, 21–36, 1963.
[11] Sewell, M., A method of post-buckling analysis, *J. Mech. Phys. Solids*, 17, 219, 1969.
[12] Tennyson, R. C., Booton, M. and Chan, K. H., Buckling of short cylinders under combined loading, *J. Appl. Mech.*, ASME, 45, 574–578, 1978.
[13] Thompson, J. M. T., and Hunt, G. W., *A General Theory of Elastic Stability*, Wiley, London, 1973.
[14] Walker, A. C., Segal, Y., and McCall, S., The buckling of thin-walled ring-stiffened steel shells, in *Buckling of Shells*, E. Ramm, Ed., Springer-Verlag, Berlin, 1982.
[15] Weingarten, V. I., and Seide, P., Elastic stability of thin-walled cylindrical and conical shells under combined external pressure and axial compression, *AIAA J.*, 3(5), 913–920, 1965.
[16] Yamaki, N., *Elastic Stability of Circular Cylindrical Shells*, North-Holland, Amsterdam, 1984.

CHAPTER
FOURTEEN

MODE INTERACTION

14.1 INTRODUCTION

In previous chapters we performed the analysis at a critical state under the assumption that there was only one distinct mode shape at the level of the lowest critical load. In other words, at the critical state the system had only one direction of instability. In mathematical terms, the eigenvectors were associated with different eigenvalues, and at the level of the first eigenvalue no two of them were coincident or nearly coincident. For example, the first critical state of an isotropic column is given by a critical load and a half sine mode shape. The second critical load requires a load four times higher than the value of the first one, with a mode shape given by a two half sine mode shape. No interaction is expected between modes as distant as those two, and each mode should be considered as a distinct bifurcation, isolated from the rest.

But it is not possible to accept such a hypothesis in all cases. There are many practical problems in structural engineering for which several mode shapes, identified from eigenvector analysis, are associated with the same or nearly the same eigenvalue.

Why do we care about coincidence of modes at a given load level? There are several reasons for concern:

- The analysis should be modified, because there are two or more directions of instability at the first critical state. The new formulation is discussed in this chapter.
- The interaction of coincident modes generates new postcritical equilibrium paths in the system, different from those that occur when the modes are treated in isolation. This is illustrated by means of examples at the end of this chapter.

- In many problems two modes that produce stable bifurcations when they are considered in isolation induce a new postcritical path when combined, which is unstable. In other words, coupling and interaction between coincident modes may generate unstable behavior. Again, examples at the end of the chapter illustrate this.
- The sensitivity of a problem with interacting modes changes with respect to a similar problem without interaction. This is shown in Chapter 15, where interactions between modes are shown to lead to severe imperfection sensitivity of the lowest critical load.

Research into mode interaction in buckling analysis started about 25 years ago. Pioneers in the field of mode interaction in systems defined by a finite number of unknowns have been Chilver [10, 11], Supple [35, 36, 37], van der Neut [41], Koiter [18], Thompson and Hunt [39], Tvergaard [40], and Reis and Roorda [29–32].

Systems with coincidence or near coincidence of critical loads for different modes are illustrated in section 14.2, where differences are made between unavoidable coincidences and coincidences by design. There are new modes that emerge in the postcritical path due to interaction, and this is formulated in section 14.3. The total potential energy W is written in section 14.4, taking into account uncoupled and coupled modes. The new mode shapes that emerge are calculated in section 14.5, while the mode amplitudes are investigated in section 14.6. The analysis is restricted in section 14.7 to the important case of only two interacting modes, with special emphasis on modes that are doubly symmetric. Examples of mode interaction are reported in sections 14.8 to 14.10.

In section 14.8 we explore a simple column problem with two degrees of freedom, in which coincidence between two modes occurs because of the design chosen. This is called the Augusti column, and the energy becomes doubly symmetric, leading to a first-order interaction problem. The two uncoupled paths are stable, but the coupled path that emerges from the critical point when the two modes are coincident is unstable. When the modes are not exactly coincident, but are sufficiently close to each other, we have a tertiary path from a secondary bifurcation. Our second example (section 14.9) is a thin-walled column with the possibility of changing one of the critical loads by modifications in the design. Again, the two main modes are stable. Mode interaction is investigated using three modes in the analysis, with first-order interaction. The coupled path for coincident modes is again shown to be unstable. Finally, the cylindrical shell under axial load is presented in section 14.10.

14.2 EXAMPLES OF SYSTEMS WITH MODE INTERACTION

There are two possibilities to have coincident modes in a structure: either this coincidence is unavoidable and is built into the nature of the structure (as in some shells), or else it is induced by design (as in stiffened plates, thin-walled columns, etc.).

To illustrate mode interaction let us consider first the bifurcation behavior of a complete spherical shell under external pressure. Bifurcation states for this case are computed from a linear fundamental path, and the results show that there are many

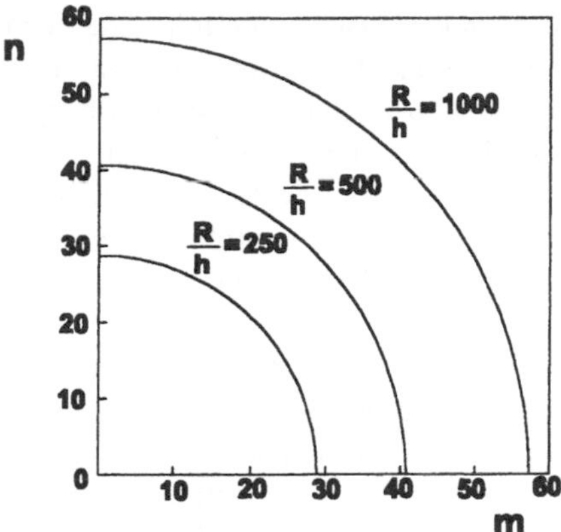

Figure 14.1 Koiter's circle for a spherical shell under external pressure for Poisson ratio 0.3.

modes for which the critical load is the same. This coincident value is the lowest critical pressure of the sphere, so that it occurs for the most relevant critical state.

The modes associated with the lowest critical pressure are plotted in Figure 14.1. It is very interesting to notice that the coincident modes in the sphere describe a circle in the plane of the mode shapes in two principal directions. This circle is named after its discoverer, Koiter [18].

Clearly, this coincidence of critical loads for different modes was not a matter of "luck," because it would also occur if we take different dimensions (thickness, radius, etc.) or different material properties (modulus of elasticity, Poisson's ratio, etc.). What changes is the actual number of modes that coincide, not the fact that there is coincidence of modes.

Further examples of many coincident modes at the lowest critical load are found in the cylindrical shell under axial load. A typical plot that illustrates the modes associated with the lowest bifurcation load is shown in Figure 14.2. The drawing

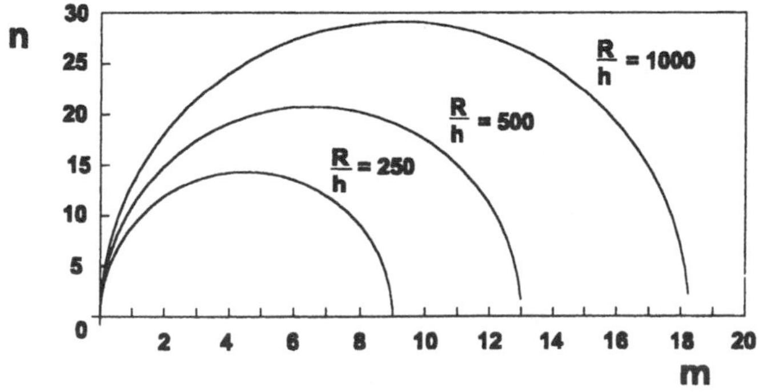

Figure 14.2 Koiter's circle for a cylindrical shell under axial load.

described by the coincident modes is again a circle. The cylindrical shell is similar to the spherical shell also in the sense that there is no way to avoid the coincidence of modes: it is built into the system of load and shell. Changes in the design, such as modifications in the radius/thickness ratio, do not change the nature of the coinciding modes but only the number of modes affected. Coincidence among modes does not entirely depend on the geometry, but on a combination of geometry and loading. For instance, the cylindrical shell under lateral pressure does not show coincident critical states as when we load it axially.

A different situation arises in other structural systems. Let us consider a thin-walled column, with an I-shaped cross section, under axial compression. The column displays global bifurcation modes, which are a function of $1/L^2$, where L is the length of the column. In this case the cross section is displaced sideways but without distortion of the shape. Another form of buckling may also occur, which involves considerable change in the cross section of the column, with rotation of the flanges and bending of the web. This is identified as local buckling and does not require lateral displacements of the column as in the global bifurcation mode. If we modify the length of the column, the local bifurcation load does not change: what changes is the number of waves in which the cross section distorts. Figure 14.3 shows a plot of the critical load versus the length of the column to illustrate that in long columns the first critical state is a bifurcation in a global mode, while short columns display bifurcation in local modes. We can identify a set of geometric parameters (of both the cross section and the length for a given set of material properties) for which the two

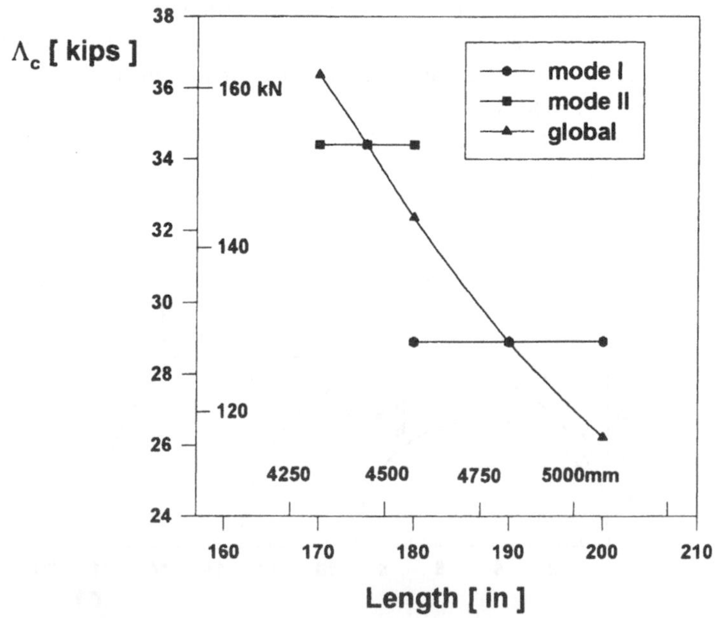

Figure 14.3 Critical load versus length of a column.

curves intersect; i.e., the two modes occur for the same value of critical load. If we plot the local versus global mode for a specific design, as in Figures 14.1 and 14.2, we would only find one point instead of a circle.

In the thin-walled column, the coincidence of modes is not present in all problems under axial load, but only in those in which the dimensions are chosen so as to achieve coincidence. Other examples of coincident modes by design are the reticulated column under axial load, discussed, for example, in [39]. In the stiffened plate, again we find local and global modes that can occur for the same load level (see, for example, [42]).

All the coincident modes discussed up to now are bifurcations. However, we can also have coincidence between a bifurcation mode and a limit-point mode. This case is discussed in detail in [13]. The problem arises in a circular arch discussed in Chapter 6, where we saw that for a design in which

$$\left(\frac{\pi}{2}\right)^4 \frac{I}{A}\frac{1}{R^2\phi^4} = 0.074, \tag{14.1}$$

bifurcation into two half wave modes becomes coincident with the limit point load.

Interaction between two or more modes can occur not only when their associated critical loads are identical, but also in cases when they are sufficiently close to each other. Of course, the interaction between modes is weaker as they become far apart, as in any nonlinear system.

In summary, if we do not pay enough attention to this point in the design of a structural component, we may end up having unintended interactions between nearly coincident modes.

14.3 NEW MODES EMERGING FROM INTERACTION

14.3.1 Modes Without Interaction

Let us consider the analysis in terms of the W-formulation. In Chapter 13 the energy was written in terms of incremental coordinates q_i with respect to an equilibrium path

$$W\,[q_i, \lambda] = W_1 + W_2 + W_3 + W_4$$

with

$$W_2 = \frac{1}{2}(A_{ij} + \Lambda\; A_{ijk}\; \bar{Q}_k)\; q_i q_j,$$

where $\bar{Q}_k$ is the displacement associated with a unit value of the load parameter Λ along a linear fundamental path. Here, Λ is a negative parameter, unlike in [31], where it is taken as positive.

The condition of critical stability was shown in Chapter 13 to reduce to

$$W_{ij}\; x_j|^c = [A_{ij} + \Lambda\; A_{ijk}\; \bar{Q}_k]\; x_j = 0. \tag{14.2}$$

The solution of (14.2) leads to a set of N eigenvalues, which we will denote as ${}^s\Lambda$. In this chapter, a superscript to the left of a variable identifies the mode to which one refers, because there is more than one active mode relevant in the analysis. The associated eigenvectors are denoted by ${}^s\mathbf{x}$, for $s = 1, \ldots, N$. These modes are orthogonal, in the sense that the scalar product of any two of them vanishes, i.e.,

$$ {}^s x_j \cdot {}^t x_j = 0 \qquad \text{for } s \neq t. \tag{14.3} $$

To investigate mode interaction we will use a different normalization of modes from that adopted in previous chapters. The new normalization is given by the condition

$$ \left[W'_{ij}(q_i = 0) \right] {}^s x_j {}^t x_i = \delta_{st} $$

or else

$$ [A_{ijk} \bar{Q}_k] \; {}^s x_j \; {}^t x_i = \delta_{st}, \tag{14.4} $$

where δ_{st} is the Kronecker delta.

Notice that if we employ (14.4) in (14.2) we get the new condition

$$ \frac{1}{\Lambda_c} A_{ij} \left({}^s x_j\right) \left({}^t x_i\right) = -\delta_{st}. \tag{14.5} $$

The analysis in Chapter 13 was carried out using perturbations, so that for each isolated mode one could write an expansion in terms of a perturbation parameter s in the form

$$ {}^t q_i = \left[{}^t q_i^{(1)c} \right] s + \frac{1}{2!} \left[{}^t q_i^{(2)c} \right] s^2 + \cdots \tag{14.6} $$

with

$$ \left[{}^t q_i^{(j)c} \right] \equiv \left[\frac{d^j ({}^t q_i)}{ds^j} \right]^c . $$

The perturbation parameter was chosen as one of the generalized coordinates of the problem, say, $s = q_1$.

14.3.2 First-Order Interaction

Next, we investigate the case in which there is a set of M eigenvalues that are relatively close or even coincident. Furthermore, we assume that the remaining eigenvalues occur at significantly higher values of Λ or that they do not influence the coupling between modes. We shall refer to the first M eigenvalues ${}^s\Lambda$ (for $s = 1, \ldots, M$) and their associated eigenvectors as "active," in the sense that they play a crucial role in the interactive buckling process. The remaining $(N - M)$ eigenvalues, ${}^s\Lambda$ (for $s = M+1, \ldots, N$), are "passive," and are not considered as important in the analysis.

As a first approximation, one can investigate a linear combination of the "active" modes to reflect mode interaction. This first-order approximation is given by

$$ q_i^I = {}^s x_i \, \xi_s, \tag{14.7} $$

where ξ_s are modal amplitudes (i.e., scalar quantities), the coupled modes are identified by ${}^s x_i$ with $s = 1, \ldots, M$, and the degrees of freedom are $i = 1, \ldots, N$.

- Equation (14.7) is the complete solution of the problem if we include all the modes in the analysis, i.e., $M = N$. But this is extremely expensive, because the analysis includes the solution of nonlinear equations, as is shown in this chapter, and inclusion of all modes is seldom done in practical problems.
- Equation (14.7) is a very good approximation if all the relevant modes that determine the postcritical behavior of the coupled problem are included.
- Sometimes we do not know a priori which modes should be included for coupling, and this may lead to incorrect solutions of the problem. For example, if not all relevant modes are included in (14.7), then the analysis may show that there is no coupling between the selected modes. This would be a limitation of the analysis.

Let us study ways to improve the solution when only a few modes are included. This leads to second-order terms in the mode coupling analysis.

14.3.3 Second-Order Interaction

To improve the solution we have to include new mode shapes, but they should not be a linear combination of the chosen modes, because that would have been taken care of by the linear interaction, equation (14.7). Thus, we include quadratic terms with new coupled modes.

The interaction between the ${}^k x_i$ modes creates new coupled modes, which will be denoted as $\left[{}^{kl} x_i\right]$ arising from the interaction between modes ${}^k x_i$ and ${}^l x_i$. A quadratic combination of modes may be written in the form

$$q_i^{II} = {}^{st} x_i \, \xi_s \, \xi_t, \tag{14.8}$$

where again ξ_s and ξ_t are modal amplitudes (i.e., scalars), the coupled modes are identified by ${}^{st} x_i$ with $s, t = 1, \ldots, M$, and the degrees of freedom are $i = 1, \ldots, N$. Notice that q_i^{II} is a function of the load parameter Λ, and in (14.8) the amplitude ξ_s is the function of the load.

Higher-order interaction terms could also be used in the analysis, including, for example, interactions between three modes at a time.

If we restrict the analysis to a quadratic combination, the initial postcritical path q_i including linear and quadratic combinations may be described as

$$q_i = q_i^I + q_i^{II} \tag{14.9}$$

or else

$$q_i = {}^s x_i \, \xi_s + {}^{st} x_i \, \xi_s \, \xi_t. \tag{14.10}$$

This combination may be found in the works of Reis [29] and Reis and Roorda [31].

Remarks on the coupled modes follow:

- For a specific value of t, ${}^{t}x_i$ is a vector of dimension N (an isolated eigenvector), while $\left[{}^{st}x_i\right]$ is a matrix of dimension $N \times M$, which reflects the interaction between mode s and all the M other active modes.
- If we set s and t to specific values, then $\left[{}^{st}x_i\right]$ is a vector of dimension $N \times 1$.
- The coupled modes are orthogonal to the isolated modes

$$A_{ijk}\bar{Q}_k \, {}^{u}x_j \, {}^{st}x_i = 0. \tag{14.11}$$

- Another property of the coupled modes is that they are symmetric

$${}^{st}x_i = {}^{ts}x_i. \tag{14.12}$$

- Notice that the modes considered need not be simultaneous; that is, we do not require that the associated eigenvalues be exactly coincident.

To summarize, in this section we only wrote the displacement fields in the post-buckling path in terms of the isolated modes that interact with (${}^{s}x_i$), and new coupled modes that enter into the analysis (${}^{st}x_i$). The isolated modes are computed using an eigenvalue analysis as in Chapter 13, but the coupled modes are unknown at the moment, in the sense that we do not even know what shape they represent. To follow the path we employ scalar coordinates ξ_s, which could also be seen as participating factors of the modes. These are also unknown at the moment.

Our task in this chapter is to indicate how ${}^{st}x_i$ and ξ_s can be computed in the analysis, and to do that we investigate the equilibrium conditions using the energy W.

14.4 THE TOTAL POTENTIAL ENERGY W

We start by substituting the postcritical displacements of (14.9) into the energy W and expanding

$$\begin{aligned} W\left[q_i^I + q_i^{II}, \Lambda\right] &= \frac{1}{2!}(A_{ij} + \Lambda A_{ijk}\bar{Q}_k)(q_i^I q_j^I + q_i^{II} q_j^{II} + 2q_i^I q_j^{II}) \\ &+ \frac{1}{3!}(A_{ijk} + \Lambda A_{ijkl}\bar{Q}_l)(q_i^I q_j^I q_k^I + 3q_i^{II} q_j^I q_k^I + 3q_i^{II} q_j^{II} q_k^I + q_i^{II} q_j^{II} q_k^{II}) \\ &+ \frac{1}{4!}(A_{ijkl})(q_i^I q_j^I q_k^I q_l^I + 4q_i^{II} q_j^I q_k^I q_l^I + 6q_i^{II} q_j^{II} q_k^I q_l^I + 4q_i^{II} q_j^{II} q_k^{II} q_l^I \\ &+ q_i^{II} q_j^{II} q_k^{II} q_l^{II}). \end{aligned} \tag{14.13}$$

Here the indices i, j, k, l refer to all the degrees of freedom of the system. We do not specify in (14.13) how many modes interact.

Substituting (14.7) and (14.8) into W yields the energy in terms of the coupled modes and the modal amplitudes

$$W = W[{}^{st}x_j, \xi_s, \Lambda].$$

Next we neglect terms of fifth and sixth order in ξ_s and retain only up to fourth-order terms. The functional W reduces to

$$W = \frac{1}{2!}(A_{ij} + \Lambda A_{ijk}\bar{Q}_k)(q_i^I q_j^I + q_i^{II} q_j^{II} + 2q_i^I q_j^{II})$$

$$+\frac{1}{3!}(A_{ijk} + \Lambda A_{ijkl}\bar{Q}_l)(q_i^I q_j^I q_k^I + 3q_i^{II} q_j^I q_k^I)$$

$$+\frac{1}{4!}(A_{ijkl})(q_i^I q_j^I q_k^I q_l^I).$$

We can recognize that some terms in W are zero due to the condition of orthogonality between modes, equation (14.4). First, the following term is zero:

$$(A_{ij} + \Lambda\ A_{ijk}\ \bar{Q}_k)\ q_i^I\ q_j^{II} = 0. \tag{14.14}$$

Next, the following two terms are equal:

$$\frac{1}{2}(A_{ij} + \Lambda\ A_{ijk}\ \bar{Q}_k)\ q_i^I\ q_j^I = \frac{1}{2}\delta_{st}\ \xi_s\ \xi_t\ (\Lambda - {}^s\Lambda_c), \tag{14.15}$$

where ${}^s\Lambda_c$ is the critical load for the isolated mode ${}^s\mathbf{x}$. The proof of the conditions (14.14) and (14.15) is found as problems solved at the end of this chapter.

Finally, the energy W can be written as

$$W = \frac{1}{2}\delta_{st}\xi_s\xi_t(\Lambda - {}^s\Lambda_c)$$

$$+\frac{1}{2}(A_{ij} + \Lambda A_{ijk}\bar{Q}_k)\,{}^{st}x_i\,{}^{uv}x_j\ \xi_s\xi_t\xi_u\xi_v$$

$$+\frac{1}{3!}(A_{ijk} + \Lambda A_{ijkl}\bar{Q}_l)\,{}^s x_i\,{}^t x_j\,{}^u x_k\ \xi_s\xi_t\xi_u$$

$$+\frac{1}{2}(A_{ijk} + \Lambda A_{ijkl}\bar{Q}_l)\,{}^{st}x_i\,{}^u x_j\,{}^v x_k\ \xi_s\xi_t\xi_u\xi_v$$

$$+\frac{1}{4!}(A_{ijkl})\,{}^s x_i\,{}^t x_j\,{}^u x_k\,{}^v x_l\ \xi_s\xi_t\xi_u\xi_v. \tag{14.16}$$

Thus, W is a function of ξ_s and ${}^{st}x_i$. Notice that all the terms depend on the amplitude ξ_s, but the first, third, and fifth terms are independent of the coupled mode ${}^{st}x_i$.

14.5 EVALUATION OF COUPLED MODE SHAPES

14.5.1 A Condition for Coupled Modes

The modes ${}^{st}x_i$ arising from coupling between isolated modes also define paths of equilibrium in the $\Lambda - q_i$ space. This section follows the presentation of [28], and to obtain the mode shapes ${}^{st}x_i$ of the coupled modes, we set to zero the first variation of W with respect to the ${}^{st}x_i$ for constant values of the amplitude parameter ξ_s.

Thus, the starting condition is

$$\frac{\partial W}{\partial({}^{st}x_i)} = 0 \tag{14.17}$$

$$= \left[\left(A_{ij} + \Lambda A_{ijk}\bar{Q}_k\right) {}^{uv}x_j + \frac{1}{2}\left(A_{ijk} + \Lambda A_{ijkl}\bar{Q}_l\right) {}^{u}x_j\, {}^{v}x_k\right] \xi_s\xi_t\xi_u\xi_v.$$

From the above equation, one can conclude that

$$[A_{ij} + \Lambda A_{ijk}\bar{Q}_k]\, {}^{uv}x_j = -\frac{1}{2}(A_{ijk} + \Lambda A_{ijkl}\bar{Q}_l)\, {}^{u}x_j\, {}^{v}x_k$$

or else

$$\hat{W}_{ij}\, {}^{uv}x_j = -\frac{1}{2}\hat{W}_{ijk}\, {}^{u}x_j\, {}^{v}x_k. \tag{14.18}$$

The member on the right side is clearly nonzero; this shows that $\left[{}^{uv}x_j\right]$ are not eigenvalues of the "tangent matrix" $[A_{ij}+\Lambda A_{ijk}\bar{Q}_k]$. This is correct, since we already calculated all the eigenvectors and denoted them by ${}^{s}\mathbf{x}$.

Let us look into the number of systems to be solved in (14.18). If there are M interacting modes, then there will be $\left[\left(M + M^2\right)/2\right]$ systems to obtain the coupled modes.

Notice that the coupled modes are a function of Λ, i.e.,

$${}^{uv}x_j = {}^{uv}x_j\,(\Lambda)\,.$$

A solution of the complete problem without further assumptions is shown [28] and in Problem 14.3 at the end of this chapter. However, this is somewhat cumbersome, and we show a simplified approach next.

14.5.2 Solution of Coupled Modes

A simplification in the analysis can be obtained if the interacting modes are close to each other or if they are coincident. In this case it is possible to assume that the coupled mode shape $\left[{}^{uv}x_j\right]$ remains the same between the two critical loads. Under this condition, the derivative of (14.18) with respect to the load Λ leads to

$$[A_{ijk}\bar{Q}_k]\, {}^{uv}x_j = -\frac{1}{2}(A_{ijkl}\bar{Q}_l)\, {}^{u}x_j\, {}^{v}x_k \tag{14.19}$$

or else

$$\hat{W}'_{ij}\,{}^{uv}x_j = -\frac{1}{2}\hat{W}'_{ijk}\,{}^{u}x_j\,{}^{v}x_k.$$

The matrix $\left[\hat{W}'_{ij}\right]$ is not singular, so that there must be a unique solution to the coupled mode shape $\left[{}^{uv}x_j\right]$.

14.6 EVALUATION OF MODE AMPLITUDES

We now return to (14.16) for the energy W. Since the coupled modes ${}^{uv}x_j$ are now known from the solution of (14.19), the complete path can be traced by computation of the mode amplitudes ξ_s. In other words, we have constructed a basis of vectors (the modes ${}^{s}x_j$ and the coupled modes ${}^{st}x_j$) to express the equilibrium condition, and we now seek to find how these amplitudes change with the evolving path.

14.6.1 Solution of Coupled Mode Amplitudes

At this stage, there are still M unknowns: the values of ξ_s. At an equilibrium state, W must be stationary with respect to all possible variations of the generalized coordinates q_i, but according to (14.10) the variations are written in terms of the ξ_s.

Thus,

$$\frac{\partial W}{\partial \xi_s}[\xi_t, \Lambda] = W_s = 0, \qquad s = 1, \ldots, M, \tag{14.20}$$

or else, from (14.16),

$$\begin{aligned} W_s = {} & \left[\left(\Lambda - {}^{s}\Lambda\right)\delta_{st}\right]\xi_t \\ & + \left[\frac{1}{2}(A_{ijk} + \Lambda\ A_{ijkl}\bar{Q}_l)\,{}^{s}x_i\,{}^{t}x_j\,{}^{u}x_k\right]\xi_s\xi_t + [2(A_{ij} + \Lambda\ A_{ijk}\bar{Q}_k)\,{}^{st}x_i\,{}^{uv}x_j \\ & + 2(A_{ijk} + \Lambda\ A_{ijkl}\bar{Q}_l)\,{}^{st}x_i\,{}^{u}x_j\,{}^{v}x_k + \frac{1}{6}A_{ijkl}\,{}^{s}x_i\,{}^{t}x_j\,{}^{u}x_k\,{}^{v}x_l]\xi_t\xi_u\xi_v \\ & = 0. \end{aligned} \tag{14.21}$$

Alternatively, we can write (14.21) in the compact form

$$\left[\left(\Lambda - {}^{s}\Lambda\right)\delta_{st}\right]\xi_t + \alpha_{stu}\xi_s\xi_t + \beta_{stuv}\xi_t\xi_u\xi_v = 0 \tag{14.22}$$

with the coefficients α_{stu} and β_{stuv} given by the bracketed expressions in (14.21).

We can employ any method of solution of nonlinear algebraic equations, including Newton–Raphson and perturbation techniques, to obtain ξ_t from (14.22) for each load level Λ. The number of cubic equations to be solved is M; these are scalar simultaneous equations.

Table 14.1 shows the number of systems that have to be solved for a different number of interacting modes M. In the second column we indicate the number of

Table 14.1

Number of interacting modes ${}^s x_j$	Coupled modes to be identified ${}^{uv} x_j$	Simultaneous cubic equations ξ_t
2	3	2
3	6	3
4	10	4
5	15	5

coupled modes that need to be computed for which one has to solve a system of linear equations of dimension $N \times N$. The third column shows how many cubic equations have to be solved for the computation of the amplitudes ξ_t.

14.6.2 An Algorithm for the Evaluation of Coupled Mode Paths

The above formulation can be summarized in the form of an algorithm as follows:

1. Compute the linear fundamental path.
2. Evaluate critical states (eigenvalues, eigenvectors, and displacements at the critical states).
3. Normalize eigenvectors as

$$[A_{ijk}\bar{Q}_k]\ {}^s x_j\ {}^t x_i = \delta_{st}.$$

4. Evaluate coupled mode shapes from

$$[A_{ijk}\bar{Q}_k]\,{}^{uv} x_j = -\frac{1}{2}(A_{ijkl}\bar{Q}_l)\,{}^u x_j\ {}^v x_k.$$

5. Evaluate coefficients

$$\alpha_{stu} = \left[\frac{1}{2}(A_{ijk} + \Lambda\ A_{ijkl}\bar{Q}_l)\,{}^s x_i\ {}^t x_j\ {}^u x_k\right],$$

$$\beta_{stuv} = \Big[2(A_{ij} + \Lambda\ A_{ijk}\bar{Q}_k)\,{}^{st} x_i\ {}^{uv} x_j$$

$$+2(A_{ijk} + \Lambda\ A_{ijkl}\bar{Q}_l)\,{}^{st} x_i\ {}^u x_j\ {}^v x_k + \frac{1}{6}A_{ijkl}\ {}^s x_i\ {}^t x_j\ {}^u x_k\ {}^v x_l\Big].$$

6. Solve ξ_t from the cubic equations

$$\left[(\Lambda -{}^s\Lambda)\,\delta_{st}\right]\xi_t + \alpha_{stu}\xi_u\xi_t + \beta_{stuv}\xi_t\xi_u\xi_v = 0.$$

7. Postcritical coupled path is

$$q_i = {}^s x_i\ \xi_s + {}^{st} x_i\ \xi_s\ \xi_t.$$

14.7 TWO INTERACTING MODES

14.7.1 Interaction between Two Nonsymmetric Modes

This is a very common problem often found in structures. The analysis can be carried out as in the general case by setting the number of interacting modes $M = 2$. However, simpler explicit expressions are possible because of the small number of modes involved. The analysis in this section includes linear and quadratic interaction terms.

The displacements in the postcritical path are

$$q_i = \left({}^1x_i\right)\xi_1 + \left({}^2x_i\right)\xi_2 + \left[{}^{11}x_i\right]\xi_1^2 + 2\left[{}^{12}x_i\right]\xi_1\xi_2 + \left[{}^{22}x_i\right]\xi_2^2. \qquad (14.23)$$

The three vectors ${}^{uv}x_i$ are evaluated from (14.18). For interaction between modes 1 and 2 we have to solve

$$[A_{ijk}\,\bar{Q}_k]\left[{}^{12}x_j\right] = -\frac{1}{2}(A_{ijkl}\,\bar{Q}_l)\left({}^1x_j\right)\left({}^2x_k\right).$$

A similar procedure is followed to solve the coupling modes $\left[{}^{11}x_i\right]$ and $\left[{}^{22}x_i\right]$

$$[A_{ijk}\,\bar{Q}_k]\left[{}^{11}x_j\right] = -\frac{1}{2}(A_{ijkl}\,\bar{Q}_l)\left({}^1x_j\right)\left({}^1x_k\right),$$

$$[A_{ijk}\,\bar{Q}_k]\left[{}^{22}x_j\right] = -\frac{1}{2}(A_{ijkl}\,\bar{Q}_l)\left({}^2x_j\right)\left({}^2x_k\right).$$

The amplitudes ξ_s are calculated from (14.21). Let us write (14.21) in the form

$$(\Lambda - {}^s\Lambda_c)\delta_{st}\xi_t + \alpha^{stu}\xi_t\xi_u + \beta^{stuv}\xi_t\xi_u\xi_v = 0 \qquad (14.24)$$

for $s, t, u, v = 1, 2$.

Expanding the above equation we get the two conditions

$$\begin{aligned}&(\Lambda - {}^1\Lambda_c)\xi_1 + \alpha^{111}\xi_2^2 + 2\alpha^{112}\xi_1\xi_2 + \alpha^{122}\xi_2^2 \\ &+\beta^{1111}\xi_1^3 + 3\beta^{1122}\xi_1\xi_2^2 + 3\beta^{1112}\xi_1^2\xi_2 + \beta^{1222}\xi_2^3 \\ &= 0,\end{aligned} \qquad (14.25)$$

$$\begin{aligned}&(\Lambda - {}^2\Lambda_c)\xi_2 + \alpha^{211}\xi_1^2 + 2\alpha^{212}\xi_1\xi_2 + \alpha^{222}\xi_2^2 \\ &+\beta^{2111}\xi_1^3 + 3\beta^{2221}\xi_1\xi_2^2 + 3\beta^{2211}\xi_1^2\xi_2 + \beta^{2222}\xi_2^3 \\ &= 0.\end{aligned} \qquad (14.26)$$

There are six α coefficients and eight β coefficients in (14.24). For example,

$$\alpha^{122} = \frac{1}{2}(A_{ijk} + \Lambda A_{ijkl}\bar{Q}_l)\,{}^1x_i\,{}^2x_j\,{}^2x_k,$$

$$\beta^{2111} = 2(A_{ij} + \Lambda A_{ijk}\bar{Q}_k)\left[{}^{21}x_i\right]{}^{11}x_j$$

$$+2(A_{ijk} + \Lambda A_{ijkl}\bar{Q}_l)\left[{}^{21}x_i\right]{}^1x_j\,{}^1x_k + \frac{1}{6}(A_{ijkl})\,{}^2x_i\,{}^1x_j\,{}^1x_k\,{}^1x_l.$$

Thus, we have two cubic equations, (14.25) and (14.26), in two unknowns, ξ_1 and ξ_2. The solutions are a function of Λ through the coefficients of the equations. But we stop the analysis here because we are interested in an even simpler case of symmetric systems.

14.7.2 Interaction between Two Symmetric Modes

There are many practical cases in which the modes in competition are symmetric. We say that a system is singly symmetric if the energy is symmetric with respect to one of the buckling mode amplitudes [29]. For example,

$$W[\xi_1, \xi_2, \Lambda] = W[\xi_1, -\xi_2, \Lambda] \tag{14.27}$$

indicates symmetry with respect to the amplitude ξ_1.

Doubly symmetric systems, on the other hand, are characterized by

$$W[\xi_1, \xi_2, \Lambda] = W[-\xi_1, -\xi_2, \Lambda], \tag{14.28}$$

so that both modes are symmetric. It has been shown in [31] that in doubly symmetric systems the cubic terms in ξ_s in (14.16) vanish. The values of $\beta^{stuv} = 0$ and the coupled mode shapes $\left[{}^{uv}x_j\right]$ are zero; thus, (14.18) will not be used in the analysis. We go back to (14.21), or (14.24) for two modes, and set ${}^{uv}x_j = 0$. The only remaining unknowns are ξ_1 and ξ_2, and they can be calculated from (14.24).

The displacements in the postcritical path are given by

$$q_i = {}^1x_i\,\xi_1 + {}^2x_i\,\xi_2. \tag{14.29}$$

The energy W results in

$$W[\xi_s, \Lambda] = \frac{1}{2!}\delta_{st}(\Lambda - {}^s\Lambda_c)\xi_s\xi_t + \frac{1}{4!}(A_{ijkl}\,{}^sx_i\,{}^tx_j\,{}^ux_k\,{}^vx_l)\xi_s\xi_t\xi_u\xi_v$$

with $i, j, k, l = 1, \ldots, N$; and $s, t, u, v = 1, 2$. However, not all the terms in W are nonzero: Of the quadratic terms, only three remain in this case. Thus, the expanded version of W leads to

$$W[\xi_1, \xi_2, \Lambda] = \frac{1}{2}(\Lambda - {}^1\Lambda_c)\xi_1^2 + \left(c_{11}\xi_1^4 + c_{22}\xi_2^4 + 2c_{12}\xi_1^2\xi_2^2\right). \tag{14.30}$$

The coefficients c_{ij} are calculated from

$$c_{11} = \frac{1}{4!}(A_{ijkl}\,{}^1x_i\,{}^1x_j\,{}^1x_k\,{}^1x_l), \tag{14.31}$$

$$c_{22} = \frac{1}{4!}(A_{ijkl}\,{}^2x_i\,{}^2x_j\,{}^2x_k\,{}^2x_l),$$

$$c_{12} = \frac{3}{4!}(A_{ijkl}\,{}^1x_i\,{}^1x_j\,{}^2x_k\,{}^2x_l).$$

The relation between the c_{ij} coefficients and the curvature of the postcritical path is discussed in Problem 14.4. The energy W is now defined in a space of three dimensions in terms of Λ, ξ_1, and ξ_2.

Equilibrium conditions. The equilibrium conditions are obtained from

$$\frac{\partial W}{\partial \xi_s} = 0, \tag{14.32}$$

where $s = 1, 2$. From the above we have the explicit form

$$(\Lambda - {}^1\Lambda_c)\xi_1 + 4(c_{11}\xi_1^3 + c_{12}\xi_1\xi_2^2) = 0,$$

$$(\Lambda - {}^2\Lambda_c)\xi_2 + 4(c_{22}\xi_2^3 + c_{12}\xi_1^2\xi_2) = 0.$$

We can rewrite these equations as

$$[(\Lambda - {}^1\Lambda_c) + 4(c_{11}\xi_1^2 + c_{12}\xi_2^2)]\,\xi_1 = 0, \tag{14.33}$$

$$[(\Lambda - {}^2\Lambda_c) + 4(c_{12}\xi_1^2 + c_{22}\xi_2^2)]\,\xi_2 = 0.$$

Let us investigate the solutions of the above cubic equations:

a) A possible solution is $\xi_1^F = \xi_2^F = 0$ and it corresponds to the fundamental path.

b) Away from the fundamental path, $\xi_1 \neq 0$, $\xi_2 = 0$ leads to the uncoupled postcritical path related to the eigenvector ${}^1\mathbf{x}$,

$$\Lambda = \left({}^1\Lambda_c\right) - 4(c_{11}\xi_1^2), \tag{14.34}$$

$$\xi_2 = 0,$$

and for $\xi_1 = 0$, $\xi_2 \neq 0$ we get

$$\Lambda = \left({}^2\Lambda_c\right) - 4(c_{22}\xi_2^2), \tag{14.35}$$

$$\xi_1 = 0.$$

These are the equations of the postbuckling uncoupled paths. Notice that Λ are negative values, so that stable paths are associated with $c_{11} > 0$, $c_{22} > 0$.

c) If we solve the two equilibrium equations but for $\xi_1 \neq 0$, $\xi_2 \neq 0$ we obtain a system of two quadratic equations leading to the tertiary paths:

$$(\Lambda - {}^1\Lambda_c) + 4(c_{11}\xi_1^2 + c_{12}\xi_2^2) = 0, \tag{14.36}$$

$$(\Lambda - {}^2\Lambda_c) + 4(c_{12}\xi_1^2 + c_{22}\xi_2^2) = 0.$$

The simultaneous solution of these two equations leads to the explicit form

$$\xi_1^2 = \frac{1}{4}\left[\frac{c_{12}(\Lambda - {}^2\Lambda_c) - c_{22}(\Lambda - {}^1\Lambda_c)}{\det(c_{ij})}\right], \tag{14.37}$$

$$\xi_2^2 = \frac{1}{4}\left[\frac{c_{11}(\Lambda - {}^2\Lambda_c) - c_{12}(\Lambda - {}^1\Lambda_c)}{\det(c_{ij})}\right].$$

The coefficient c_{12} is crucial in the interaction problem. If $c_{12} = 0$, then equations (14.37) reduce to the uncoupled paths.

Conditions of critical stability. The condition

$$\det\left[\frac{\partial^2 W}{\partial\xi_s\partial\xi_t}\right] = \det\left[W_{st}\right] = 0 \tag{14.38}$$

is satisfied at a critical state. Notice that this is a general condition of critical stability and should be valid for the primary bifurcations from the primary path as well as for secondary bifurcations along the secondary path.

We first calculate the components of W_{st} in the form

$$W_{11} = (\Lambda - {}^1\Lambda_c) + 12\, c_{11}\xi_1^2 + 4\, c_{12}\xi_2^2,$$

$$W_{22} = (\Lambda - {}^2\Lambda_c) + 12\, c_{22}\xi_2^2 + 4\, c_{12}\xi_1^2,$$

$$W_{12} = W_{21} = 8\, c_{12}\xi_1\xi_2.$$

The determinant is given by

$$\det[W_{st}] = (\Lambda - {}^1\Lambda_c + 12\, c_{11}\xi_1^2 + 4\, c_{12}\xi_2^2)$$
$$\times\, (\Lambda - {}^2\Lambda_c + 12\, c_{22}\xi_2^2 + 4\, c_{12}\xi_1^2) - 64\, c_{12}^2\xi_1^2\xi_2^2$$
$$= 0.$$

Expanding the determinant, we get

$$(\Lambda - {}^1\Lambda_c)(\Lambda - {}^2\Lambda_c) + 12(\Lambda - {}^1\Lambda_c)c_{22}\xi_2^2 + 12c_{11}(\Lambda - {}^2\Lambda_c)\xi_1^2$$
$$+4(\Lambda - {}^1\Lambda_c)c_{12}\xi_1^2 + 4c_{12}(\Lambda - {}^2\Lambda_c)\xi_2^2$$
$$+48c_{11}c_{12}\xi_1^4 + 48c_{22}c_{12}\xi_2^4 - 48c_{12}^2\xi_1^2\xi_2^2 + 144c_{11}c_{22}\xi_1^2\xi_2^2$$
$$= 0$$

or else, in the more compact form,

$$(\Lambda - {}^1\Lambda_c)(\Lambda - {}^2\Lambda_c) + 4\xi_2^2[(\Lambda - {}^2\Lambda_c)\, c_{12} + 3(\Lambda - {}^1\Lambda_c)\, c_{22}] \quad (14.39)$$
$$+4\xi_1^2[(\Lambda - {}^1\Lambda_c)\, c_{12} + 3(\Lambda - {}^2\Lambda_c)\, c_{11}]$$
$$+48[c_{11}c_{12}\xi_1^4 + c_{22}c_{12}\xi_2^4 + (3c_{11}c_{22} - c_{12}^2)\xi_1^2\xi_2^2]$$
$$= 0.$$

This equation governs the critical state of the problem. At the first bifurcation point from the fundamental path we have $\Lambda = {}^1\Lambda_c$ and $\xi_1 = \xi_2 = 0$, and the equation becomes an identity. The same occurs for the bifurcation state $\Lambda = {}^2\Lambda_c$. Furthermore, (14.39) provides new information for secondary bifurcations, and this is discussed in the following section.

Secondary bifurcation points. Along the secondary path given by

$$(\Lambda - {}^1\Lambda_c) = -4c_{11}\xi_1^2,$$
$$\xi_2 = 0,$$

we investigate the occurrence of critical states. For the above equilibrium paths, the determinant reduces to

$$(\Lambda - {}^1\Lambda_c)(\Lambda - {}^2\Lambda_c) + (\Lambda - {}^1\Lambda_c)^2\frac{c_{12}}{c_{11}}$$
$$-3(\Lambda - {}^1\Lambda_c)(\Lambda - {}^2\Lambda_c) + 3(\Lambda - {}^1\Lambda_c)^2\frac{c_{12}}{c_{11}}$$
$$= 0$$

or else

$$c_{11}(\Lambda - {}^1\Lambda_c)(\Lambda - {}^2\Lambda_c) + (\Lambda - {}^1\Lambda_c)^2 c_{12} = 0$$

or

$$-(\Lambda - {}^2\Lambda_c)c_{11} + (\Lambda - {}^1\Lambda_c)c_{12} = 0.$$

Finally, a secondary bifurcation point is given by

$${}^s\Lambda_c = \frac{c_{12}\left({}^1\Lambda_c\right) - c_{11}\left({}^2\Lambda_c\right)}{c_{12} - c_{11}}. \tag{14.40}$$

Using (14.40) we can locate the critical load. The displacement at this state can be computed as

$$({}^s\xi_1)^2 = \frac{{}^s\Lambda_c - {}^1\Lambda_c}{4c_{11}}, \qquad {}^s\xi_1 = 0. \tag{14.41}$$

There are two critical secondary states at the same load level, one on each branch of the secondary path.

Remarks

- It is interesting to notice that for coincident critical states ${}^2\Lambda_c = {}^1\Lambda_c$, we get ${}^s\Lambda_c = {}^1\Lambda_c$ and ${}^s\xi_1 = 0$; that is, the secondary bifurcation states occur at the same load as the primary coincident states.
- Depending on the nature of the paths, the secondary bifurcations may appear on the second path associated with ${}^2\Lambda_c$. In this case,

$${}^s\Lambda_c - {}^2\Lambda_c = \frac{c_{22}}{c_{22} + c_{12}}({}^2\Lambda_c - {}^1\Lambda_c). \tag{14.42}$$

We stop the formulation at this point, but it is important to visualize the possibilities of coupled paths for this class of doubly symmetric systems. Figure 14.4 is taken from a work by Supple [37]: In case (a) there is branching from the upper path without affecting the lower secondary path. Case (b) shows branching from the lower path, leading to a descending path. This has serious consequences for the structure. In case (c) there is no interaction. Finally, in case (d) the tertiary path is an ellipse.

Several theorems have been formulated by Supple for doubly symmetric systems with just two active degrees of freedom:

Theorem 14.1. "When one or both uncoupled equilibrium paths are falling, the coupled paths branching from the lower uncoupled path will always be falling."

Theorem 14.2. "When one or both uncoupled equilibrium paths are rising, the coupled paths branching from the upper uncoupled path will always be rising."

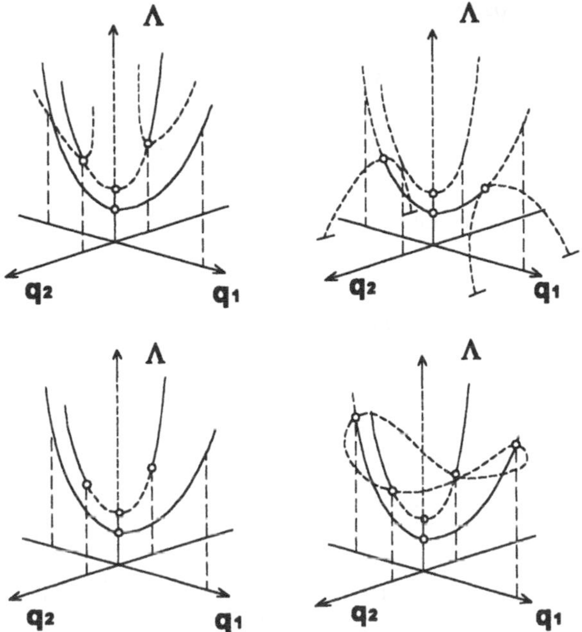

Figure 14.4 Forms of doubly symmetric buckling, according to Supple.

These and other theorems are discussed in [37] in the context of the diagonal energy formulation.

14.7.3 An Algorithm for Coupled Paths in Two-Mode Interaction of Doubly Symmetric Systems

For doubly symmetric systems, the general algorithm reduces to the following:

1. Compute the fundamental path.
2. Evaluate critical states.
3. Normalize eigenvectors and construct energy W.
4. Verify modes are doubly symmetric.
5. Evaluate coefficients

$$c_{11} = \frac{1}{4!}(A_{ijkl}\,{}^1x_i\,{}^1x_j\,{}^1x_k\,{}^1x_l),$$

$$c_{22} = \frac{1}{4!}(A_{ijkl}\,{}^2x_i\,{}^2x_j\,{}^2x_k\,{}^2x_l),$$

$$c_{12} = \frac{6}{4!}(A_{ijkl}\,{}^1x_i\,{}^1x_j\,{}^2x_k\,{}^2x_l).$$

6. Obtain secondary paths

$$\Lambda = {}^1\Lambda_c - 4(c_{11}\xi_1^2),$$

$$\Lambda = {}^2\Lambda_c - 4(c_{22}\xi_2^2).$$

7. Obtain secondary bifurcations

$${}^s\xi_1 = \pm\frac{1}{2}\sqrt{\frac{{}^2\Lambda_c - {}^1\Lambda_c}{c_{11} + c_{12}}}.$$

8. Obtain tertiary path

$$\xi_1^2 = \frac{1}{4}\left[\frac{c_{12}(\Lambda - {}^2\Lambda_c) - c_{22}(\Lambda - {}^1\Lambda_c)}{\det(c_{ij})}\right],$$

$$\xi_2^2 = \frac{1}{4}\left[\frac{c_{11}(\Lambda - {}^2\Lambda_c) - c_{12}(\Lambda - {}^1\Lambda_c)}{\det(c_{ij})}\right].$$

14.8 THE AUGUSTI COLUMN

In 1964 Augusti presented a column problem with two degrees of freedom, which displayed interactions between modes [1]. This has been reproduced many times in the literature to illustrate a simple case with complex coupling behavior. In this section we discuss the Augusti column with reference to the theoretical framework presented in this chapter and employ the two-mode analysis for the doubly symmetric system.

The total potential energy. The structure and load are shown in Figure 14.5: This is a rigid bar with a pin support at the base and two rotational springs in perpendicular planes. The stiffness of the rotational springs are K_1 and K_2; the rotations of the column in each plane are Q_1 and Q_2, as indicated in Figure 14.5, and the load is P. The vertical deflection of the top of the column is Δ.

The total potential energy results in [1]

$$V = \frac{1}{2}\left(K_1 Q_1^2 + K_2 Q_2^2\right) - P\Delta, \tag{14.43}$$

where P and Δ are positive downward, as indicated in Figure 14.5.

We investigate the general case in which the two springs are not identical and bear the relation

$$K_2 = \tau K_1. \tag{14.44}$$

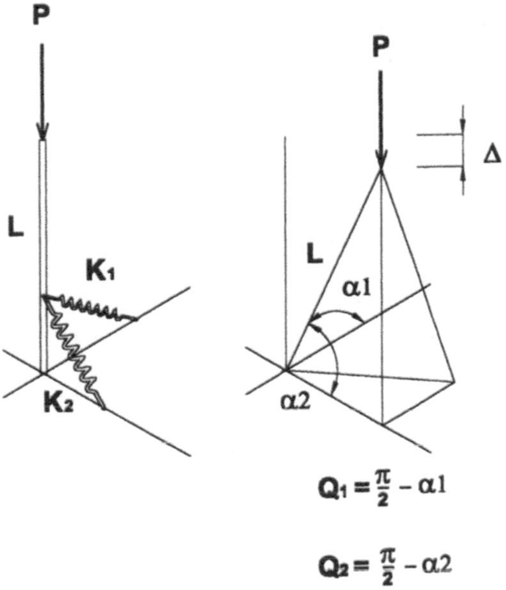

Figure 14.5 The Augusti column.

It is possible to write the kinematic equations for Δ by means of geometric considerations as

$$\Delta = L\left[\frac{1}{2}\left(Q_1^2 + Q_2^2\right) - \frac{1}{24}\left(Q_1^4 + Q_2^4 - 6Q_1^2Q_2^2\right)\right]$$

leading to the energy in the form

$$V\left[Q_1, Q_2, \Lambda\right] = K_1\left\{\frac{1}{2}\left(Q_1^2 + \tau Q_2^2\right) - \Lambda\left[\frac{1}{2}\left(Q_1^2 + Q_2^2\right) - \frac{1}{24}\left(Q_1^4 + Q_2^4 - 6Q_1^2Q_2^2\right)\right]\right\},$$

where

$$\Lambda = \frac{PL}{K_1} \geq 0.$$

Expansion of the energy yields

$$V\left[Q_1, Q_2, \Lambda, \tau\right] = \frac{1}{2!}\left(A_{11}Q_1^2 + A_{22}Q_2^2\right) + \frac{1}{4!}\left(A_{1111}Q_1^4 + 6A_{1122}Q_1^2Q_2^2 + A_{2222}Q_2^4\right), \quad (14.45)$$

where the linear and cubic coefficients A_i and A_{ijk} are all zero, and the nonzero terms are

$$A_{11} = K_1(1-\Lambda), \qquad A_{22} = K_1(\tau - \Lambda), \tag{14.46}$$
$$A_{1111} = A_{2222} = K_1\Lambda, \qquad A_{1122} = -K_1\Lambda.$$

The energy depends not only on the generalized coordinates and the load, but also on the constitutive parameter τ. For $\tau = 1$ the two springs are identical (and it will be shown that the critical loads are also identical), while for values of τ close to unity the two springs are similar but not identical (in which case the associated critical loads are close but not coincident). Problems in which the nature of the behavior changes with a design parameter (as would be the case of τ in the Augusti column) are investigated by means of what is called design sensitivity analysis, and this subject is treated in detail in the next chapter.

Equilibrium conditions. The equilibrium conditions take the form

$$V_1 = \left(A_{11} + \frac{1}{6}A_{1111}Q_1^2 + \frac{1}{2}A_{1122}Q_2^2\right)Q_1 = 0, \tag{14.47}$$
$$V_2 = \left(A_{22} + \frac{1}{6}A_{2222}Q_2^2 + \frac{1}{2}A_{1122}Q_1^2\right)Q_2 = 0.$$

The fundamental path is trivial, i.e.,

$$Q_1^F = 0, \qquad Q_2^F = 0, \tag{14.48}$$

is a solution of (14.47). Thus, it is not necessary to distinguish between the forms of energy V and W.

Clearly, there are more possibilities to satisfy the equilibrium conditions of (14.47). For $Q_2 = 0$ but $Q_1 \neq 0$ we can use $V_1 = 0$ and obtain a secondary path given by

$$A_{11} + \frac{1}{6}A_{1111}Q_1^2 = 0 \tag{14.49}$$

from which

$$\Lambda = 1 + \frac{1}{6}Q_1^2. \tag{14.50}$$

Similarly, for $Q_1 = 0$ but $Q_2 \neq 0$ we can use $V_2 = 0$ and obtain another secondary path

$$A_{22} + \frac{1}{6}A_{2222}Q_2^2 = 0 \tag{14.51}$$

and the following results

$$\Lambda = \tau\left(1 + \frac{1}{6}Q_2^2\right). \tag{14.52}$$

We know that these are secondary paths because they cross the primary path at some point. But there is a new path in which neither generalized coordinate is zero and satisfies the condition (14.47), leading to

$$Q_1^2 = 3\frac{6(1-\tau) - (3+\tau)(Q_2^2 + \frac{1}{9}Q_2^4)}{(Q_2^2 - 3)(3+\tau)}, \tag{14.53}$$

$$\Lambda = \frac{3+\tau}{4}\left(1 - \frac{1}{3}Q_2^2\right). \tag{14.54}$$

This is not a secondary path, because it has no points in common with the trivial primary path, so it must be a tertiary path (or perhaps more than one).

The equilibrium paths in terms of the original coordinates have been plotted in Figure 14.6 for a specific value of $\tau = 1.1$. The lowest uncoupled path occurs in the plane $\Lambda - Q_1$ and is shown in Figure 14.6.a to be stable and rising. The second critical state is associated with Q_2 and is rising, as illustrated in Figure 14.6.b. The tertiary path has displacement components in both coordinates and is descending, and is thus unstable. It intersects the lowest uncoupled path at approximately $Q_1 = 0.4$ and does not have any point in common with the second uncoupled mode. A plot of Q_1 versus Q_2 displays a nonlinear relation (a hyperbola) showing that the coupled path is three dimensional (not contained in a plane). For this perfect system, the maximum load that the system can reach is the secondary bifurcation at approximately $\Lambda = 1.05$.

Next, in Figure 14.7 we consider the results for $\tau = 1$, for which the two modes are coincident. The secondary bifurcations also coincide with the primary bifurcations; thus four paths emerge from the critical state $\Lambda = 1$. The uncoupled secondary paths have the same curvature, and the coupled paths also have the same curvature between them. The tertiary paths are contained in planes at $\pm 45°$ from the coordinate planes (Figure 14.7.c). The maximum load that the system can reach is the coincident critical state $\Lambda = 1$.

For this very simple two-degrees-of-freedom problem it was possible to compute all the equilibrium paths in closed form. However, we continue with the analysis to show the energy transformation in terms of uncoupled modes.

Stability conditions. The second derivatives of the energy result in

$$V_{11} = W_{11} = A_{11} + \frac{1}{2}A_{1111}Q_1^2 + \frac{1}{2}A_{1122}Q_2^2, \tag{14.55}$$

$$V_{22} = W_{22} = A_{22} + \frac{1}{2}A_{2222}Q_2^2 + \frac{1}{2}A_{1122}Q_1^2,$$

$$V_{12} = W_{12} = \frac{2}{3}A_{1122}Q_1Q_2.$$

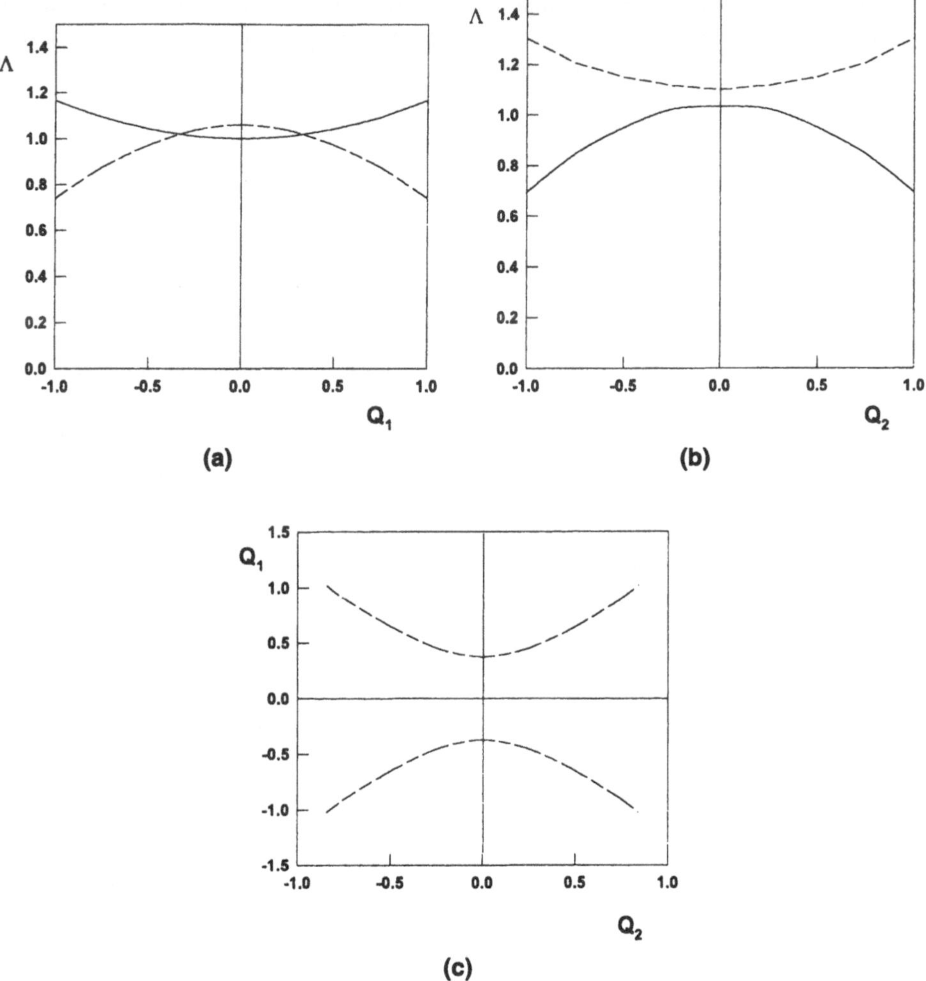

Figure 14.6 The Augusti column: Coupled paths for noncoincident modes for $\tau = 1.1$.

The critical states along the fundamental path can be evaluated using (14.55) as

$$\det(V_{ij}\,[Q_1 = 0, Q_2 = 0, \Lambda]) = 0,$$

from which we get

$$\det\begin{bmatrix} A_{11} & 0 \\ 0 & A_{22} \end{bmatrix} = 0.$$

A critical state arises when $A_{11} = 0$, leading to

$$^{1}\Lambda_c = 1, \qquad {}^{1}x_j = \{1 \quad 0\}. \tag{14.56}$$

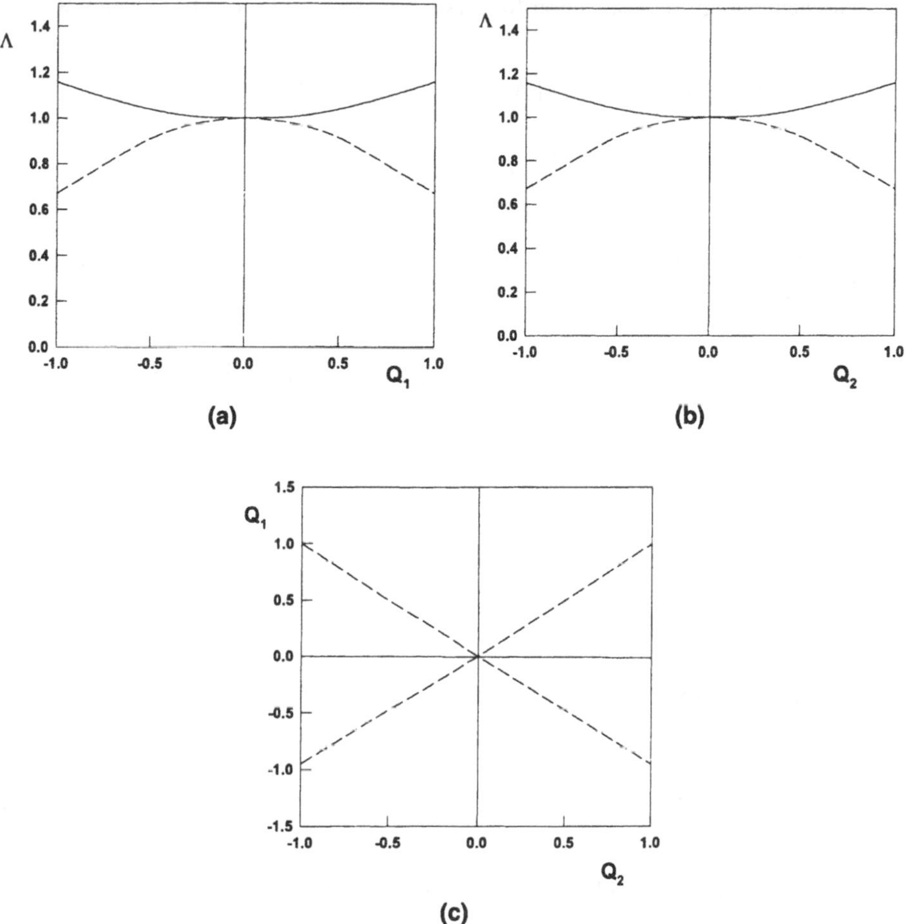

Figure 14.7 The Augusti column: Coupled paths for coincident modes for $\tau = 1.0$.

Another condition can be used: $A_{22} = 0$ and yields

$$^2\Lambda_c = \tau, \qquad\qquad ^2x_j = \{0 \quad 1\}. \tag{14.57}$$

Notice that for $\tau = 1$ (i.e., the rotational springs have the same stiffness) the two modes become coincident, and $^1\Lambda_c =^2 \Lambda_c$. We continue the analysis to investigate coupling between the two modes described by (14.56) and (14.57) and assume $\tau \geq 1$ so that $\left|^2\Lambda_c\right| \geq \left|^1\Lambda_c\right|$.

Transformation of the energy. The displacement field in the coupled mode results in

$$q_j = \xi_1 \left(^1x_j\right) + \xi_2 \left(^2x_j\right). \tag{14.58}$$

The new energy representation follows from substitution of (14.58) into W:

$$W[\xi_1, \xi_2, \Lambda] = -\frac{1}{2}\left(\Lambda - {}^1\Lambda_c\right)\xi_1^2 - \frac{1}{2}\left(\Lambda - {}^2\Lambda_c\right)\xi_2^2$$

$$+\frac{1}{4!}\left[A_{1111}\xi_1^4 + 6A_{1122}\xi_1^2\xi_2^2 + A_{2222}\xi_2^4\right].$$

Because all powers in the mode amplitudes are even, the energy results in a doubly symmetric form. We write

$$W[\xi_1, \xi_2, \Lambda] = -\frac{1}{2}\left(\Lambda - {}^1\Lambda_c\right)\xi_1^2 - \frac{1}{2}\left(\Lambda - {}^2\Lambda_c\right)\xi_2^2 \tag{14.59}$$

$$+\left(c_{11}\xi_1^4 + 2c_{12}\xi_1^2\xi_2^2 + c_{22}\xi_2^4\right)$$

with

$$c_{11} = c_{22} = \frac{A_{1111}}{24} = \frac{1}{24}K_1\Lambda, \qquad c_{12} = \frac{A_{1122}}{8} = -\frac{1}{8}K_1\Lambda. \tag{14.60}$$

Secondary bifurcations and tertiary paths. We can write the lowest secondary path in the plane $\Lambda - \xi_1$ as

$$\Lambda = {}^1\Lambda + 4c_{11}\xi_1^2 = {}^1\Lambda + \frac{A_{1111}}{6}\xi_1^2. \tag{14.61}$$

The other secondary path develops in the plane $\Lambda - \xi_2$:

$$\Lambda = {}^2\Lambda + 4c_{22}\xi_2^2 = {}^2\Lambda + \frac{A_{2222}}{6}\xi_2^2. \tag{14.62}$$

To obtain the secondary bifurcations along the uncoupled secondary path, we evaluate derivatives of W in the form

$$W_{11} = -\left(\Lambda - {}^1\Lambda_c\right) + 12c_{11}\xi_1^2 + 4c_{12}\xi_2^2,$$

$$W_{22} = -\left(\Lambda - {}^2\Lambda_c\right) + 12c_{22}\xi_2^2 + 4c_{12}\xi_1^2,$$

$$W_{12} = 4c_{12}\xi_1\xi_2$$

and compute

$$\det\left(W_{ij}\right) = 0$$

together with (14.61). This yields

$$\xi_1^c = \pm\frac{1}{2}\sqrt{\frac{\left({}^2\Lambda - {}^1\Lambda_c\right)}{c_{11} + c_{22}}}$$

or else

$$\xi_1^c = \pm\frac{1}{2}\sqrt{\frac{\tau - 1}{6\left(A_{1111} + 6A_{1122}\right)}}. \tag{14.63}$$

There are two roots in (14.63); this means that there are two critical states along the secondary path. Such states are symmetrical, as shown in Figure 14.6.

The tertiary paths can be obtained from the equilibrium conditions in exact form:

$$W_1\left[\xi_1, \xi_2, \Lambda\right] = -\left(\Lambda - {}^1\Lambda_c\right) + 4\left(c_{11}\xi_1^2 + c_{12}\xi_2^2\right) = 0, \tag{14.64}$$

$$W_2\left[\xi_1, \xi_2, \Lambda\right] = -\left(\Lambda - {}^2\Lambda_c\right) + 4\left(c_{22}\xi_2^2 + c_{12}\xi_1^2\right) = 0,$$

where $\xi_1 \neq 0$ and $\xi_2 \neq 0$.

The tertiary path has been computed in Figure 14.6 and descends from the secondary bifurcation. The important consequence of mode coupling in this case is that new paths emerge that are unstable, although the secondary paths were stable. The maximum load that the system can attain is given by the secondary bifurcation load. We shall see in Chapter 15 that this system is highly sensitive to initial imperfections, so that such a load cannot be reached and the column fails at a smaller load.

Next we consider the case in which $\tau = 0$. The secondary bifurcations occur for $\xi_1^c = 0$, that is, the primary critical point. Not only do the two primary bifurcations occur for the same load level, but also the primary bifurcation and the secondary bifurcations are coincident.

The maximum load that can be obtained from this column is the first critical load. However, imperfections reduce the load even further, and this is discussed in Chapter 15.

14.9 A THIN-WALLED COLUMN

Thin-walled columns under axial load have interactive buckling only by design. The basic problem is interaction between global modes, involving lateral displacements like an Euler mode and local modes with short waves in the longitudinal direction. The two modes can be set to be coincident by appropriate selection of length and cross-sectional properties of the column. Years ago such a design with coincident modes was thought to be a good choice, because there was no material employed to make one mode stronger than the other, and an optimization of this problem leads to a coupled mode solution.

Examples of interactive buckling in I-section columns under axial load have been presented in the literature for isotropic materials. A simplified version of the problem was presented by van der Neut [41] in which the entire buckling resistance of the cross section was provided by the flanges. Each of the two flanges can display local buckling independently of the other, and this leads to a strong interaction problem. A more refined model of interaction of a local and a global mode was considered in [34, 6, 7] by assuming a local mode (similar to mode 2 in this section) and a global mode, and by means of a second-order field they found a new mode containing only displacements in the flanges (mode 3 in this section). However, in [33] it was found that the previous solution had an error of unknown magnitude; thus it was proposed to include the mode arising from the second-order field as one of the principal modes participating in the analysis. Hence, the problem solved in [33] was interaction between a global, a primary local, and a secondary local mode, in which only first-order fields were included. The two-mode interaction of a primary and a secondary local mode was studied in [5] for a composite column to show that if the modes are doubly symmetric, then the second-order contributions vanish and the first-order interaction leads to stable coupling of the modes, which is incorrect.

The symmetric bending of an I-beam was studied in [15] again using a first-order analysis with three modes and also included the influence of second-order fields. Columns made of composite materials were investigated by the author and coworkers in [14, 28], which form the basis of this section. Other cross sections have been considered in [21, 22, 23].

Geometry and displacements. A simplified analysis can be performed in this case using a Ritz procedure. The displacement field of the web is represented from [14] in terms of an axial shortening q_1, a rotation of the flanges q_3, a lateral displacement of

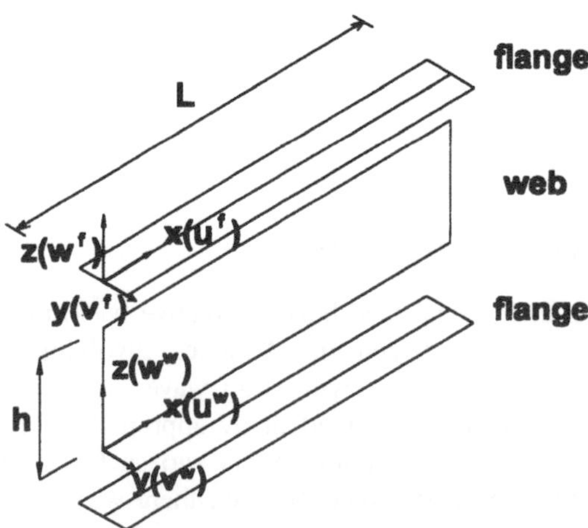

Figure 14.8 Thin-walled column: Convention for geometry and displacements in the flanges and in the web.

the column in a global mode q_2, and a bending of the flanges q_4. These degrees of freedom are represented in Figures 14.8 and 14.10, and the displacements result in

$$w^{web} = q_2 \sin\left(\frac{\pi x}{L}\right) + h\left(\frac{y}{h} - \frac{y^2}{h^2}\right)\left(q_3 - \frac{q_3^3}{6}\right)\sin\left(\frac{n\pi x}{L}\right), \tag{14.65}$$

$$u^{web} = q_1\left(\frac{x}{L}\right), \qquad v^{web} = -\frac{\nu}{L}\left|y - \frac{h}{2}\right| q_1.$$

For the flanges we employ the following representation:

$$u^{flange} = \left(\frac{x}{L}\right)q_1 - y\frac{\pi}{L}\cos\left(\frac{\pi x}{L}\right)q_2, \tag{14.66}$$

$$v^{flange} = \sin\left(\frac{\pi x}{L}\right)q_2 - \frac{1}{2}y\sin^2\left(\frac{n\pi x}{L}\right)q_3^2 - \nu\left[\frac{y}{L}q_1 + \frac{1}{2}\left(\frac{\pi y}{L}\right)^2\sin\left(\frac{\pi x}{L}\right)q_2\right],$$

$$w^{flange} = y\sin\left(\frac{n\pi x}{L}\right)\left(q_3 - \frac{q_3^3}{6}\right) - \frac{h}{12}\sin^2\left(\frac{n\pi x}{L}\right)\left(q_3^2 - \frac{q_3^4}{3}\right) + \left(\frac{2y}{b}\right)\sin\left(\frac{n\pi x}{L}\right)q_4.$$

q4 δ δ q5 q2

Displacements of the flange

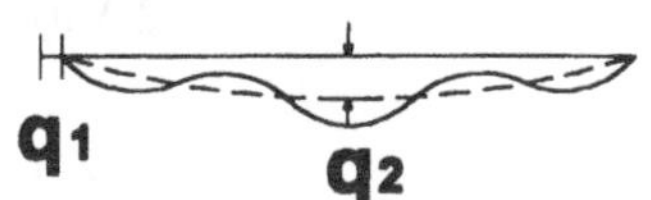

Displacements of the center of the web

Figure 14.9 Local and global modes considered. Reprinted from [14] with permission of Technomic Publishing Co., Inc.

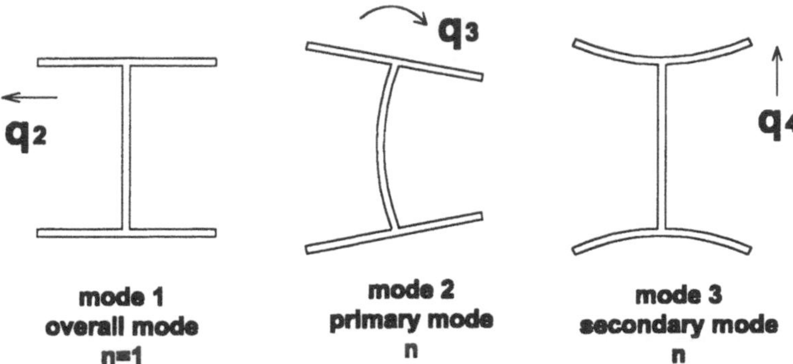

Figure 14.10 Degrees of freedom for the displacement field assumed in the thin-walled column. Reprinted from [14] with permission of Technomic Publishing Co., Inc.

Details of the derivation of the above field may be found in [14] and include the conditions of compatibility of displacements along the joints between flanges and web, compatibility of rotations, and transverse expansion of the flanges and the web. Poisson's modulus ν is present in the displacement definition because of the transverse expansion of the column. The generalized coordinate q_1 is necessary to model the primary path, in which end shortening occurs. The coordinate q_2 is the lateral displacement and models a global buckling mode. The coordinate q_3 is associated with a local buckling mode. Finally, it is necessary to include another mode q_4 into the analysis to allow for interaction. A similar analysis is discussed in [15]. The cross-sectional deflections for the global and local modes considered are shown in Figures 14.9 and 14.10. If only q_1, q_2, and q_3 are included in the model, it has been shown that interaction leading to a descending path is not modeled [5].

A study concerning what terms are required in the nonlinear analysis was done in [7]. The kinematic expressions become

$$\varepsilon_{11} = \frac{\partial u}{\partial x} + \frac{1}{2}\left(\frac{\partial v}{\partial x}\right)^2 + \frac{1}{2}\left(\frac{\partial w}{\partial x}\right)^2, \tag{14.67}$$

$$\varepsilon_{22} = \frac{\partial v}{\partial y} + \frac{1}{2}\left(\frac{\partial u}{\partial y}\right)^2 + \frac{1}{2}\left(\frac{\partial w}{\partial y}\right)^2,$$

$$\varepsilon_{12} = \frac{1}{2}\left(\frac{\partial u}{\partial y} + \frac{\partial v}{\partial x} + \frac{\partial w}{\partial x}\frac{\partial w}{\partial y}\right),$$

$$\chi_{11} = -\frac{\partial^2 w}{\partial x^2}, \tag{14.68}$$

$$\chi_{22} = -\frac{\partial^2 w}{\partial y^2},$$

$$\chi_{12} = -\frac{\partial^2 w}{\partial x \partial y}.$$

For a composite laminated plate, the constitutive equations are given by classical lamination theory.

Total potential energy. The strain energy of the web can be written as the sum of the membrane and bending contributions

$$U^{web} = U^{web}_{membrane} + U^{web}_{bending}, \tag{14.69}$$

where the membrane contribution is given by

$$U^{web}_{membrane} = \frac{1}{2}\int_{x=0}^{L}\int_{y=0}^{h} (N_{11}\varepsilon_{11} + N_{22}\varepsilon_{22} + 2N_{12}\varepsilon_{12})^{web}\, dx\, dy, \tag{14.70}$$

while the bending part results in

$$U^{web}_{bending} = \frac{1}{2}\int_{x=0}^{L}\int_{y=0}^{h} (M_{11}\chi_{11} + M_{22}\chi_{22} + 2M_{12}\chi_{12})^{web}\, dx\, dy. \tag{14.71}$$

A similar derivation for the flanges leads to

$$U^{flanges} = 2\left(U^{flange}_{membrane} + U^{flange}_{bending}\right), \tag{14.72}$$

$$U^{flange}_{membrane} = \frac{1}{2}\int_{x=0}^{L}\int_{y=-b/2}^{b/2} (N_{11}\varepsilon_{11} + N_{22}\varepsilon_{22} + 2N_{12}\varepsilon_{12})^{flange}\, dx\, dy, \tag{14.73}$$

$$U^{flange}_{bending} = \frac{1}{2}\int_{x=0}^{L}\int_{y=-b/2}^{b/2} (M_{11}\chi_{11} + M_{22}\chi_{22} + 2M_{12}\chi_{12})^{flange}\, dx\, dy. \tag{14.74}$$

Finally, the total potential energy of the column is given by

$$V = U^{web} + U^{flanges} - Pq_1. \tag{14.75}$$

The explicit form of the energy in terms of energy coefficients is more complex in this problem and is not given here for the sake of brevity. However, it can be easily reproduced using a symbolic manipulator and the above equations.

Fundamental path and critical states. With the energy defined as in (14.75), one can get the equilibrium conditions and the displacements along a linear fundamental path

$$Q_j^F = \begin{Bmatrix} \bar{Q}_1 \\ 0 \\ 0 \\ 0 \end{Bmatrix}, \tag{14.76}$$

where only the first degree of freedom (i.e., the axial shortening) is nonzero. This is a nontrivial fundamental path and results in the linear form

$$q_1 = \Lambda \bar{Q}_1. \tag{14.77}$$

The critical state is given by the eigenvalue problem

$$\left(A_{ij} + \Lambda^c A_{ijk}\, \bar{Q}_k\right)^s x_j = 0. \tag{14.78}$$

Solution of (14.78) yields the following eigenvalues:

$$^1\Lambda^c = -\frac{A_{22}}{A_{221}\bar{Q}_1}, \qquad {}^2\Lambda^c = -\frac{A_{33}}{A_{331}\bar{Q}_1}, \qquad {}^3\Lambda^c = -\frac{A_{44}}{A_{441}\bar{Q}_1}. \tag{14.79}$$

All eigenvalues are associated with stable symmetric bifurcation, so that the post-buckling paths are rising with the load parameter. The first two eigenvalues are a primary local mode, a global mode, and a secondary local mode.

Following the theory developed in this chapter, we normalize the eigenvectors as

$$A_{ijk}\, \bar{Q}_k \left({}^s x_i\right)\left({}^t x_j\right) = \delta_{st}. \tag{14.80}$$

The explicit form of the three eigenvectors is

$$^1\mathbf{x} = \begin{Bmatrix} 0 \\ \left(A_{221}\bar{Q}_1\right)^{-1/2} \\ 0 \\ 0 \end{Bmatrix}, \qquad {}^2\mathbf{x} = \begin{Bmatrix} 0 \\ 0 \\ \left(A_{331}\bar{Q}_1\right)^{-1/2} \\ 0 \end{Bmatrix}, \tag{14.81}$$

$$^3\mathbf{x} = \begin{Bmatrix} 0 \\ 0 \\ 0 \\ \left(A_{441}\bar{Q}_1\right)^{-1/2} \end{Bmatrix}. \tag{14.82}$$

Notice that in this case the eigenvectors have the same direction of the generalized coordinates (similar to the Augusti column); that is, ξ_1 is coincident with q_3, ξ_2 is the same as q_2, and ξ_3 is coincident with q_4.

Coupled postcritical paths. The coupled postcritical path is written as a linear combination of three modes to perform a first-order interaction analysis

$$q_i = \left({}^1x\right)\xi_1 + \left({}^2x\right)\xi_2 + \left({}^3x\right)\xi_3. \tag{14.83}$$

The total potential energy can now be written in terms of the mode amplitudes

$$W[\xi_1\ \xi_2\ \xi_3\ \Lambda] = \frac{1}{2}\left[\left(\Lambda - {}^{1}\Lambda^c\right)\xi_1^2 + \left(\Lambda - {}^{2}\Lambda^c\right)\xi_2^2 + \left(\Lambda - {}^{3}\Lambda^c\right)\xi_3^2\right] \quad (14.84)$$

$$+\left(\alpha_{123}\xi_1\ \xi_2\ \xi_3 + \alpha_{223}\ \xi_2^2\ \xi_3\right)$$

$$+(\beta_{1111}\xi_1^4 + \beta_{2222}\xi_2^4 + \beta_{3333}\xi_3^4$$

$$+\beta_{1122}\xi_1^2\xi_2^2 + \beta_{1133}\xi_1^2\xi_3^2 + \beta_{2233}\xi_2^2\xi_3^2), \quad (14.85)$$

where

$$\alpha_{123} = A_{234}\left({}^{1}x_2\right)\left({}^{2}x_3\right)\left({}^{3}x_4\right), \quad (14.86)$$

$$\alpha_{223} = \frac{1}{2}\left(A_{334} + \Lambda A_{1334}\bar{Q}_1\right)\left({}^{2}x_3\right)^2\left({}^{3}x_4\right), \quad (14.87)$$

and

$$\beta_{1111} = \frac{1}{24}A_{2222}\left({}^{1}x_2\right)^4, \qquad \beta_{2222} = \frac{1}{24}A_{3333}\left({}^{2}x_3\right)^4, \quad (14.88)$$

$$\beta_{3333} = \frac{1}{24}A_{4444}\left({}^{3}x_4\right)^4, \qquad \beta_{1122} = \frac{1}{4}A_{2233}\left({}^{1}x_2\right)^2\left({}^{2}x_3\right)^2,$$

$$\beta_{2233} = \frac{1}{4}A_{3344}\left({}^{3}x_4\right)^2\left({}^{2}x_3\right)^2, \qquad \beta_{1133} = \frac{1}{4}A_{2244}\left({}^{1}x_2\right)^2\left({}^{3}x_4\right)^2.$$

The only coefficient that depends on the load parameter is α_{223}.

Next we examine the properties of symmetry of the energy W and find that

$$W[-\xi_1,\ -\xi_2,\ \xi_3,\ \Lambda] = W[\xi_1,\ \xi_2,\ \xi_3,\ \Lambda], \quad (14.89)$$

$$W[-\xi_1,\ \xi_2,\ -\xi_3,\ \Lambda] \neq W[\xi_1,\ \xi_2,\ \xi_3,\ \Lambda],$$

$$W[\xi_1,\ -\xi_2,\ -\xi_3,\ \Lambda] \neq W[\xi_1,\ \xi_2,\ \xi_3,\ \Lambda].$$

This means that the energy is not symmetric with respect to all modes, but only with respect to the first two modes.

The three equilibrium conditions are next written as

$$W_1[\xi_1,\ \xi_2,\ \xi_3,\ \Lambda] = \left[\left(\Lambda - {}^{1}\Lambda^c\right)\xi_1\right] + (b_{123}\ \xi_2\ \xi_3)$$

$$+\left(4c_{1111}\xi_1^3 + 2c_{1122}\xi_1\xi_2^2 + 2c_{1133}\xi_1\xi_3^2\right)$$

$$= 0,$$

$$W_2\,[\xi_1\ \xi_2\ \xi_3\ \Lambda] = \left[\left(\Lambda -^2 \Lambda^c\right)\xi_2\right] + (b_{123}\ \xi_1\ \xi_3 + 2b_{223}\xi_2\ \xi_3) \qquad (14.90)$$

$$+\left(4c_{2222}\xi_2^3 + 2c_{1122}\xi_1^2\xi_2 + 2c_{2233}\xi_2\xi_3^2\right)$$

$$= 0,$$

$$W_3\,[\xi_1\ \xi_2\ \xi_3\ \Lambda] = \left[\left(\Lambda -^3 \Lambda^c\right)\xi_3\right] + \left(b_{123}\ \xi_1\ \xi_2 + b_{223}\xi_2^2\right)$$

$$+\left(4c_{3333}\xi_3^3 + 2c_{1133}\xi_1^2\xi_3 + 2c_{2233}\xi_2^2\xi_3\right)$$

$$= 0.$$

A quadratic approximation of (14.90) can be obtained by neglecting cubic terms and results in the form

$$W_1\,[\xi_1\ \xi_2\ \xi_3\ \Lambda] = \left[\left(\Lambda -^1 \Lambda^c\right)\xi_1\right] + (\alpha_{123}\ \xi_2\ \xi_3) = 0, \qquad (14.91)$$

$$W_2\,[\xi_1\ \xi_2\ \xi_3\ \Lambda] = \left[\left(\Lambda -^2 \Lambda^c\right)\xi_2\right] + (\alpha_{123}\ \xi_1\ \xi_3 + 2\alpha_{223}\xi_2\ \xi_3) = 0,$$

$$W_3\,[\xi_1\ \xi_2\ \xi_3\ \Lambda] = \left[\left(\Lambda -^3 \Lambda^c\right)\xi_3\right] + \left(b_{123}\ \xi_1\ \xi_2 + \alpha_{223}\xi_2^2\right) = 0.$$

Equation (14.91) can be solved explicitly, and details are given in [14].

The coupled path results as a consequence of the interaction between two local modes in the form

$$q_j^{coupled} = \begin{Bmatrix} 0 \\ 0 \\ \left({}^2x_3\right)\xi_2 \\ \left({}^3x_4\right)\xi_3 \end{Bmatrix}. \qquad (14.92)$$

Numerical results. To illustrate the behavior of an I-section composite column, a specific geometry with $b = h$ and $L = 16.67b$ is reported in [14]. The analytical results computed from the expressions of this section have been checked by a finite element analysis based on the theory of this chapter and show good agreement. One difference in the model is that the finite element discretization allows for "amplitude modulation"; that is, the amplitude of the displacements is allowed to decrease near the edges of the column. A second source of difference is the boundary conditions: simply supported conditions are satisfied by the present simplified model, while in a finite element solution it is possible to have deformations away from the Bernoulli assumptions.

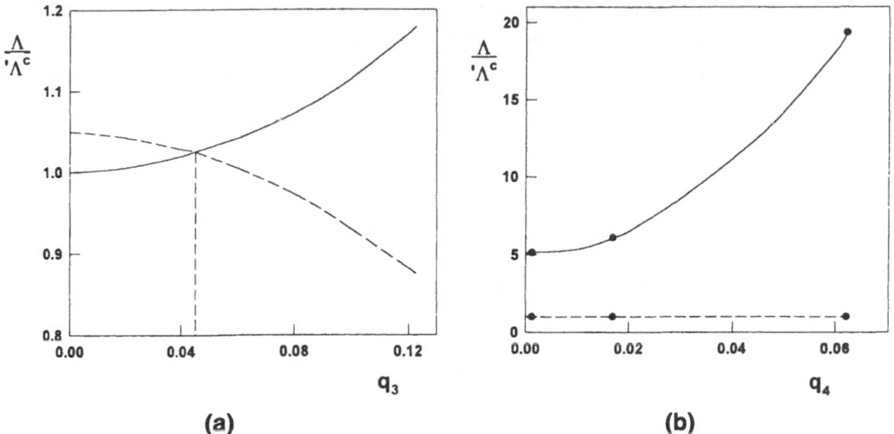

Figure 14.11 Secondary and tertiary paths for an I-section with $b = h$. (a) Coupled path and primary local mode; (b) coupled path and secondary local mode. Reprinted from [14] with permission of Technomic Publishing Co., Inc.

The length is chosen so as to achieve interaction between modes. Figure 14.11.a shows the uncoupled rising equilibrium path and the projection of the coupled path in terms of the amplitude of the primary local mode. Similar to what we observed in the Augusti column, we find that for small rotations in the uncoupled mode the rising secondary path is below the falling tertiary path. A secondary bifurcation along the stable path occurs for a load Λ slightly above ${}^1\Lambda_c$ and $q_3 = 0.046$. The tertiary path dominates the scene for larger rotations.

The behavior in terms of the secondary local mode q_4 is reflected in Figure 14.11.b. The drawing is similar to the Augusti column in the sense that the secondary uncoupled and the tertiary paths do not intersect. The tertiary path seems flat in the $\Lambda - q_4$ plane simply because of the scale.

The material properties are very important in the interaction process of the I-column. This can also be used to avoid the occurrence of interaction for a given column length. This is one of the advantages of composite materials, and would not be possible in homogeneous sections, for which the only solution is to change the length to avoid mode interaction.

14.10 THE CYLINDRICAL SHELL UNDER AXIAL LOAD

Under axial load, the cylindrical shell has many modes for which the critical load is the same. This led to the so-called Koiter circle, of which many modes may be involved in interactions at the same time. This has been a challenging problem for over 50 years, and new contributions to understand the complex behavior of such structural system are still being published.

What we report in this section are results obtained from [3, 2]. The kinematic relations employed are the same discussed in Chapter 13 and attributed to Donnell.

The critical states were computed using the following shapes, as in Chapter 13:

$$w\,(n, m) = hq_{nm} \cos\left(\frac{ny}{R}\right) \sin\left(\frac{m\pi x}{L}\right),$$

where n is the number of half waves in the circumferential direction, while m is the number of half waves in the meridional direction of the cylinder.

We mentioned before that the lowest critical load is given by

$$N^c = \frac{Eh^2}{cR} \qquad \text{with} \qquad c = \sqrt{3\left(1 - \nu^2\right)}.$$

But there are many modes for which this is the critical load. Those modes satisfy the condition known as Koiter's circle

$$\alpha^2 + \beta^2 - \alpha = 0,$$

where

$$\alpha^2 = \frac{Rh}{2cL^2} m^2, \qquad \beta^2 = \frac{h}{2cR} n^2.$$

It is interesting to notice that Koiter's circle contains one axisymmetric mode, given by

$$w\,(0, m) = hq_{0m} \sin\left(\frac{m\pi x}{L}\right).$$

For example, a metal cylindrical shell with $R/h = 500$, $L/R = 1$, and $h = 1$ has 13 coincident modes. One of them contains 13 half waves in the meridional direction and zero in the circumferential direction, that is, an axisymmetric mode. The other 12 modes are nonsymmetric in the sense that they also involve circumferential waves.

Mode coupling. Which modes do we combine in an interaction analysis? One answer is to combine those modes of the Koiter's circle. Another answer is to combine all modes below a certain critical load level. Otherwise, we could consider all modes included in the analysis.

There are limitations in terms of the number of modes that can be considered for interaction, since they lead to nonlinear equations. We would like to keep this number as small as possible, and to do that we need to identify which modes lead to strong interactions. This is a difficult question and requires some physical interpretation about how the interaction between stable modes produces unstable paths.

A criteria to identify the modes included in the analysis of interaction is as follows: We investigate a certain mode $w(n, m)$ and evaluate the stress resultants that provide equilibrium. Then we search for modes that produce a similar stress distribution, so that they can cancel the contribution of the first mode considered when they couple with opposite signs.

Donnell [12] was perhaps the first to identify a coupling mechanism between a generic mode in the circle, let us say $w(n, m)$, and a symmetric mode $w(0, 2m)$ with twice the number of axial waves. Thus, he considered

$$w = \xi_1 \cos\left(\frac{n}{R}y\right) \sin\left(\frac{m\pi}{L}x\right) + \xi_2 \cos\left(2\frac{m\pi}{L}x\right).$$

Due to the waviness of the mode $w(n, m)$ there are some circles in the circumference that increase their length, while a decrease occurs in others (Figure 14.12). This generates a pattern of tension and compression in the circumferential stress resultant N_{22}. A second mode $w(0, 2m)$ induces again a pattern of hoop stress resultants that cancel those required by the shell for equilibrium. There is an erosion in the capacity of the shell to provide equilibrium. This mechanism has been further discussed by Batista and Croll [4] and extended to other load cases in the cylindrical shell.

Four modes were considered in [3]

$$w = w(n, m) + w(0, 2m) + w(2n, 0) + w(2n, 2m)$$

or else

$$w = \xi_1 \cos\left(\frac{n}{R}y\right) \cos\left(\frac{m\pi}{L}x\right) + \xi_2 \cos\left(2\frac{m\pi}{L}x\right) \tag{14.93}$$

$$+\xi_3 \cos\left(2\frac{n}{R}y\right) + \xi_4 \cos\left(2\frac{n}{R}y\right) \cos\left(2\frac{m\pi}{L}x\right),$$

where the amplitudes ξ_k are normalized with respect to the thickness of the shell.

Notice that not all modes in (14.93) are present in Koiter's circle.

The plots of the coupled paths become more complicated as the number of interacting modes increases. In the present case there are four interacting modes, so that

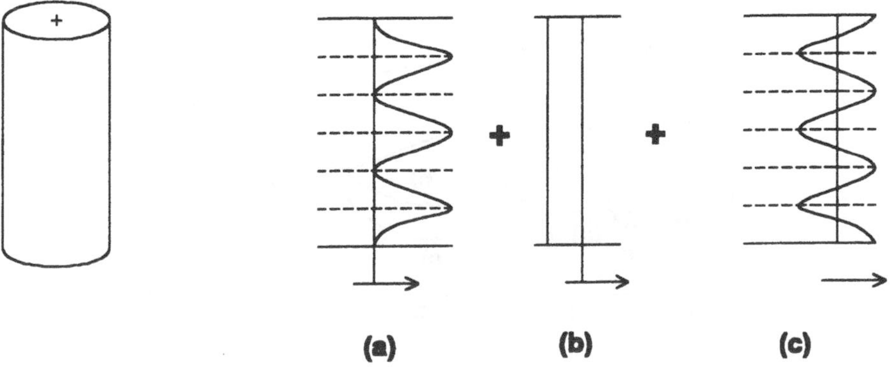

Figure 14.12 Physical interpretation of modes that couple in a cylindrical shell under axial load (a) nonlinear hoop tension; (b) linear constant tension field; (c) linear hoop stress in mode $m = 2j$, $n = 0$. From [3].

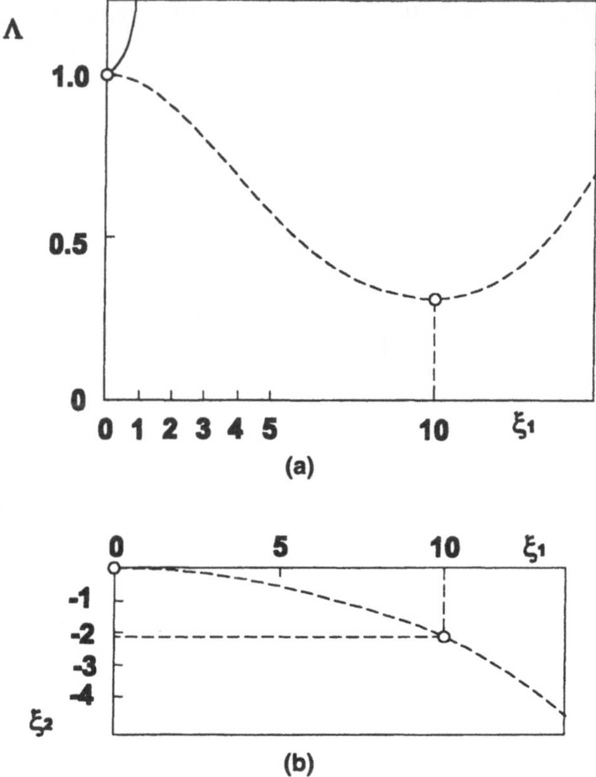

Figure 14.13 Coupled mode in a metal cylinder under axial load, with $R/h = 1000$, $L/R = 1$. The basic mode considered has $m = 1$ and $n = 13$. (a) Load versus amplitude in mode 1, (b) equilibrium in terms of modes 1 and 2. From [3].

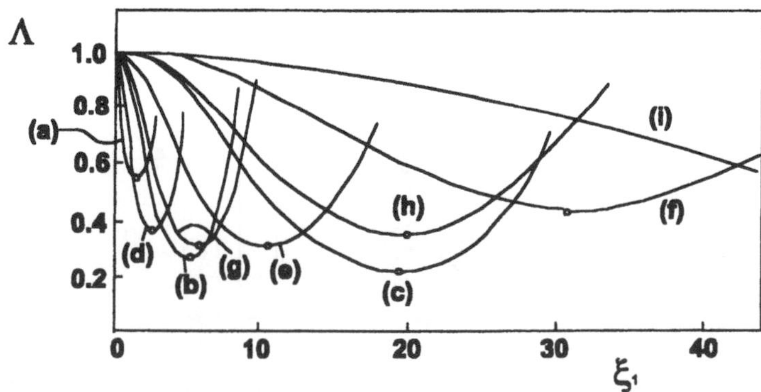

Figure 14.14 Coupled mode in a metal cylinder under axial load, with $R/h = 200$ to 5000, $L/R = 0.25$ to 4. The basic mode considered has $m = 1$ and $n = 13$. Each curve is identified in Table 14.2. From [3].

Table 14.2 Identification of cases studied in Figure 14.14

Curve	*R/h*	*L/R*
(*a*)	200	1/4
(*b*)	200	1
(*c*)	200	4
(*d*)	1000	1/4
(*e*)	1000	1
(*f*)	1000	4
(*g*)	5000	1/4
(*h*)	5000	1
(*i*)	5000	4

only projections of equilibrium paths can be represented. In Figure 14.13.a we show the isolated mode $w(13, 1)$ for a shell with $R/h = 1000$, $L/R = 1$, $\nu = 0.3$. The isolated modes have a stable behavior; however, the coupled modes become unstable and descend as ξ_1 increases. There is a minimum value below $\Lambda = 0.4\ {}^1\Lambda_c$. For $\xi_1 > 10$ the coupled path rises. The same path is also shown in Figure 14.13.b in the coordinate plane $\xi_1 - \xi_2$, the two modes producing strong interaction. Clearly the path has components in both modes, and the coupling is stronger as deformations increase. But the largest deflections occur in the mode ξ_1. For example, at the minimum load the deflections are $\xi_1 = 10$ and $\xi_2 = 2.2$.

How the geometry of the shell influences the coupled path is illustrated in Figure 14.14. The geometry of each case studied is indicated in Table 14.2. Thinner shells have a lower minimum postcritical coupled state; and this minimum is also lower for small values of the ratio L/R (short cylinders).

14.11 FINAL REMARKS

In this chapter we presented an approach in which a few modes are included in the interactive analysis, and the rest are taken as passive. However, there is no guarantee that the passive modes are not essential to model the coupling behavior, and because of that the analysis is improved by means of a search of new modes emerging from the interaction of the active modes.

The new mode $({}^{st}x_i)$ discussed in section 14.5 can also be written as a linear combination of **all** the eigenvectors of the problem, but in most cases it is a combination of only **a few** of them. The coupled mode could even be just one of the eigenvectors that was not included in the linear part of the analysis. Thus, the quadratic analysis can be seen as a way to improve a linear interaction analysis, by means of identification of the set of passive modes that should be considered as active. Notice that if all the relevant active modes have been properly identified, the analysis can be carried out by means of linear interaction analysis, i.e., (14.7).

We discussed just one approach to investigate mode interaction in the stability of structure. An alternative approach is presented in [39, 16], where the generalized coordinates of the problem are split into active and passive coordinates. A clear difference is that in this chapter we employed active modes, while in the approach of [39, 16] what are active or passive are the generalized coordinates. Elimination of passive coordinates is a key point in this second analysis: They are expanded in terms of the active coordinates using perturbation techniques and with the help of equilibrium conditions. Then the results are substituted back into the potential energy W of the system. The new transformed energy depends only on active coordinates.

In all algorithms of solution for interactive buckling, the nonlinear equations to be solved lead to a number of paths. Bounds to the number of solutions may be found using the theorem of Bezout, and this is discussed, for example, in [37]. Structures with a small number of degrees of freedom may have a large number of possible paths. For example, a structure with 2 degrees of freedom has a maximum number of postbuckling paths of 4 (as in the Augusti column); if the degrees of freedom are 3, the possible paths are between 7 and 13, while if there are 6 active degrees of freedom, the maximum number of paths could perhaps be up to 63.

Instead of using a theoretical formulation based on the energy approach, as in the present chapter and many other works in this field, one can employ a numerical approach [8]. The structural system may be modeled by means of finite elements with nonlinear kinematic relations, and the equilibrium paths followed using continuation techniques. There are, however, difficulties in such an approach, as mentioned in [37]:

> The existence of a multiplicity of equilibrium paths at or near the critical load level has far reaching consequences for numerical non-linear stability analysis. Numerical methods which converge to an equilibrium solution from an approximately predicted solution using an iterative process may experience difficulty when there are many equilibrium solutions in the region of the initial guess. We may say that such situations would be ill-conditioned for numerical investigation of this type.

For a structure having several buckling modes that are linearly independent, conventional optimization of this structure yields a solution in which the critical load due to each buckling mode is the same. In this way, there is a compound (not distinct) bifurcation buckling problem via an optimization of the design. Such an optimum solution would not consider the possibility of mode interaction and has been repeatedly called ''the naive approach'' to optimum design in buckling. This illustrates that the optimum of a problem may generate a solution of higher complexity than the original problem itself. In this case the optimization applied to problems with a singularity of order one in the second derivatives of the energy yields another problem with singularity of order two or higher.

The list of problems with interaction between competing modes in structural mechanics exceeds by far the few cases considered in this chapter. Most notably, the spherical shell under external pressure has a behavior similar to the cylindrical shell under axial load. Cylindrical shells with axial stiffeners, under axial load, have been investigated, for example, in [9, 17]. Stiffened panels were considered in [20]. Simply supported channels under uniform compression are studied in [27]. A column under special simplified conditions was reported in [41], while other column problems are treated in [19, 38].

The concepts of mode interaction originated in the field of stability of structures, as discussed in this chapter, but there are many fields in science where this is also relevant. A sociological study of catastrophic failures in high-tech industries [26], such as nuclear power, petrochemical, and aerospace industries, was reported some time ago. In such industries the transformation system is nonlinear and displays interactions as in our buckling mode interaction analysis. Again, we find the same concepts by which interactions can occur because of design of the plants, or because it is inherent to the process (as in nuclear power plants). "Complex interactions are those in which one component can interact with one or more other components outside the normal production sequence, either by design or not by design" [26, p. 77]. The above study shows a number of examples of accidents in innovative as well as in traditional industries where this is valid. The overall conclusions about interaction in this case are negative: "Given that the very nature of the transformation system requires non-linear interactions and even tighter coupling, the chances of system accidents no doubt will increase" [26, p. 121].

14.12 PROBLEMS

Review questions. (*a*) Discuss different possibilities to have coincident modes. (*b*) Is mode coincidence the same as mode interaction? (*c*) What are the consequences of mode interaction? (*d*) Give examples of systems that exhibit interaction between modes. (*e*) Describe how mode interaction occurs in a spherical shell. (*f*) Why do new modes emerge from interaction of two modes? (*g*) What are passive and active modes? (*h*) Explain the differences between first- and second-order interaction. (*i*) When are modes considered symmetric? (*j*) Describe cases in which mode interaction does not produce a reduction in the maximum load. (*k*) Explain what consequences may have mode interaction on an analysis via continuation methods.

Problem 14.1 (Theory). Show that (14.14) is valid:

$$(A_{ij} + \Lambda A_{ijk}\bar{Q}_k)q_i^I q_j^{II} = 0.$$

Problem 14.2 (Theory). Show that (14.15) is valid:

$$\frac{1}{2}(A_{ij} + \Lambda A_{ijk}\bar{Q}_k)q_i^I q_j^I = \frac{1}{2}\delta_{st}\xi_s\xi_t({}^s\Lambda_c - \Lambda).$$

Problem 14.3 (Theory) [28]. Calculate the coupled modes ${}^{st}x_j$ when the eigenvalues ${}^s\Lambda$ and ${}^t\Lambda$ are not coincident.

Problem 14.4 (Theory). Are the coefficients c_{ij} the same as the curvatures $\Lambda^{(2)c}$ of the postbuckling path?

Problem 14.5 (Column on elastic foundation). Consider the column on elastic foundation investigated in section 7.9. For $k = 3.75$ evaluate the new paths that emerge as a consequence of interaction.

Problem 14.6 (Column on elastic foundation). Calculate Problem 14.5 for $k = 4$.

Problem 14.7 (Column on elastic foundation). Calculate Problem 14.5 for $k = 4.2$.

Problem 14.8 (Two-bar truss) [31]. A truss formed by two bars was studied in Problem 5.5. Evaluate the new paths emerging from interaction. Does interaction produce a reduction in the maximum load that the system can reach?

14.13 BIBLIOGRAPHY

[1] Augusti, G., Stabilita di strutture elastiche elementari in presenza di grandi spostamenti, *Atti Accad. Sci. Fis. Mat. di Napoli*, Ser. 3era, 4(5), 1964 (in Italian).

[2] Antonini, R. C., *Influencia da Interacao entre modos e imperfecoes na Flambagem de Cascas Cilindricas Axialmente Comprimidas*, M. Sc. thesis, Universidade Federal de Rio do Janeiro, Rio de Janeiro, 1981 (in Portuguese).

[3] Batista, R. C., *Estabilidade Elastica de Sistemas Mecanicos Estruturais*, III Escola de Matematica Aplicada, Laboratorio de Computacao Cientifica, Rio de Janeiro, 1982 (in Portuguese).

[4] Batista, R. C., and Croll, J. G. A., Explicit lower bounds for the buckling of axially loaded cylinders, *Internat. J. Mech. Sci.*, 23(6), 331–343, 1981.

[5] Barbero, E. J., Raftoyiannis, I., and Godoy, L. A., Mode interaction in FRP columns, in *Mechanics of Composite Materials: Non-Linear Effects*, M. H. Hyer, Ed., AMD, vol. 159, American Society of Mechanical Engineers, New York, 1993, 9–18.

[6] Benito, R. and Sridharan, S., Interactive buckling with finite strips, *Internat. J. Numer. Meth. Engrg.*, 21, 145–161, 1985.

[7] Benito, R., and Sridharan, S., Mode interaction in thin-walled structural members, *J. Structural Mech., ASCE*, 12(4), 517–542, 1985.

[8] Butterworth, J. W., Numerical post-buckling analysis, Chapter 8, in *Structural Instability*, W. Supple, Ed., IPC Press, Guilford, England, 1973, 111–123.

[9] Byskov, E., and Hutchinson, J., Mode interaction in axially stiffened cylindrical shells, *AIAA J.*, 15(7), 941–948, 1977.

[10] Chilver, A. H., Coupled modes of elastic buckling, *J. Mech. Phys. Solids*, 15, 15–28, 1967.

[11] Chilver, A. H., and Johns, K. C., Multiple path generation at coincident branching points, *Internat. J. Mech. Sci.*, 13, 899–910, 1971.

[12] Donnell, L. H., A new theory for the buckling of thin cylinders under axial compression and bending, *Trans. ASME*, 56, 795, 1934.

[13] El Naschie, M. S., *Stress, Stability and Chaos in Structural Engineering: An Energy Approach*, McGraw-Hill, New York, 1990.

[14] Godoy, L. A., Barbero, E. J., and Raftoyiannis, I., Interactive buckling analysis of fiber-reinforced thin-walled columns, *J. Composite Materials*, 29(5), 591–613, 1995.

[15] Goltermann, P., and Mollmann, H., Interactive buckling of thin-walled beams: II Applications, *Internat. J. Solids Structures*, 25(7), 729–749, 1989.

[16] Hunt, G. W., An algorithm for the non-linear analysis of compound bifurcation, *Proc. Roy. Soc.*, 300, 443–471, 1981.

[17] Justino, M. R., *Estabilidade Local de Paneis Cilindricos Enrijecidos*, M.Sc. thesis, Federal University of Rio de Janeiro, Brazil, 1982 (in Portuguese).

[18] Koiter, W. T., *General Theory of Mode Interaction in Stiffened Plate and Shell Structures*, Report 590, Delft University of Technology, Delft, Holland, 1976.

[19] Koiter, W. T., and Kuiken, G., *The interaction between local buckling and overall buckling on the behavior of built-up columns*, Report 447, Laboratory of Engineering Mechanics, Delft University, Delft, Holland, 1971.

[20] Koiter, W. T., and Pignataro, M., *A General Theory for the Interaction between Local and Overall Buckling of Stiffened Panels*, Report 556, Dept. Mechanical Engineering, Delft University of Technology, Delft, Holland, 1976.

[21] Kolakowski, Z., Mode interaction in thin-walled trapezoidal column under uniform compression, *Internat. J. Thin Walled Structures*, 5, 329–342, 1987.

[22] Kolakowski, Z., Some thoughts on mode interaction in thin-walled columns under uniform compression, *Internat. J. Thin-Walled Structures*, 7, 23–35, 1989.

[23] Kolakowski, Z., Interactive buckling of thin-walled beam-columns with open and closed cross sections, *Internat. J. Thin-Walled Structures*, 15, 159–183, 1993.

[24] Maskaant, R., *Interactive Buckling of Biaxially Loaded Elastic Plate Structures*, Ph.D. thesis, University of Waterloo, Waterloo, Canada, 1989.

[25] Mollmann, H., and Goltermann, P., Interactive buckling of thin-walled beams: I Theory, *Internat. J. Solids Structures*, 25(7), 715–728, 1989.

[26] Perrow, C., *Normal Accidents: Living with High-Risk Technologies*, Basic Books (Harper Collins), New York, 1984.

[27] Pignataro, M., Luongo, A., and Rizzi, N., On the effect of the local-overall interaction on the post-buckling of uniformly compressed channels, *Internat. J. Thin-Walled Structures*, 3, 293–321, 1985.

[28] Raftoyiannis, I., Godoy, L. A., and Barbero, E. J., Buckling mode interaction in composite plate assemblies, *Appl. Mech. Rev.*, 48(11), Part 2, 52–60, 1995.

[29] Reis, A. J., *Interactive Buckling in Elastic Structures*, Ph.D. thesis, University of Waterloo, Waterloo, Canada, 1977.

[30] Reis, A. J., Interactive buckling in thin-walled structures, Chapter 7, in *Developments in Thin Walled Structures*, vol. 3, J. Rhodes and A. C. Walker, Eds., Elsevier, Oxford, 1987, 237–279.

[31] Reis, A. J., and Roorda, J., Post-buckling behavior under mode interaction, *J. Engrg. Mech., ASCE*, 105(4), 1979, 609–621.

[32] Roorda, J., and Reis, A. J., Nonlinear interactive buckling: Sensitivity and Optimality, *J. Structural Mech.*, 5(2), 207–232, 1977.

[33] Sridharan, S., and Ali, A., An improved interactive buckling analysis of thin-walled columns having doubly symmetric cross sections, *Internat. J. Solids Structures*, 22(4), 429–443, 1986.

[34] Sridharan, S., and Benito, R., Columns: Static and dynamic interaction buckling, *J. Engrg. Mech., ASCE*, 110(1), 49–65, 1984.

[35] Supple, W., Coupled branching configurations in the elastic buckling of symmetric structural systems, *Internat. J. Mech. Sci.*, 9, 97–112, 1967.

[36] Supple, M. J., On the branching of equilibrium paths, *Proc. Roy. Soc. London, Ser. A*, 315, 499–518, 1970.

[37] Supple, W., Coupled buckling modes of structures, Chapter 3, in *Structural Instability*, W. Supple, Ed., IPC Press, Guildford, England, 1973.

[38] Svenson, S., and Croll, J. G. A., Interaction between local and overall buckling, *Internat. J. Mech. Sci.*, 17, 307–321, 1975.

[39] Thompson, J. M. T., and Hunt, G. W., *A General Theory of Elastic Stability*, Wiley, London, 1973.

[40] Tvergaard, V., Buckling behavior of plate and shell structures, *Proc. 14th IUTAM Congress*, Delft, Holland, 1976.

[41] van der Neut, A., The interaction of local buckling and column failure of thin-walled compression members, *Proc. XII Int. Congress on Applied Mechanics, IUTAM*, Springer-Verlag, Berlin, 1969, 389–399.

[42] Walker, A. C., Interactive buckling of structural components, *Sci. Progress*, 62, 579–597, 1975.

CHAPTER

FIFTEEN

SENSITIVITY OF SYSTEMS WITH MODE INTERACTION

15.1 INTRODUCTION

This chapter is a continuation of Chapter 14, with the addition of imperfections on the analysis and behavior. We restrict our attention to interaction between bifurcation states, and this makes the analysis similar to what was seen in Chapter 10.

There are many imperfection shapes that can be taken into account; however, the solution is much simpler if we restrict our attention to imperfections with the shape of the modes competing during interaction. This is a very important class of imperfections, since the structure is most affected in its weakest directions, i.e., the interacting modes.

In most cases, the new terms in the total potential energy can be written in the form

$$\bar{W} = -\Lambda \xi_j \bar{\xi}_j, \tag{15.1}$$

where $\bar{\xi}_j$ are the amplitudes of the imperfections in the isolated mode ξ_j.

We saw before that there are two main steps in nonlinear interaction between modes: first, the evaluation of coupled mode shapes, as given by $\left({}^{st}x_j\right)$; and, second, the evaluation of the amplitudes of displacements ξ_j. We notice that the new term $\bar{W}$ does not contain the second-order shape $\left({}^{st}x_j\right)$. The mode shapes are then unaffected by the presence of imperfections, since

$$\frac{\partial \bar{W}}{\partial^{st} x_j} = 0. \tag{15.2}$$

The equilibrium conditions of Chapter 14 change, and there is a new term due to the imperfections

$$-\Lambda\bar{\xi}_j + \left[\left(\Lambda -^{s} \Lambda\right)\delta_{st}\right]\xi_t + \alpha_{stu}\xi_s\xi_t + \beta_{stuv}\xi_t\xi_u\xi_v = 0. \tag{15.3}$$

The solution of this problem is no longer a bifurcation, due to the contribution of a new constant term. The simplest way to solve the problem is to include the imperfection as part of a nonlinear analysis and carry out the computations of the nonlinear equilibrium paths using the best available tools. In some cases it is possible to employ analytical tools, but it is often necessary to use numerical techniques.

The Augusti column is explored in section 15.2 using an analytical solution. Section 15.3 deals with experimental work done for a thin-walled box column, in which imperfections were identified by means of a Southwell plot. Other problems investigated in this section are the spherical shell (section 15.4) and the cylinder under axial load (section 15.5).

15.2 THE AUGUSTI COLUMN

This is a classical example employed in the technical literature to illustrate imperfection sensitivity in systems with mode interaction. The basic problem was presented in Chapter 14, and will be again investigated to understand the influence of an imperfection in the geometry.

The geometry and load of the problem are shown in Figure 14.5. The imperfection is reflected in a new term of the form

$$\begin{aligned}\bar{V} &= -\bar{\xi}Q_1K_1 - \bar{\xi}Q_2K_2 \\ &= -\bar{\xi}K_1\left(Q_1 + \tau Q_2\right) \\ &= -\bar{\xi}\left(B_1Q_1 + B_2Q_2\right).\end{aligned} \tag{15.4}$$

The equilibrium conditions in the original coordinates result in

$$\begin{aligned}V_1 &= \left(A_{11} + \frac{1}{6}A_{1111}Q_1^2 + \frac{1}{2}A_{1122}Q_2^2\right)Q_1 + B_1\bar{\xi} = 0, \\ V_2 &= \left(A_{22} + \frac{1}{6}A_{2222}Q_2^2 + \frac{1}{2}A_{1122}Q_1^2\right)Q_2 + B_2\bar{\xi} = 0.\end{aligned} \tag{15.5}$$

This system does not admit a trivial solution to the fundamental path, and a nonlinear analysis is now required to compute the new path starting at the unloaded state. We have to remember from Chapter 14 that the load parameter is inside the A_{ij} and A_{ijkl} coefficients in this problem.

From $V_1 = 0$ we get the load $\Lambda = PL/K_1$ along the fundamental path

$$\Lambda^F = \frac{6(Q_1 - \bar{\xi})}{Q_1(3Q_2^2 - Q_1^2 + 6)}. \tag{15.6}$$

Next we substitute Λ in $V_2 = 0$ and get a relation between Q_1 and Q_2 along the fundamental path

$$\begin{aligned} V_2 &= (6\bar{\xi}\tau)\, Q_1 + (-6\bar{\xi})\, Q_2 + 6(1-\tau)\, Q_1 Q_2 \\ &\quad + \left[Q_1^3(-\bar{\xi}\tau) + Q_1^2 Q_2(-3\bar{\xi}) + Q_1 Q_2^2(3\bar{\xi}\tau) + Q_2^3(\bar{\xi})\right] \\ &= 0. \end{aligned}$$

The last bracket collects cubic terms in the displacements. If we restrict our attention to a quadratic solution, then we get

$$Q_1 = \frac{\bar{\xi} Q_2}{\bar{\xi}\tau + Q_2(1-\tau)}. \tag{15.7}$$

Coincident modes. For $\tau = 1$, (15.7) yields

$$Q_1 = Q_2, \tag{15.8}$$

meaning that the path evolves along planes at 45° from each axis of displacements.

In this case we can make use of the condition (15.8) to investigate the coupled path. The equilibrium equation becomes

$$V_1 = A_{11} Q_1 + \frac{1}{6} A_{1111} Q_1^3 + B_1 \bar{\xi} = 0. \tag{15.9}$$

The equilibrium paths for coincident modes, $\tau = 1$, have been evaluated in Figure 15.1 and are shown projected on the plane $Q_1 - \Lambda$. The plot shows the influence of different imperfection amplitudes ranging from $\bar{\xi} = 0$ to $\bar{\xi} = 0.06$. The imperfection has an effect on the coupled path similar to imperfections in unstable symmetric bifurcations, discussed in Chapter 15. Notice the presence of a complementary path for each imperfection amplitude.

The complete path in three dimensions is shown in Figure 15.2. Coupling between the two modes in the Augusti column (in which isolated modes are imperfection insensitive) yields an imperfection-sensitive equilibrium path whenever imperfections are included in the analysis.

The second derivative is in this case

$$V_{11} = A_{11} + \frac{1}{2} A_{1111} Q_1^2. \tag{15.10}$$

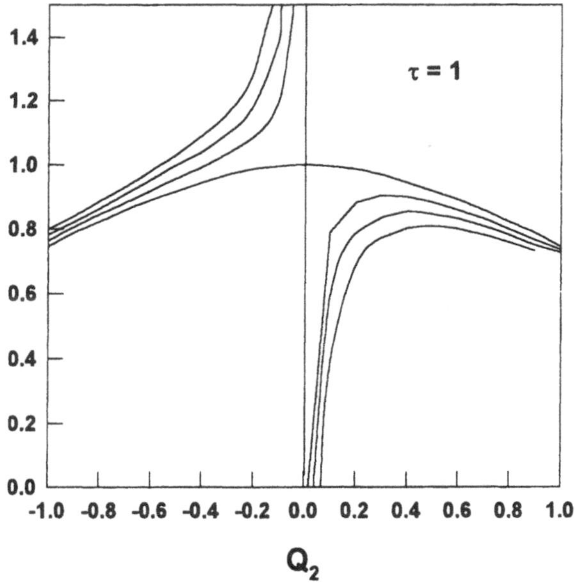

Figure 15.1 The Augusti column. Equilibrium paths for coincident modes.

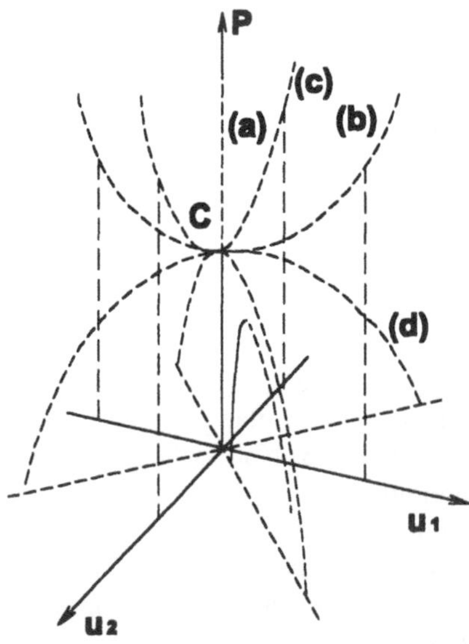

Figure 15.2 Coupled paths for the Augusti column, seen in three dimensions: (a) fundamental path, (b) secondary path, (c) secondary path, (d) tertiary paths.

Simultaneous solution of the two equations (15.9) and (15.10) yields the imperfection sensitivity in the form

$$\Lambda^M = \Lambda^c \left\{ 1 - \left[\left(\frac{3}{8} \right)^{2/3} 4^{2/3} \right] \bar{\xi}^{2/3} \right\}. \tag{15.11}$$

The sensitivity has the imperfection amplitude $\bar{\xi}$ to an exponent 2/3. This is again a 2/3 power law, as in unstable symmetric bifurcations.

Noncoincident modes. Solution of the equilibrium condition for $\tau \neq 1$ is shown in Figure 15.3 and represents a more complex path, which is fully three dimensional.

Here the path approaches the secondary and tertiary paths and loses stability close to the secondary bifurcation state. Sensitivity is now given by

$$\Lambda^M = \Lambda^c \left[1 - \left(\frac{3}{8} \right)^{2/3} \bar{\xi}^{2/3} \right]. \tag{15.12}$$

However, the coefficient in (15.11) is larger due to the factor $4^{2/3} = 2.52$. Thus, coalescence of modes has the effect of increasing the imperfection sensitivity of the problem.

The Augusti column is a rather academic problem; however, it shows the complexity of the behavior that one might expect to find in real engineering applications. More practical cases are discussed in the next sections.

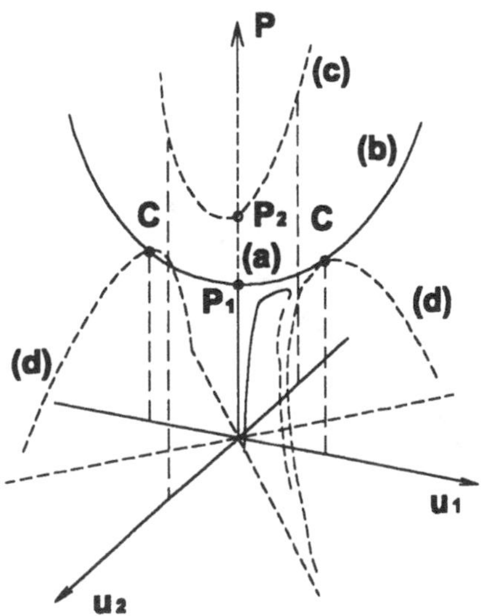

Figure 15.3 Equilibrium paths for the Augusti column for noncoincident modes. (a) Fundamental path, (b) secondary path, (c) secondary path, (d) tertiary paths.

15.3 EXPERIMENTS WITH A BOX COLUMN

A most interesting experimental investigation on the interactive buckling of aluminum box columns was carried out by Reis [8]. The specific column reported has a square thin-walled cross section, with thickness and length dimensions ($b/t = 46.748$ and $L/b = 30.472$, with $b = 76.2mm$) chosen so that local and overall buckling occur for the same load level ($P = 55.7kN$). This load is 33% below the elastic proportional limit stress of aluminium from experimental evaluation, so that the phenomenon was expected to occur in the form of elastic buckling. The column was tested with simply supported boundary conditions and under axial load.

The midspan deflection and the axial load have been used in Figure 15.4 to draw a Southwell plot of the test. The results are very close to a straight line, the slope of which is identified as the critical load. This load is only 77% of the theoretical load computed from an analysis of isolated modes ($42.889kN$). The displacement at $w/P = 0$ is the imperfection in the global buckling mode (a half-wave imperfection along the length) with a small value ($\xi/L = 0.00056$).

For this imperfection the tests show sensitivity of 23%; i.e., mode interaction and imperfections reduce the maximum load to only 0.77 of the expected value under conditions of no interaction.

The same tests were modelled using a finite strip analysis, and a 4% agreement was obtained in the maximum load [2].

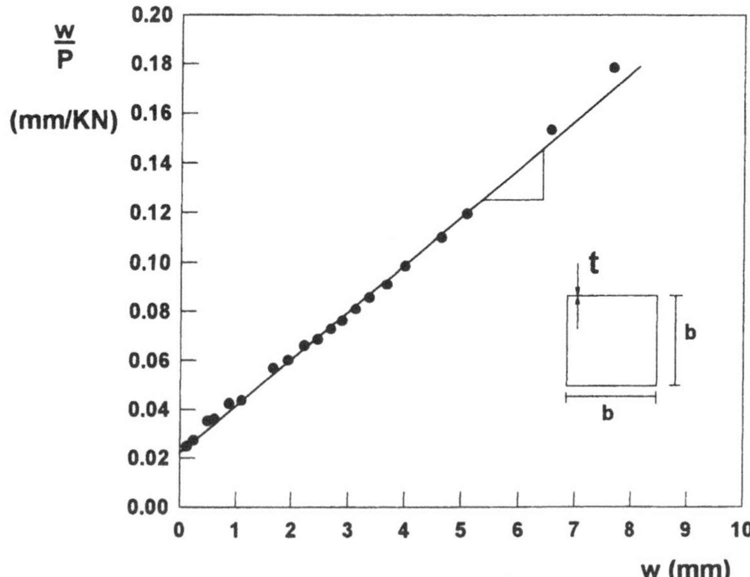

Figure 15.4 Southwell plot for a box column tested by Reis. Reprinted from [8] with permission from Elsevier Science.

15.4 PRESSURIZED SPHERICAL SHELL

Sensitivity of the buckling load in pressurized spherical shells with imperfections is a problem that does not occur because of a specific design (as was the case in thin-walled columns, plates with stiffeners, etc.), but it occurs in **all** thin-walled spherical shells. There is no way to escape this problem of interaction among several modes, and imperfection sensitivity, as presented in Chapter 10, is not adequate because the sphere never has isolated modes of buckling.

In this section we follow the presentation of Hutchinson [4], although it has been adapted to the notation of this book. A section of a complete sphere is considered, and it is treated as a shallow shell. This is a simplification of the problem, but it is reasonable if the shallow section considered is several times the length of the dominant buckling waves.

The buckling modc can be represented in the form

$$w = q_j \cos\left(\frac{mx}{R}\right)\cos\left(\frac{ny}{R}\right), \tag{15.13}$$

where q_j is the generalized coordinate in the mode defined by the integers m and n. We saw in Chapter 14 that there are many modes for which the buckling load is the same, and they satisfy the condition of Koiter's circle as

$$m^2 + n^2 = \alpha^2 \qquad \text{where} \qquad \alpha^4 = 12\left(1-\nu^2\right)\left(\frac{R}{h}\right)^2. \tag{15.14}$$

An interesting feature of the sphere is that the nonlinear equations for the amplitude ξ when we include all the modes of Koiter's circle uncouple into separate sets of equations. These sets are associated with the interaction between two or three of the critical modes.

Interaction between two modes. Interaction between two modes occurs when one of the modes has $m = 0$ or $n = 0$. Let us consider interaction between the modes ${}^1w$ and ${}^2w$:

$$ {}^1w = \xi_1 h \cos\left(\frac{\alpha x}{R}\right), \tag{15.15}$$

$$ {}^2w = \xi_2 h \sin\left(\frac{\alpha x}{2R}\right)\sin\left(\frac{\sqrt{3}}{2}\frac{\alpha y}{R}\right).$$

The equilibrium equations of interaction (15.3) are written in terms of the amplitudes in the coupling modes, i.e., ξ_1 and ξ_2,

$$\left(1-\frac{p}{p_c}\right)\xi_1 - \frac{9}{32}c\xi_2^2 = \frac{p}{p_c}\bar{\xi}_1, \tag{15.16}$$

$$\left(1-\frac{p}{p_c}\right)\xi_2 - \frac{9}{8}c\xi_2\xi_1 = \frac{p}{p_c}\bar{\xi}_2,$$

where the pressure is p, the imperfection amplitudes are $\bar{\xi}_1$ and $\bar{\xi}_2$, and $c = \sqrt{3\left(1 - \nu^2\right)}$. Two cases are considered in [4]: First, if the imperfection has components in $\xi_1 > 0$ and $\xi_2 = 0$, then the fundamental path takes the form

$$\xi_1 = \left(1 - \frac{p}{p_c}\right)^{-1} \frac{p}{p_c} \bar{\xi}_1. \tag{15.17}$$

The maximum pressure p^M is obtained from

$$\left(\frac{p^M}{p^c}\right)^2 - \left(2 + \frac{9}{8} c \bar{\xi}_1\right) \left(\frac{p^M}{p^c}\right) + 1 = 0. \tag{15.18}$$

Second, if the imperfection has components in ξ_2 and $\xi_1 = 0$, then the imperfection sensitivity curve takes the form

$$\left(\frac{p^M}{p^c}\right)^2 - \left(2 + \frac{27\sqrt{3}}{32} c \left|\bar{\xi}_2\right|\right) \left(\frac{p^M}{p^c}\right) + 1 = 0. \tag{15.19}$$

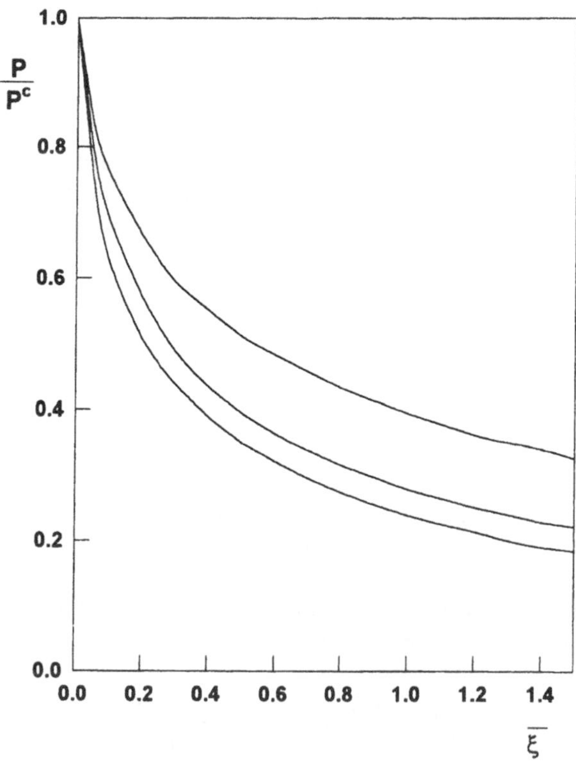

Figure 15.5 Imperfection sensitivity in the spherical shell: (a) imperfection with $\xi_1 > 0$ and $\xi_2 = 0$; (b) imperfection with ξ_2 and $\xi_1 = 0$; (c) three-mode case with $\bar{\xi}_1 = 0$, $\bar{\xi}_2 = 0$, and $\bar{\xi}_3 > 0$.

The above imperfection sensitivity relations, (15.18) and (15.19), have been plotted in Figure 15.5. It is clear that the second case, i.e., imperfection with ξ_2 and $\xi_1 = 0$, has a more detrimental effect on the maximum load. Both cases show an initial drop in the load for very small values of $\bar{\xi}$: For example, if the imperfection is of the order of 10% of the thickness of the shell, the load drops to about 65% of the theoretical value for the perfect shell. For an imperfection of the order of the thickness, on the other hand, the load is only 25%–30% of the theoretical value. This is a very severe imperfection sensitivity, so that spherical shells should be designed taking this effect into account.

Interaction of three modes. If none of the modes in the interaction has $m = 0$ or $n = 0$, then we have three interacting modes. Remember that in this analysis the interacting modes were taken from Koiter's circle, so that

$$
\begin{aligned}
{}^1w &= \xi_1 h \cos\left(m_1 \frac{x}{R}\right) \cos\left(n_1 \frac{y}{R}\right), &\quad \text{where} &\quad (m_1)^2 + (n_1)^2 = \alpha^2; \\
{}^2w &= \xi_2 h \sin\left(m_2 \frac{x}{R}\right) \sin\left(n_2 \frac{y}{R}\right), &\quad \text{where} &\quad (m_2)^2 + (n_2)^2 = \alpha^2; \\
{}^3w &= \xi_3 h \sin\left(m_3 \frac{x}{R}\right) \sin\left(n_3 \frac{y}{R}\right), &\quad \text{where} &\quad (m_3)^2 + (n_3)^2 = \alpha^2.
\end{aligned} \tag{15.20}
$$

The quadratic equilibrium equations become

$$
\begin{aligned}
\left(1 - \frac{p}{p_c}\right)\xi_1 - \frac{9}{16}c\xi_2\xi_3 &= \frac{p}{p_c}\bar{\xi}_1, \\
\left(1 - \frac{p}{p_c}\right)\xi_2 - \frac{9}{16}c\xi_3\xi_1 &= \frac{p}{p_c}\bar{\xi}_2, \\
\left(1 - \frac{p}{p_c}\right)\xi_3 - \frac{9}{16}c\xi_2\xi_1 &= \frac{p}{p_c}\bar{\xi}_3.
\end{aligned} \tag{15.21}
$$

Consider the case $\bar{\xi}_1 = 0$, $\bar{\xi}_2 = 0$, and $\bar{\xi}_3 > 0$; then

$$
\xi_3 = \left(1 - \frac{p}{p_c}\right)^{-1} \frac{p}{p_c}\bar{\xi}_3. \tag{15.22}
$$

Imperfection sensitivity takes the form

$$
\left(\frac{p^M}{p^c}\right)^2 - \left(2 + \frac{9}{16}c\bar{\xi}_3\right)\left(\frac{p^M}{p^c}\right) + 1 = 0. \tag{15.23}
$$

The results are also plotted in Figure 15.5, and can now be compared with the case of interaction between two modes. It can be seen that the curve for this three-mode case lies above the other two, showing that it is less sensitive to imperfections.

In all the above results, only one modal imperfection was allowed in each case. Shells in which the imperfection is more general than the one presented in [4] would have a different imperfection sensitivity; however, the main features of the behavior are shown in this analysis in which the sphere displays a severe drop in the maximum value of the load that can be attained.

15.5 CYLINDRICAL SHELL UNDER AXIAL COMPRESSION

The nonlinear interaction among modes in the cylinder is only second to that occurring in spherical shells reviewed in the last section. There are many studies on this problem, but only some simple models are discussed here, following [1]. The study uses the shell equations due to Donnell.

Consider a model of the cylinder as in Chapter 14, and we now include the influence of imperfections. Geometric imperfections are assumed in the form of the buckling modes that interact, and several cases have been studied.

Similar to what was done in Chapter 14, we consider four modes in the interaction studies

$$w = \xi_1 \cos\left(\frac{n}{R}y\right)\cos\left(\frac{m\pi}{L}x\right) + \xi_2 \cos\left(2\frac{m\pi}{L}x\right) \tag{15.24}$$
$$+\xi_3 \cos\left(2\frac{n}{R}y\right) + \xi_4 \cos\left(2\frac{n}{R}y\right)\cos\left(2\frac{m\pi}{L}x\right).$$

Geometric imperfections also take the same form, i.e.,

$$\bar{w} = \bar{\xi}_1 \cos\left(\frac{n}{R}y\right)\cos\left(\frac{m\pi}{L}x\right) + \bar{\xi}_2 \cos\left(2\frac{m\pi}{L}x\right). \tag{15.25}$$

Substitution of (15.24) and (15.25) into the nonlinear equations of this problem leads to a nonlinear problem of equilibrium. This has been solved in [1] for a shell with $R/h = 1000$, $\nu = 0.3$, and $L/R = 1$. Plots of equilibrium curves are reproduced in Figure 15.6, taken from [1]. The first mode in the interaction, with the amplitude ξ_1, has wave numbers $m = 1$ and $n = 13$.

The results show the maximum load is reduced from the theoretical value from isolated mode analysis, due to the presence of imperfections. This induces a drop in the maximum load, which depends on the amplitude of $\bar{\xi}_i$. For very small imperfections in both modes ($\bar{\xi}_1 = 0.05$, $\bar{\xi}_2 = -0.05$), we have a drop in Λ^M of about 10%. The case with the same imperfections amplified by a factor of 10 ($\bar{\xi}_1 = 0.5$, $\bar{\xi}_2 = -0.5$) leads to a value of $\Lambda^M = 0.5\Lambda^c$. As the imperfection grows, the curve of equilibrium is more nonlinear, and for large imperfections there is no maximum in the plot, but the displacements grow to large values with increasing load.

Other studies in [1] include the influence of imperfections in modes $\bar{\xi}_3$ and $\bar{\xi}_4$. For example, the imperfection with $\bar{\xi}_j = 0.1$ for $j = 1, 2, 3, 4$, leads to $\Lambda^M = 0.8\Lambda^c$. The modifications in Λ^M due to such imperfections ($\bar{\xi}_3$ and $\bar{\xi}_4$) are not substantially different from cases in which they are neglected.

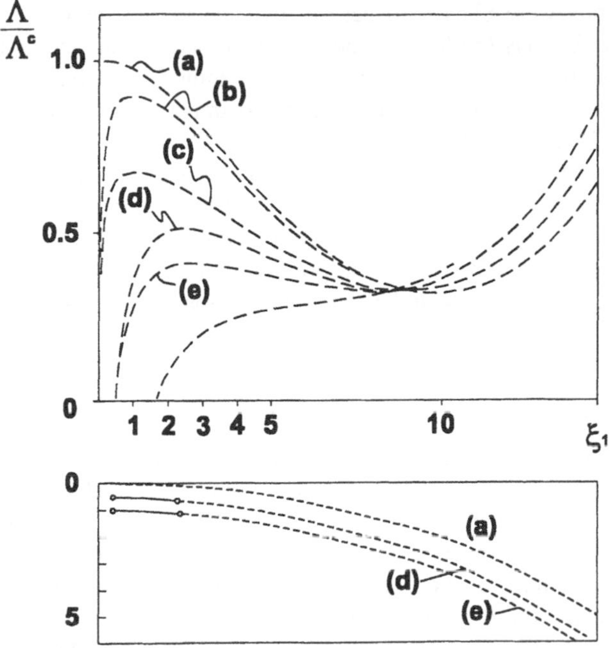

Figure 15.6 Equilibrium paths for the cylindrical shell under axial load: (a) perfect shell; (b) $\bar{\xi}_1 = 0.05,\ \bar{\xi}_2 = -0.05$; (c) $\bar{\xi}_1 = 0.05,\ \bar{\xi}_2 = -0.5$; (d) $\bar{\xi}_1 = 0.5,\ \bar{\xi}_2 = -0.5$; (e) $\bar{\xi}_1 = 0.5,\ \bar{\xi}_2 = -1.0$.

Imperfection sensitivity of the cylinder was studied by Koiter [5]. For axisymmetric imperfections (those identified by $\bar{\xi}_2$ in the present section), the relation between Λ^M and $\bar{\xi}_2$ is given by

$$\left(\frac{\Lambda^M}{\Lambda^c}\right)^2 - \left(2 + \frac{3}{2}c\bar{\xi}_2\right)\left(\frac{\Lambda^M}{\Lambda^c}\right) + 1 = 0. \tag{15.26}$$

For the same values of Poisson's ratio, the sensitivity of the cylinder is similar to the case of the sphere. This is left to the reader as a problem at the end of the chapter.

15.6 FINAL REMARKS

Koiter and Pignataro [6] studied the influence of imperfections on the nonlinear interaction in stiffened panels. The structure considered consists of a flat plate and stiffeners built up from steel flat plate strips. The interaction investigated occurs between local and overall modes. Mode interaction and imperfections in the stiffened plate account for a reduction in the load-bearing capacity of about 10%.

Imperfections were chosen to act on just one of the modes. The influence of combined imperfections (both local and overall imperfections acting simultaneously) was not part of the study. "Our reason for this omission is that no disastrous consequences are to be feared from such combined imperfections, worse than those for imperfections of one type" [6]. The stiffened panel is extremely sensitive to very small imperfections (of the order of a few percent of the thickness), but this sensitivity is reduced for the larger imperfections that are found in practical aeronautical structures. Following previous experience, such as [6], it seems reasonable to focus attention on the influence of modal imperfections in problems of mode interaction.

Finally, design sensitivity under mode interaction has been the subject of [9, 3].

15.7 PROBLEMS

Review questions. (*a*) Compare the imperfection sensitivity under ode interaction with that of isolated modes. (*b*) Explain why coupled mode shapes are not modified by imperfections. (*c*) What reduces most the maximum load in the Augusti column: coincidence or near coincidence of modes? (*d*) What are the levels of reduction in the maximum load that can be expected in a square box column due to interaction between local and overall modes? (*e*) For a spherical shell under external pressure there are many coincident modes. Explain why only 2 or 3 of them interact.

Problem 15.1. Compare the imperfection sensitivity of the pressurized spherical shell obtained in this chapter with Figure 15.1. Discuss what differences you find between such experimental results and the theoretical ones obtained here.

Problem 15.2. Plot the imperfection sensitivity of the Augusti column.

Problem 15.3. Plot the imperfection sensitivity curve for a cylindrical shell under axial compression, according to (15.26). Compare this with the case of the sphere under external pressure.

Problem 15.4. Find the imperfection sensitivity of an I-section column under axial load, already discussed in Chapter 14.

15.8 BIBLIOGRAPHY

[1] Batista, R. C., *Estabilidade Elastica de Sistemas Mecanicos Estruturais*, Laboratorio de Computacao Cientifica, LNCC-CNPq, Rio de Janeiro, 1982.

[2] Benito, B., and Sridharan, S., Mode interaction in thin-walled structural members, *J. Engrg. Mech.*, ASCE, 12(4), 145–161, 1985.

[3] Godoy, L. A., and Almanzar, L., Design sensitivity under buckling mode interaction in *Engineering Mechanics: A Force for the XXI Century*, H. Murakami and J. E. Lucco, Eds., ASCE, Reston, VA, 1998.

[4] Hutchinson, J. W., Imperfection sensitivity of externally pressurized spherical shells, *J. Appl. Mech.*, 34(March), 49–55, 1967.

[5] Koiter, W. T., Elastic stability and postbuckling behavior, *Proc. Symposium on Non-Linear Problems*, R. E. Langer, Ed., University of Wisconsin Press, Madison, 1963, 257.

[6] Koiter, W. T., and Pignataro, M., *A General Theory for the Interaction Between Local and Overall Buckling of Stiffened Panels*, Delft University of Technology, Report 76-83, April, 1976.
[7] Maaskant, R., *Interactive Buckling of Biaxially Loaded Elastic Plate Structures*, Ph.D. Thesis, University of Waterloo, Waterloo, 1989.
[8] Reis, A. J., Interactive buckling in thin-walled structures, in *Developments in Thin-Walled Structures*, vol. 3, J. Rhodes and A. C. Walker, Eds., Elsevier Science, London, 1987, 237–279.
[9] Roorda, J., and Reis, A. J., Nonlinear interactive buckling: Sensitivity and optimality, *J. Structural Mech.*, 5(2), 207–232, 1977.

APPENDIX

A

DISPLACEMENTS, STRAINS, AND STRESSES

In this appendix we review some elementary concepts about the variables involved in the computation of the total potential energy. A reader with basic background in the theory of elasticity could perhaps skip this section.

A.1 GEOMETRY

We start by considering a general three-dimensional solid with a volume $\mathcal{V}$ and a surface $\mathcal{S}$ containing the volume, as shown in Figure A.1.

The geometry of a solid body is described by the positions of its points with respect to an arbitrary origin O in space. Vector $\mathbf{r}$ defines the position of a generic point A, and the Cartesian components of $\mathbf{r}$ may be written as

$$\mathbf{r} = x_i \mathbf{t_i}, \quad i = 1, 2, 3, \tag{A.1}$$

where $\mathbf{t_i}$ are unit vectors in the Cartesian reference system, and x_i are the components of $\mathbf{r}$ (see Figure A.2). Index notation (often attributed to Einstein) is employed throughout this book, in which repeated indices imply summation. For example, (A.1) can be expanded as

$$\mathbf{r} = x_1 \mathbf{t_1} + x_2 \mathbf{t_2} + x_3 \mathbf{t_3}.$$

It will be necessary to identify the direction (say, μ) of a segment in the body. For that, we employ the direction μ in vector form of unit value, given by

$$\mu = \mu_i \mathbf{t_i}.$$

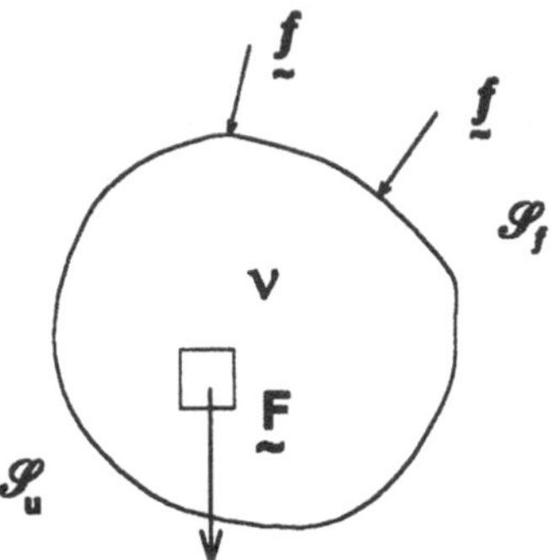

Figure A.1 Notation for a solid three-dimensional body.

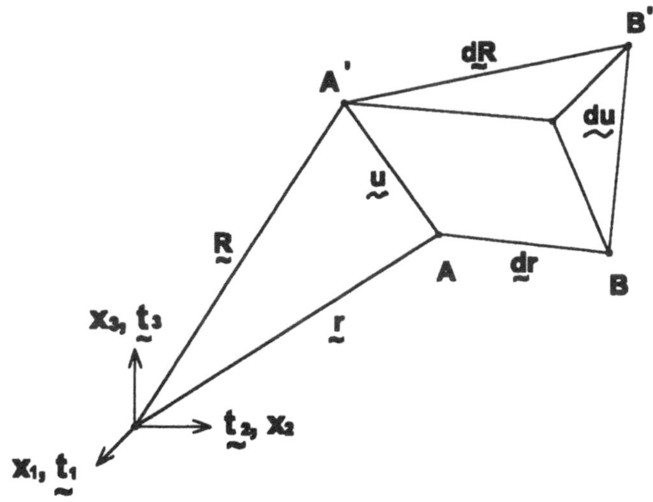

Figure A.2 Displacements of a point A and a segment AB.

A.2 FORCES

In the basic problem of mechanics, a system of forces is applied to the body: They may include gravitational loads on part or on the complete volume, surface loads applied on part of the boundary, and other types of loads, such as thermal, electrical, etc. These loads are increased from an initial value (usually zero) to a given value by means of control parameters. In the developments presented in this book, a single control parameter (Λ) is considered for the loads, so that all the loads acting on the body are assumed to increase at the same rate. Of course, many engineering problems are such that several independent loading rates affect different components of the load system, but for the sake of simplicity of the presentation we start from the single load parameter controlling the forces on the body.

Volume forces **F** are denoted as

$$\mathbf{F} = \Lambda F_i \mathbf{t_i} \tag{A.2}$$

and act on the volume $\mathcal{V}$. The surface forces **f** are written as

$$\mathbf{f} = \Lambda f_i \mathbf{t_i} \tag{A.3}$$

and act on the part of the boundary denoted by $\mathcal{S}_f$ in Figure A.1.

The rate at which we increase the loads is assumed to be "slow," in the sense that there are no inertia effects so that we only consider static actions. Furthermore, we restrict our attention to conservative loads.

A.3 DISPLACEMENTS

Under the action of forces, the particles of the body undergo displacements $\mathbf{u}(\mathbf{x})$ that define a new configuration indicated by the position vector $\mathbf{R}(\mathbf{r}, \mathbf{u})$. This new configuration, illustrated in Figure A.1, may be expressed as

$$\mathbf{R} = \mathbf{r} + \mathbf{u} \tag{A.4}$$

and

$$\mathbf{dR} = \mathbf{dr} + \mathbf{du}. \tag{A.5}$$

The scalar modulus dS of $\mathbf{dR}$ is obtained as

$$dS^2 = \mathbf{dR} \cdot \mathbf{dR} = (\mathbf{dr} + \mathbf{du}) \cdot (\mathbf{dr} + \mathbf{du}).$$

The modulus dS^2 can be expanded in terms of Cartesian components in the form

$$dS^2 = ds^2 + \left(\frac{\partial u_i}{\partial x_j} + \frac{\partial u_j}{\partial x_i} + \frac{\partial u_m}{\partial x_i} \frac{\partial u_m}{\partial x_j} \right) dx_i \, dx_j, \tag{A.6}$$

where $ds^2 = \mathbf{dr} \cdot \mathbf{dr}$ is the scalar modulus of the undeformed vector $\mathbf{dr}$.

A.4 STRAINS

The deformations of the body at a given point are summarized using the information of the strain tensor ε_{ij} defined as

$$\varepsilon_{ij} = \frac{1}{2} \left(\frac{\partial u_i}{\partial x_j} + \frac{\partial u_j}{\partial x_i} + \frac{\partial u_m}{\partial x_i} \frac{\partial u_m}{\partial x_j} \right). \tag{A.7}$$

Equations such as (A.7) are usually known as "kinematic relations." Notice that index i (and also j) is a free index and indicates that the number of equations are written in compact form. Thus, (A.7) represents 9 conditions.

It may be shown that the ε_{ij} are the components of a second-order symmetric tensor, known as the strain tensor of Green–Lagrange. Other strain tensors are employed in the literature but will not be reviewed here.

The strain tensor ε_{ij} has a linear and a quadratic part in terms of u_i, and they will be denoted by

$$\varepsilon_{ij} = \varepsilon'_{ij} + \varepsilon''_{ij}, \tag{A.8}$$

where

$$\varepsilon'_{ij} = \frac{1}{2}\left(\frac{\partial u_i}{\partial x_j} + \frac{\partial u_j}{\partial x_i}\right),$$

$$\varepsilon''_{ij} = \frac{1}{2}\frac{\partial u_m}{\partial x_j}\frac{\partial u_m}{\partial x_i}.$$

Some measures of changes in length, angle, and volume at a generic point of the body are of special interest. Segment ds, originally oriented in the direction $\boldsymbol{\mu}$, changes its length to dS but also changes its orientation in space.

We define the longitudinal strain ϵ_μ to measure the change in length of a segment that was initially in the direction $\boldsymbol{\mu}$,

$$\epsilon_\nu = \frac{dS - ds}{ds}. \tag{A.9}$$

In terms of displacement components, ϵ_μ results in

$$\epsilon_\mu = \left[1 - \left(\frac{\partial u_i}{\partial x_j} + \frac{\partial u_j}{\partial x_i} + \frac{\partial u_m}{\partial x_i}\frac{\partial u_m}{\partial x_j}\right)\mu_i\mu_j\right]^{1/2} - 1 \tag{A.10}$$

or else

$$\epsilon_\mu = (1 - 2\varepsilon_{ij}\mu_i\mu_j)^{1/2} - 1. \tag{A.11}$$

The change in angle between two segments originally in directions $\boldsymbol{\eta}$ and $\boldsymbol{\mu}$ may be reflected by an angular strain. With reference to Figure A.3, we consider that $\boldsymbol{\eta}$ and $\boldsymbol{\mu}$ are perpendicular in the original unloaded configuration, i.e.,

$$\boldsymbol{\eta} \cdot \boldsymbol{\mu} = 0.$$

The angle following deformation is denoted by α. Then we can define a new angle $\varphi_{\mu\eta} = \pi/2 - \alpha$ and write

$$\sin\varphi_{\mu\eta} = \frac{2\varepsilon_{ij}\mu_i\eta_j}{(1+\epsilon_\mu)(1+\epsilon_\eta)}. \tag{A.12}$$

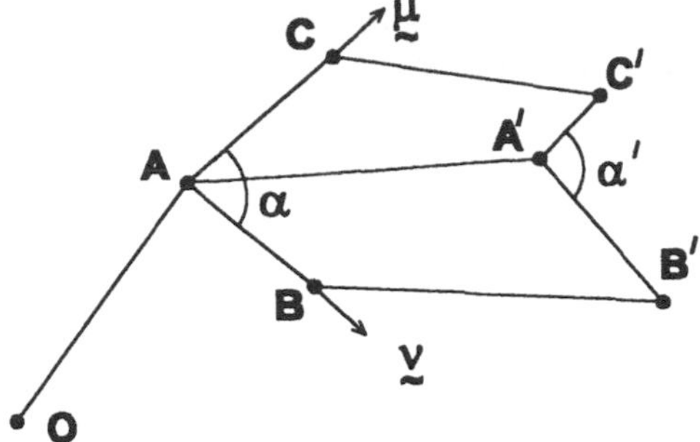

Figure A.3 Angle deformations.

Finally, a volumetric strain Δ is defined by the change in volume of a sphere surrounding a point

$$\Delta = \frac{d\mathcal{V} - d\mathcal{V}_0}{d\mathcal{V}_0}. \tag{A.13}$$

As a first approximation, it is assumed that the original spherical volume $d\mathcal{V}_0$ deforms into an ellipsoidal volume $d\mathcal{V}$, and it may be shown that

$$\Delta = (1 + \epsilon_\mu)(1 + \epsilon_\eta)(1 + \epsilon_\rho) - 1, \tag{A.14}$$

where η, μ, and ρ are orthogonal directions before deformation.

A set of relations has been established, as illustrated in Figure A.4: the strains ϵ_μ, $\varphi_{\mu\eta}$, and Δ may be completely defined in terms of the strain tensor ε_{ij}; and ε_{ij} is defined in terms of the displacement components u_i. Clearly, the specific strains are a nonlinear function of the tensor ε_{ij}: for example, a square root is present in ϵ_μ, and a more complicated dependence is present in $\varphi_{\mu\eta}$ and Δ. Furthermore, the relation between ε_{ij} and u_i is again nonlinear.

A theory that includes all the above nonlinearity is known as a "large strain theory." Although accurate and general, its use is rather cumbersome, and some simplifications have been developed.

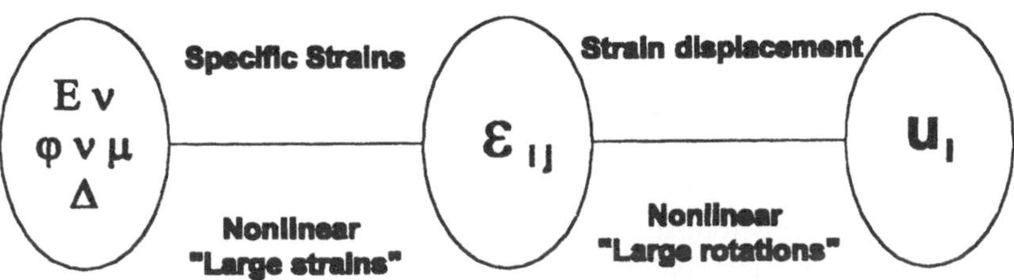

Figure A.4 Relations between specific strains, strains, and displacements.

Let us concentrate on the relations between ϵ_μ, $\varphi_{\mu\eta}$, and Δ with ε_{ij}. Under the assumptions that $\sin\varphi_{\nu\mu} \approx \varphi_{\nu\mu}$ and $(\epsilon_\mu)^2 \ll \epsilon_\mu$, it is possible to obtain

$$\epsilon_\mu \approx \varepsilon_{ij}\mu_i\mu_j,$$

$$\varphi_{\mu\eta} \approx 2\varepsilon_{ij}\mu_i\eta_j,$$

$$\Delta \approx \varepsilon_{11} + \varepsilon_{22} + \varepsilon_{33},$$

with $\varepsilon_{ij} = \varepsilon'_{ij} + \varepsilon''_{ij}$.

Such a theory is said to consider "small strains and large rotations." Notice that ϵ_μ, $\varphi_{\nu\mu}$, and Δ are now linear functions of ε_{ij} but are nonlinear functions of u_i.

Finally, a third theory may be obtained by neglecting ε''_{ij} in comparison with ε'_{ij}. This is a completely linear theory, in the sense that ϵ_μ, $\varphi_{\nu\mu}$, and Δ become linear functions of u_i. Such a theory is called "small displacements" or "small rotations" or "linear" theory.

It is shown in Chapter 4 that a linear theory of elasticity (neglecting ε''_{ij}) is not adequate to perform stability analysis. A large strain theory, on the other hand, is not necessary for most engineering applications, so that we continue our analysis in the context of the theory of small strains and large rotations and at the initial stages of buckling. Furthermore, the main variables in the analysis will not be the specific strains ϵ_μ, $\varphi_{\nu\mu}$, Δ (which may be entirely calculated from ε_{ij}) but the strain tensor ε_{ij} and the displacement vector u_i.

A.5 STRESSES

Next, the usual definition of stress vector in the sense of Cauchy is considered by looking at a plane in the body with a normal vector μ. As the area surrounding the point tends to zero, the resultant of the internal forces acting on the plane define the vector σ_μ. Components of σ_μ may be found as the projection of σ_μ on the normal μ and on the plane considered (Figure A.5)

$$\sigma_\mu = \sigma_{\mu\mu}\boldsymbol{\mu} + \sigma_{\mu s}\mathbf{s}. \tag{A.15}$$

A second-order symmetric tensor may be defined from equilibrium considerations as σ_{ij}, and related to σ_μ by

$$\sigma_\mu = \sigma_{ij}\mu_i\mathbf{t_j} \tag{A.16}$$

with $\sigma_{ij} = \sigma_{ji}$. Thus, σ_μ is linear in σ_{ij} and independent of the deformations of the body.

In theories of finite elasticity, another definition of the stress tensor is required, which depends on the deformations (such as the Piola–Kirchhoff tensors). Whenever there are large strains, the equilibrium equations should be written in the deformed configuration. This occurs, for example, for large displacements along the postbuckling equilibrium path.

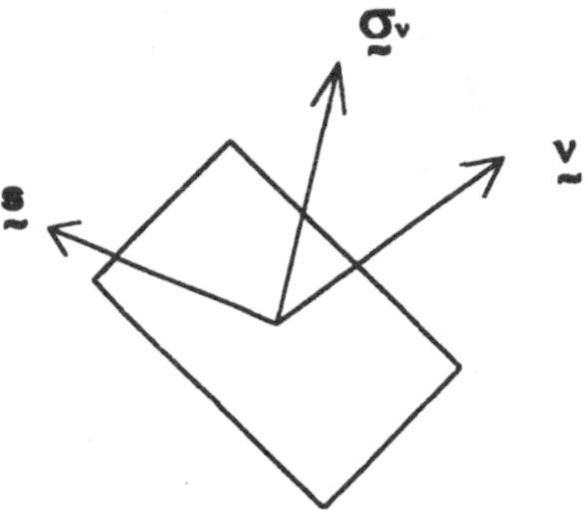

Figure A.5 Stress vector.

However, for intermediate situations, it is acceptable to employ the nonlinear tensor of strains of (A.7) with σ_{ij}, and perform the necessary integrations in the volume of the solid before deformations. Such intermediate situations occur in the vicinity of bifurcations and are typical of the initial postcritical behavior studied in this book.

A.6 CONSTITUTIVE RELATIONS

In linear elasticity the isotropic constitutive equations linking stresses σ_{ij} and strains ε_{ij} reduce to

$$\sigma_{ij} = \frac{E}{1+\nu}\left(\varepsilon_{ij} + \frac{\nu}{1-2\nu}\delta_{ij}\varepsilon_{mm}\right), \qquad \text{(A.17)}$$

where the scalar moduli E and ν are the usual modulus of elasticity and Poisson's ratio, and δ_{ij} is the Kronecker delta ($\delta_{ij} = 1$ if $i = j$, and $\delta_{ij} = 0$ if $i \neq j$).

A few problems in this book refer to orthotropic materials, and the appropriate constitutive equations will be defined as part of the example.

A.7 BOUNDARY CONDITIONS

To have a boundary value problem we need to define a complete set of boundary conditions on $\mathcal{S}$. In kinematic boundary conditions only displacements are involved, such as

$$u_i - \bar{u}_i = 0 \quad \text{on } \mathcal{S}_u, \qquad \text{(A.18)}$$

where $\bar{u}_i$ are known values of displacements, and $\mathcal{S}_u$ is the part of the boundary where displacements are prescribed.

Mechanical boundary conditions involve known boundary forces $\bar{f}_i$ and stresses on the part of the boundary denoted by $\mathcal{S}_f$. From equilibrium, they satisfy

$$\sigma_{ij}\mu_i - \bar{f}_j = 0 \quad \text{on } \mathcal{S}_f. \qquad \text{(A.19)}$$

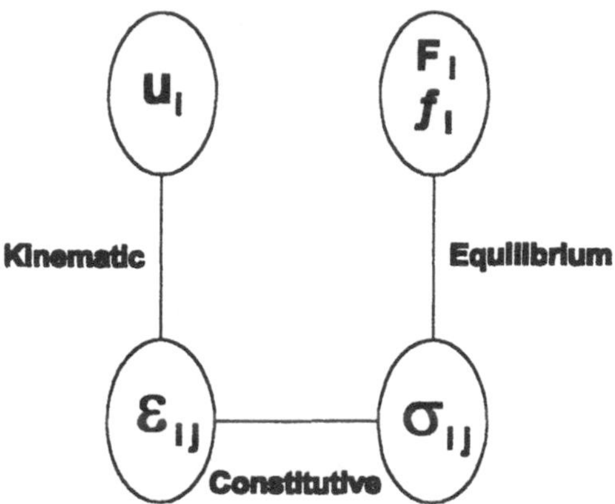

Figure A.6 Relations between different fields in elasticity.

Mixed boundary conditions can also occur at a point on the boundary, so that some components of displacements and some components of force are prescribed at the same part of the boundary.

A.8 SUMMARY

Figure A.6 shows a schematic relation between the variables defining the problem in three-dimensional elasticity. A broader picture involving a complete set of variables and a similar analysis for many fields of mechanics may be found in a work by Tonti [1].

The summary of results presented in this section is valid for three-dimensional elasticity. But in many cases it is more convenient to work with technical theories for specific applications, as when one of the dimensions of the body is small in comparison with the others. This leads to engineering theories for line and surface structural elements, such as beams, columns, rings, arches, plates, and shells, and the examples of this book refer to such engineering applications.

The specific technical nonlinear theories, in which stress resultants across the thickness rather than actual stresses are used, will not be reviewed in the following, and it is assumed that the reader is familiar with them. An excellent account of technical theories may be found in the references of Chapter 2.

A.9 BIBLIOGRAPHY

[1] Tonti, G., The reasons of the analogies in physics, in *Problem Analysis in Engineering and Science*, F. H. Branin Jr. and K. Huseyin, Eds., Academic Press, New York, 1977, 463.

APPENDIX

B

ENERGY OF BARS AND TRUSSES

B.1 ENERGY OF A FRAME MEMBER

A bar that can be part of a frame can be discretized using polynomial interpolation functions, as is usually done in structural analysis. The geometry and displacement components of a general member with length L and oriented at θ_x from the x-axis are shown in Figure 2.7.

It is assumed that the bar is extensional; i.e., the length L changes following deformation of the structure. The strain energy is given by

$$\Omega = \frac{1}{2}EA\int \varepsilon^2 dx + \frac{1}{2}EI\int \chi^2 dx, \tag{B.1}$$

where the change in curvature can be obtained from the nonlinear relation

$$\varepsilon = \frac{du}{dx} + \frac{1}{2}\left(\frac{dw}{dx}\right)^2, \qquad \chi = \frac{d^2w}{dx^2}. \tag{B.2}$$

Substitution into the strain energy leads to

$$\Omega = \frac{1}{2}\int\left\{AE\left[\frac{du}{dx} + \frac{1}{2}\left(\frac{dw}{dx}\right)^2\right]^2 + EI\left(\frac{d^2w}{dx^2}\right)^2\right\}dx. \tag{B.3}$$

The out-of-plane displacements $w(x)$ can be approximated by cubic polynomials in the local coordinate axis of the bar, i.e.,

$$w(x) = \left[1 - 3\left(\frac{x}{L}\right)^2 + 2\left(\frac{x}{L}\right)^3\right] w_A + \left[3\left(\frac{x}{L}\right)^2 - 2\left(\frac{x}{L}\right)^3\right] w_B$$

$$+ \left[\left(\frac{x}{L}\right) - 2\left(\frac{x}{L}\right)^2 + \left(\frac{x}{L}\right)^3\right] \phi_A + \left[\left(\frac{x}{L}\right)^3 - \left(\frac{x}{L}\right)^2\right] \phi_B. \tag{B.4}$$

The in-plane displacement component u is approximated by a linear polynomial

$$u(x) = u_A + \left(\frac{x}{L}\right)(u_B - u_A), \tag{B.5}$$

where u_A, w_A, and ϕ_A are the displacements and rotation at node A.

To have generalized coordinates in global coordinate axis, the nodal variables are rotated as follows

$$u_A = Q_1 \cos\theta_x + Q_2 \cos\theta_y,$$

$$w_A = -Q_1 \cos\theta_y + Q_2 \cos\theta_x,$$

$$\phi_A = Q_3,$$

$$u_B = Q_4 \cos\theta_x + Q_5 \cos\theta_y,$$

$$w_B = -Q_4 \cos\theta_y + Q_5 \cos\theta_x,$$

$$\phi_B = Q_6. \tag{B.6}$$

The strain energy has quadratic and quartic terms, while cubic terms are not present.

A simple Maple V program to carry out the computation of the energy is as follows:

```
restart : with(linalg) :
   xi := x/L :
   w := w1 * (1 - 3 * xi^2 + 2 * xi^3) + phi1 * L * (xi - 2 * xi^2 + xi^3)
        +w2 * (3 * xi^2 - 2 * xi^3) + phi2 * L * (xi^3 - xi^2);
   u := u1 + xi * (u2 - u1) :
epsilon := diff(u, x) + 1/2 * (diff(w, x))^2 :
  kappa := diff(w, x, x) :
Umembrane := 1/2 * AE * simplify(epsilon^2, x = 0..L);
 Ubending := 1/2 * IE * simplify(kappa^2, x = 0..L);
```

$$Lx := \cos(theta1) : Ly := \cos(theta2) :$$
$$u1 := Lx * Q1 + Ly * Q2 : w1 := -Ly * Q1 + Lx * Q2 :$$
$$u2 := Lx * Q4 + Ly * Q5 : w2 := -Ly * Q4 + Lx * Q5 :$$
$$phi1 := Q3 : phi2 := Q6 :$$
$$U := simplify(Umembrane + Ubending);$$

The resulting energy is a long expression, so it is left to the reader to carry out the substitutions to get an explicit form of the energy of an inclined bar.

B.2 ENERGY OF A TRUSS MEMBER

In a truss member that may buckle we assume that there is a membrane plus a bending contribution to the energy. However, unlike the case of the frame member, one should not retain rotations at the ends of the member as degrees of freedom, since they are not the same for each concurrent member at a joint. A more appropriate function for w could thus be a hierarchical interpolation in the form

$$w(x) = w_A + (w_B - w_A)\frac{x}{L} + w_c \sin\left(\frac{\pi x}{L}\right). \tag{B.7}$$

The in-plane displacement component u is approximated by a linear polynomial (B.5).

The energy can next be computed as in (B.1) to (B.3). To write the energy in global coordinates, the transformation of the degrees of freedom at the ends of the bar are required:

$$u_A = Q_1 \cos\theta_x + Q_2 \cos\theta_y,$$
$$w_A = -Q_1 \cos\theta_y + Q_2 \cos\theta_x,$$
$$u_B = Q_4 \cos\theta_x + Q_5 \cos\theta_y,$$
$$w_B = -Q_4 \cos\theta_y + Q_5 \cos\theta_x.$$

Notice that the degree of freedom w_c need not be written in global coordinates, since it is not shared by neighboring elements.

INDEX